MEN AT WORK

MEN AT WORK

APPLICATIONS OF ERGONOMICS TO
PERFORMANCE AND DESIGN

By

ROY J. SHEPHARD, M.D., PH.D.
Department of Environmental Health
School of Hygiene
University of Toronto
Toronto, Ontario, Canada

CHARLES C THOMAS • PUBLISHER
Springfield • Illinois • U.S.A.

Published and Distributed Throughout the World by
CHARLES C THOMAS • PUBLISHER
BANNERSTONE HOUSE
301-327 East Lawrence Avenue, Springfield, Illinois, U.S.A.

© *1974, by* CHARLES C THOMAS • PUBLISHER

ISBN 0-398-02965-2

Library of Congress Catalog Card Number: 73–13752

Printed in the United States of America

BB-14

Library of Congress Cataloging in Publication Data

Shephard, Roy J.
   Men at work.

   Bibliography: p.
   1. Human engineering. 2. Work—Physiological
aspects. I. Title.
T59.7.S54   612'.042   73–13752
ISBN 0–398–02965–2

PREFACE

"Men at work" was once a familiar English road sign. The unsuspecting motorist was supposed to learn from this warning that the road was under repair. Often, no men at work could be seen. But occasionally the driver might encounter a disorganized group of laborers some two or three yards beyond the warning board. The majority would be leaning on their shovels through lack of materials, management or motivation, while a few struggled mightily with poorly designed and ineffective equipment.

Here we see in miniature many of the problems reviewed in the present book. How can the vigilance of the motorist be sustained so that he reacts swiftly to a warning sign? What distance should separate the sign from the obstacle to be avoided? Can anticipation of danger be improved if the simple three-word sign is replaced by a detailed statement of highway appropriations (as in the United States and Canada) or by picturing a man aiming a shovel at a pile of rocks? What is wrong with the highway system as a whole and plans for its maintenance? Can bottlenecks in construction be eliminated? What will motivate workers to a sustained high level of activity? What limitations are imposed on performance by the cardiorespiratory system, muscular strength, neuromuscular coordination and psychological considerations? Are the men justified in complaints of fatigue and a poor working environment? And how may the design and organization of the entire system be modified to optimize performance?

"Men at work" may seem a male chauvinist's choice of title. Let me immediately reassure any ardent women's liberationist I accept that women also work. In some occupations, particularly those requiring a sensitivity to human inter-relationships or fine motor skills, I suspect that they show a greater productivity than the average male employee. Unfortunately, it is not possible to be specific, as data are generally not available. Perhaps because of social conditioning, women have been less willing than men to cooperate in tests ranging from simple measurements of leg dimensions to maximum physical effort. Such figures as we shall review stress differences of physiology and biomechanics between the two sexes. *Equal opportunity* to lift heavy weights, for example, is ergonomic nonsense. Sometimes a task can be redesigned to match the characteristics of the worker, but even in the most favorable instances, matching is incomplete. A given task is best suited to a person with specific physiological and psychological characteristics, and the majority of workers meeting such a specification will be drawn from one sex.

The material presented in this book is drawn largely from two courses offered by the author's department: a fourth-year undergraduate option entitled *The Biology of Work* and a graduate course entitled *Ergonomics*. The topics that are covered thus range widely over the physiology and psychology of work, biomechanics and ergonomics, with the unifying theme of matching design to human performance characteristics in the interests of comfort and productivity. The topics chosen have a broad interdisciplinary appeal, being relevant not only to potential ergonomists, human factors engineers and biomechanicians, but also to students of science (applied physiology, applied psychology and kinesiology), medicine (especially environmental and industrial medicine), business management, sociology, architecture and town planning. Accordingly, jargon and unnecessary technical terms have been avoided. Further, no attempt has been made to give a rigid systematic treatment of the subject. Rather, the objective has been to stimulate the interest of the student in bettering human work performance by exploring selected topics in moderate detail.

The course in Ergonomics uses the technique of team teaching, and I am therefore indebted to my colleagues for many of the ideas that are now put forward. In particular, I would like to acknowledge the contributions of Dr. Ll. Thomas (man and machines), Dr. P. Foley (vision and the special senses), Dr. J. Daniel (biomechanical fundamentals), Drs. J.R. Brown and P. Jones (lifting), Dr. C.M. Godfrey (artificial limbs), Drs. M. Wall and C. Webster (human behavior), Mr. Roy Ross (personality testing), Drs. M.G. Evans and R.F. Morrison (motivation), Dr. H. Minden (acquisition of skills), Dr. J.F. Flowers (automated teaching devices), Mr. G. Walker (industrial retraining), Mr. E.H. Hipfner (blindness), Mr. A.J. Olbrecht and Mrs. A. Csima (computer design), Dr. J. Ogilvie (simulation techniques) and Metro Traffic Control Staff (automation of traffic control). All of these people deserve much of the credit for that which is good in the sections named, although the responsibility for errors rests strictly with me. Much of the knowledge in the remainder of the book has accumulated through an amorphous process of symbiosis, useful ideas gained from visiting lecturers, textbooks, scientific papers and discussions with students being engulfed, digested and incorporated, sometimes without record of their original source. If this has led to any injustice in either acknowledgment or interpretation of facts, I hope the investigator concerned will accept my apologies and feel free to suggest amendments to any future editions of this work.

At least three other people deserve specific thanks. I refer to the secretary, Jean Karnaus, and the illustrators, John Horwood and Jadwiga Nowoslawska.

For myself, the writing of a sixth book might seem a sign of a perseverant anomaly or an insidious addiction. Nevertheless, I must confess that the preparation of this particular volume has been both an interesting and a rewarding experience. Not only have I consolidated some of my diffuse ideas on human performance and efficiency, but I have also been led to evaluate my own techniques of working—hopefully with a resultant increase in productivity. If readers share this experience and are stimulated to apply an equally critical eye to the organization and design of the society in which they live and work, then the labors of all who have contributed to the appearance of *Men at Work* will be well repaid.

Roy J. Shephard

CONTENTS

MEN AT WORK

CHAPTER 1

THE BIRTH OF ERGONOMICS

TERMS DESCRIBING HUMAN limitations of performance—biomechanics, work physiology and work psychology—need little introduction. On the other hand, *ergonomics* is a recently coined word and as yet graces few dictionaries. It bears a superficial similarity to agronomy (*Agros* = field, *Nomos* = use, *Agronomia* = land use) and economy (*Oikos* = house, *Nemein* = manage, *Oikonomia* = house management), and could reasonably be interpreted as the science of *work use* or *work management*. Certainly, the ideas of economy and economics are carried over into ergonomics—making the most of what one has, a freedom from waste, the management of affairs and resources to the best advantage, an efficient arrangement of parts, and a system for the control of production, distribution and consumption. However, the definition of ergonomics proposed by the inventors of the word was "the customs, habits or laws of work" (*Ergon* = work, *Nomos* = law). A need was felt for a term to describe the scientific study of man and his work, including not only methods of improving efficiency in the usage of both body and tools, but also the interactions between man and machines, and the various anatomical, physiological and psychological implications of man in a working environment.

The concept of ergonomics was born during the second World War, with the emergence of interdisciplinary government research laboratories dedicated to the bettering of human performance. Difficulty was encountered in compartmentalizing many military problems. Personnel were required to fly aircraft at ever greater speeds and altitudes, to penetrate steaming jungles, to survive in arctic waters and to handle efficiently such equipment as anti-aircraft guns, radar, tanks and ships. The questions posed by these requirements seemed inappropriate for solution by traditional departments of physiology, psychology and clinical medicine, and accordingly *ad hoc* teams of specialists were brought together. Not only were these teams very effective in solving military problems, but they also found satisfaction in common interests and a cross-fertilization of ideas. After the conclusion of hostilities, there was a reluctance to disband such successful and intellectually satisfying associations, and the same teams of research workers directed their interests toward the human problems of industry and the home. Man was studied in the working environment, and an attempt was made to fit the job to the worker. Domestic equipment and furniture were critically evaluated relative to the constraints of human anatomy and physiology. Problems of driving on crowded city streets and high-speed expressways were considered in the light of human performance potential. In many instances, the outcome of research was an alteration of design, whether for a tool with complex display and control panels, or a humble item of hardware such as a chair, light switch or pen.

Design is inevitably a compromise between function and appearance, and sometimes the modifications suggested by the ergonomist have violated aesthetic considerations. A value judgment is then required; how important is the proposed change to safety, productivity or comfort? The only aspect of design inherent in the original definition of ergonomics was "the underlying plan or conception," and "the organization of parts in relation to the whole and to its purpose." Nevertheless, the more practical and applied aspects of design are an important part of current ergonomics. Indeed, some ergonomists apply this criterion to weigh the qualifications of aspirant colleagues—"Has your research led to a change of design?"

An Ergonomics Research Society was founded in the United Kingdom in 1949, and now has a substantial international membership, publishing two scientific journals (*Ergonomics* and *Applied Ergonomics*). An International Ergonomics Association has also been formed, with six triennial gatherings to its credit. In North America, related organizations include the Human Factors Societies of the United States and Canada, and the small Brouha Work Physiology symposium. The ergonomics movement has attracted the interest of the Organization for European Economic Cooperation, and a specific conference on "fitting the job to the worker" was held in Zurich in 1959.

University courses in ergonomics have often appealed to a restricted segment of a given discipline. Thus, the topic has seemed more relevant to production engineers on the factory floor than to structural engineers with a penchant for (say) bridge-building. In fact, the structural engineer would do well to consider how men and machines approach and cross a bridge. Nevertheless, many prefer to remain ignorant of human factors, and perhaps for this reason ergonomics is usually taught at a postgraduate rather than an undergraduate level. Training programs are very varied from one institution to another, since the input of students and their subsequent destinations are diverse. In general, there is a need to make good deficiencies of undergraduate preparation; the engineer must master the elements of anatomy and physiology, while the student from the life sciences must increase his background in mathematics and engineering. As in any emerging discipline, there is already a trend to specialization, and the future may well see two streams of student ergonomists—the generalist who is destined for industry, and the specialist who will enter a university teaching or research career.

Ergonomics is perhaps unique in its diversity. University courses range from manufacturing methodology to the ethical, moral and social responsibilities of the design engineer. A well-balanced Institute of Ergonomics draws together personnel with expertise in anatomy, physiology, experimental psychology and sociology, engineering, applied mathematics, business administration, and even architecture and town planning. Le Corbusier, the revolutionary Swiss architect wrote "a house is a machine to live in." He pointed out that the usual starting point of building design was appearance or availability of materials, but that the proper point of departure was the man himself. Human requirements such as light, heat, humidity, air movement and the absence of excessive noise should first be specified, and then a building proposed to meet these requirements. Often, there is a failure to consider the vital factor of load, present or future. The graceful City Hall of Toronto, apparently an architectural masterpiece, does not have suf ficient elevators to cope with the demands of both tourists and office staff. The Toronto International Airport, also a graceful design, is equally a failure in terms of handling current loads (page 23).

Will automation abolish the need for ergonomists? Certainly, their functions may change. Nevertheless, the computer itself generates many human problems (page 340) and it will be increasingly necessary to study human factors in the design, construction and maintenance of robots. Some form of *machine-minding* will be unavoidable, and problems of vigilance are likely to increase rather than decrease. Many services will continue to defy automation, at least at an economic price, and reliance will continue to be placed upon man "a unique control system mass-produced by unskilled labour" (Edholm, 1967). Population growth and an increase in leisure hours will place new demands on transport systems and entertainment industries, while developments of universal health care and an aging community will call for dramatic gains in medical productivity. At the present time, no profession has any guarantee that it can operate in perpetuity, but the ergonomist seems more secure than many who work in traditional trades and professions.

Indeed, the engineer often outruns the ergonomist, leaving a trail of problems to be resolved. Thus, the need of the mountaineer for oxygen was appreciated from early physiological studies in the Andes (1910 to 1912). However, the equipment produced by the engineers was rejected on several human criteria—it was too heavy, it was unreliable, and it had no effect upon the net rate of climbing. Undaunted by this rejection, aviation engineers next developed aircraft that would operate at altitudes of 40,000 to 45,000 feet, only to see a succession of crashes attributable to oxygen lack in the pilots. Airplanes being less expendable than mountaineers, the crashes stimulated much human research in decompression chambers, and led to the development of satisfactory and reliable oxygen equipment, lightweight versions of which played a significant role in the conquest of Everest.

Aircraft designers progressively increased not only the ceiling, but also the speed and maneuverability of their aircraft, thereby outpacing the capacity of man to make the necessary physiological and psychological adjustments. Pilots became unconscious in fast dives and turns, and a further unfortunate succession of crashes led to human research in both centrifuges and aircraft, with the ultimate development of rather effective protective equipment (*anti-g* suits).

Unfortunately, ergonomic problems are not always recognized. There may be no crashes of favorite aircraft to alarm and alert an aging air marshall. Man is highly adaptable, and can develop a surprising tolerance to an awkward posture, an adverse environment and poorly designed tools. Nevertheless, careful study may reveal unnecessary fatigue, symptoms of poor health such as backache and arthritis and a general loss of working efficiency. One simple method of determining whether an ergonomic problem exists is to use a *checklist* developed by the International Ergonomics Association (Van Hattem, 1965). The initial questions explore the main features of a task, including mental and physical load, the impact of environment, the need for training, worker satisfaction and possibilities for automation. Subsequent, more detailed sections examine the physical demands of work space; mental demands in terms of vision, hearing and other senses; the legibility, grouping, positioning, accuracy, speed and conformity of dials; the suitability of controls; muscular and postural demands of the work method; the pattern of information flow; the impact of an adverse environment; and the organization of work in terms of pacing, rest pauses, overtime and shifts. Examples illustrating solutions to these various problems will be found in subsequent chapters.

CHAPTER 2

MAN IN A MACHINE'S WORLD

HISTORICAL PERSPECTIVE

IN EARLY FORMS OF SOCIETY, man and domesticated animals were the prime sources of physical energy. However, with the passing of the centuries, machines have been introduced with the objective of amplifying human ability. Until recently, amplification was conceived in physical terms—lever and pulley systems to enhance muscle power and devices to harness the natural energy of wind, sun and hydrocarbons. Now, the emphasis has changed, as computers extend and replace man's intellectual ability. The role of man has altered from energy source to supercomputer, capable of processing complex information; in the more sophisticated machines, the routine processing of data is carried out without human intervention, and the human task becomes that of designing equipment, supervising its operation, and helping it with difficult decisions such as complex pattern recognition.

Such developments have created a demand for a few highly skilled operators. Machines are continually concentrating more power and more strain upon a smaller and smaller proportion of the total population. This leads to social problems; a specialized labor force such as the airport controllers can cause a serious disruption of the national economy by a temporary withdrawal of their services, and when working the burden of frequent decisions causes them much psychosomatic illness.

It may be wondered how soon decisions will pass beyond the competence of even the most intelligent men. In fact, the human brain (page 84) still far out-classes man-made computers. This is an expression not only of the number of nerve cells in the human brain, but also of their possible range of functional inter-connections. One estimate sets the number of inter-connections at the astonishing figure of 10^{11}.

Nevertheless, modern machines can do surprising things. Many have become virtually self-controlling. This is perhaps most obvious for aircraft and rockets. During World War II, the German V-1 and V-2 missiles showed an uncanny ability to reach their intended targets, many falling within a few yards of London's principal railway stations. Despite the absence of a pilot, in-flight maneuvers conserved fuel, adjusted altitude, avoided obstacles and ensured arrival closely on target under varying weather conditions. More recently, control systems have been devised that permit aircraft such as the British *Trident* to land without human intervention; the designers complain that their only problem is to prevent the pilots from interfering with the automatic controls! Other driverless transport systems have been introduced in London (Victoria subway line) and Montreal (Expo express), while automatic car braking and steering systems have been seriously suggested as a means of reducing the carnage on modern expressways. The proper role of the human operator in an automated system is well-exemplified by the U.S. moon flights. Here, almost all operations except the crucial decision to proceed with landing were relegated to the machine. However, on one flight the unforeseen circumstance of a rough and unsuitable landing surface required temporary human intervention, with a brief period of manual control.

Prior to World War II, the philosopher could distinguish between man and machines on the basis that machines lacked purpose. Time has apparently eroded this distinction. Man is less certain that he has a purpose, while many machines show what seems quite purposive behavior. It is thus necessary to remind ourselves that the machine's behavior patterns were created by a human mind, and can in this way be related to the ultimate purpose of the universe.

At the present time, man is confronted by a wide range of powerful, fast-acting and potentially lethal pieces of equipment. Citizens of our current society are becoming increasingly cynical and afraid of machines that the engineers are creating. Too often, a gulf has developed separating designer from user, and one of the basic tenets of engineering ethics is in danger of being overlooked:

"Engineering is the art of directing the great sources of power in nature to the use and convenience of man"
(Institution of Civil Engineers, 1828).*

Is human engineering inherently evil? In most instances, it can be argued that the *status quo* needs correction. Starvation is still a significant world problem, and one potential answer is to apply the methods of ergonomics to primitive agriculture. Hand tools such as scythes and yokes have undergone some improvement in design through centuries of use, but the real solution to increased productivity is to replace man as the prime source of agricultural energy. Many of the tasks imposed upon the laborer, such as operation of water-pumping treadmills, are degrading even for men of limited intelligence and education. Only the most cynical of sociologists would wish to maintain the *status quo* in such communities. On the other hand, man/machine systems should not be hailed as a panacea for the ills of society. Individual systems should be examined critically to see (1) how their purpose

* Quoted by Thomas, 1967.

can be achieved with a minimum of damage to either operator or machine, and (2) how their design may be improved to facilitate transfer of energy, materials or information across the man/machine interface.

THE CONCEPT OF INFORMATION

Energy and materials require little definition, but it is useful to review the concept of information. Information is something that serves to reduce uncertainty; it may be *selective* or *semantic*. Selective information reduces uncertainty by a known amount, whereas semantic information is essentially qualitative. Consider a well-filled miniskirt. This can be described in simple semantic terms—"quite an eyeful." It can also be reported more scientifically as "a skirt extending from three inches below the waist to twelve inches above the midline of the knee joint, with an upper circumference of 22 inches and a maximum circumference of 36 inches." The selective approach arouses less emotion, but has a more clearly defined impact upon our uncertainty regarding both the skirt and the girl contained therein.

The ergonomist normally deals with selective information, matters that can be reduced to digits and passed between himself and the machine he controls. However, there are occasions when semantic information may serve a useful purpose; thus, if a pilot is inadvertently heading straight towards a large mountain, he may have less respect for the navigator who reports "on course 670, altitude 7000 ft, air speed 532 knots" than the stewardess who supplements her usual offer of "coffee, tea or me" with the warning "you're flying straight into that mountain."

The relative value of different items of selective information can be assessed from the extent to which they reduce uncertainty. The prediction that there will be no eclipse of the sun tomorrow has almost zero information content, since the probability of no eclipse is always close to unity. On the other hand, the prediction that an eclipse *will* occur tomorrow has a high information value.

The usual basis for the coding of selective information is the binary scale of the digital computer (page 331). The unit of information is then the *binary digit* (*bit*). This serves to reduce uncertainty by a factor of 0.5 (for example, yes or no). Let us suppose we wish to identify the position of a pawn on a chessboard; assuming it is "crowned," it can occupy any one of sixty-four possible locations, and for certain identification of its position a minimum of six *bits* of information is required (Fig. 1). Stated in mathematical terms, if there are N alternatives, the information need is $\log_2 N$. For our chessboard, $N = 2^6$ and $\log_2 N = 6$.

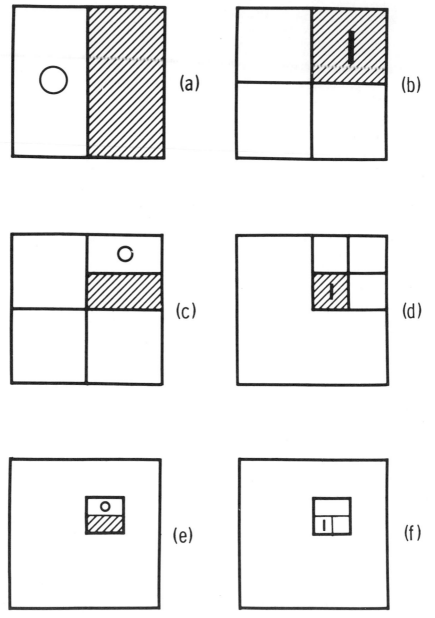

Figure 1. Identification of the position of a pawn on a chessboard, using a binary code. 0 = No, 1 = Yes. Shaded area shows uncertainty.

We may note that the position of the pawn could have been identified with rather less efficiency by checking the contents of individual squares on either a systematic or a random basis. If we had started at the top left-hand square, and moved progressively across the board as though reading a book, the location would have been discovered by using twenty-nine *bits* of information. If we had "played our hunches," then we might have used any number of bits from one to sixty-three, with an average of 31.5 in a long series of games.

Much of medical diagnosis is still a matter of playing our hunches. The experienced clinician develops the necessary intuition to operate with reasonable efficiency, but his examination is inevitably less certain than if a true binary search had been undertaken. One of the aims of teaching models (page 353) is to help in the development of such binary logic. The first step is necessarily the conversion of semantic to selective information. Picturesque terms such as chlorosis, plethora and fever are yielding place to more precise semantic comments (anemia, hypertension, raised body temperature), and these in turn are being replaced by selective and normally distributed digital variables (blood hemoglobin, systemic blood pressure, oral or rectal temperature). Unfortunately, the normal distribution of numerical readings is not well-suited to a binary search, since all values do not have an equal probability of occurrence. Let us suppose that a patient has a hemoglobin reading of 14.5 gm/100 ml of blood (Table I); he falls

TABLE I.
INFORMATION CONTENT OF BLOOD HEMOGLOBIN LEVEL (MALE PATIENT)

Reading	Information Content
15.6 gm/100 ml	Normal—almost zero information content.
14.5 gm/100 ml	Within normal limits—very low information content.
13.0 gm/100 ml	Slightly subnormal—low information content.
11.0 gm/100 ml	Clinical anemia—moderate information content.
5.0 gm/100 ml	Gross anemia—high information content.

a little below the average figure of 15.6 gm/100 ml, but is well within the anticipated range of normality. A second patient has a reading of 13.0 gm/100 ml; he is still not clinically anemic, but the physician may wish to ask a few questions regarding diet, weight loss, hemorrhage and the like. A third patient has a hemoglobin level of 11.0 gm/100 ml; he has an unequivocal anemia, and it will be necessary (1) to undertake a careful search for the cause, and (2) to initiate appropriate treatment with vitamins and iron. A fourth patient with a hemoglobin level of only 5 gm/100 ml may need emergency admission to hospital and blood transfusion.

Another very practical instance of varying information content is provided by the reporting of body temperature. A harassed mother may call her family physician, presenting purely semantic information ("Jim has a fever"). The physician urges her to convert this to a selective reading. If the figure then quoted is 99° F, the information content is so low that the doctor will replace the telephone receiver at an early opportunity. If the temperature is 100° F, he may recommend two days of bed rest. If it is 103° F, he will undoubtedly ask further questions, if 105° F he will pay a house call, and if 106° F he will immediately summon an ambulance.

Where much information must be transmitted at high speed, there is some advantage to an increased choice of selective responses. Thus a morse telegraph, essentially a binary machine (dot versus dash), transmits messages at 30 words per minute. A typist (32 choice machine) can reproduce 60 to 100 words per minute, while an experienced stenotypist (200 choice machine) can develop a speed of 250 words/minute (marginally faster than the average lecture of 180 to 200 words/minute). Although the ultimate speed of the stenotypist is high, the rate of learning is slower than for a standard typewriter, and the incidence of errors if also larger (Fig. 2).

The same type of consideration applies to the choice of alphabet—the 26 Arabic letters transmit less information than the 44 Eskimo symbols or the 3000 Chinese signs. On the other hand, almost everyone can master the Arabic alphabet, whereas many poorly educated Chinese can recognize less than a 1000 signs. Thus, with the Chinese system there is a failure of communication between the well-educated and those who are less literate.

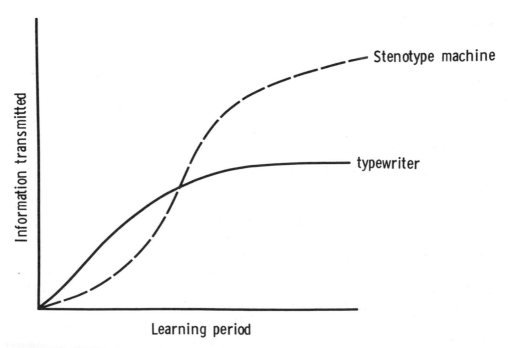

Figure 2. Learning curves for a standard typewriter and for a stenotype machine.

THE CONCEPT OF A SYSTEM

A system may be defined as a group of components connected by suitable pathways for the transmission of energy, material or information and directed to a specific purpose. Possible forms of information flow have been discussed above, although we may emphasize once again that they depend upon effective communication. Energy may take many forms, from the old-fashioned air, water and steam pressure to gas and electricity.

If a system is to function well, its purpose should be clearly defined and understood, not only by the operator, but—if we may think in anthropomorphic terms—by the various mechanical elements involved. This implies that individual components will be well matched to each other, that their several contributions to overall performance will be clearly defined, and that the designer himself will be familiar with the operation of the system, not only in theory, but also under arduous practical conditions.

Much of the research on systems design began under the pressure of wartime needs. One problem that attracted the attention of the British Radar Research Establishment at Malvern was the limited success achieved by anti-aircraft gunners. It transpired that while both the radar sets and the anti-aircraft guns were well designed as individual items of equipment, there was a difficulty in the transfer of information from the radar screen to the gunsights through the human operator; the machines had evolved independently, and no thought had been given to overall function. Application of a systems approach was very successful in meeting the challenge posed to southeastern England by the German V-1 Rockets; 98 percent of these were destroyed before reaching their targets, and at the same time there was a dramatic improvement in the proportion of conventional enemy planes destroyed by anti-aircraft fire.

In this particular example, the purpose of the system (to hit and destroy enemy aircraft and missiles) was fairly obvious. But many complex organizations have existed for many years without any attempt at a definition of their purpose. Consider the average university. Does this exist to provide an ivory tower for specialized research? Should the professors turn to mission-oriented studies that will benefit the community? Or is their prime function that of teaching undergraduates? Is there a responsibility toward part-time students, or should this aspect of education be handled through community colleges? Does the university also have a responsibility to the immediate neighborhood where it is located? (Failure to consider this last point was one significant factor in riots at Columbia University in New York City.)

Similar questions can be asked of the large metropolitan hospital. Is its proper function to treat disease? Should it carry out research, or train an enormous number of medical students and nurses? Does it have a responsibility to improve the health of the immediately surrounding area by storefront clinics and suitable preventive measures? The allocation of resources—space, personnel and money—should be contingent upon the choice of one or more of these purposes. But generally, the questions remain unresolved, and resources are allocated to those who shout loudest. Because of governmental financing policies, large numbers of students may be admitted, even to the detriment of sick patients. Surgeons compete for operating theaters, physicians for bed space, and too few of either are selected with an eye to their teaching skills. Nominal respect is paid to the community (without defining its boundaries!), but little attention may be directed to the necessary corollaries of such a concern (outpatient facilities, home-care programs, and transport for the physically disabled).

Most of us would naturally reject such a gross caricature of *our* hospital, but there is some truth in this type of criticism. Administrators have bowed to immediate pressures, without taking time to examine the purpose of the organization they control, to define its boundaries in physical and operational terms, and to consider the inevitable constraints of time, money and manpower.

Finally, let us take the example of sport. What attitude should the government adopt towards various forms of sport? Is the purpose of such activity to improve the health of the average citizen, to provide mass entertainment, or to enhance national prestige? If the first purpose is accepted, the rules need modification; baseball is more likely to fall within the scope of an average family with a small backyard if played at a slow pace, with a soft ball, and additional participation will result from the absence of a formal pitcher. Current rules are designed to produce a fast, exciting game, suited to a mammoth arena.

DEFINITION OF THE SYSTEM

The first stage in improving the design of a man/machine system is to define it. *Purpose* must be established, not only for the overall system but for individual *elements*. The latter are reviewed by the usual methods of time and motion analysis, particularly careful study of films (page 175). Dr. O. Solandt tells the story—possibly apocryphal—that when an army unit was filmed during assembly of a Bren gun, one man stood motionless throughout the procedure. Some fifty years ago, his purpose would have been to hold the horses' heads, and the loss of purpose had not been noted with mechanization of

the Bren-gun tractor. Within universities, similar vestigial functions
may persist, such as the invigilator at an *open* examination, and the
surgical clerk who pulls on an abdominal retractor, seeing nothing
and contributing nothing to the progress of an operation. It is always
healthy to ask not only what functions a man is performing, but whether
these are essential to the purpose of the system.

The *type* of system should next be clarified. Is it purely human,
purely a machine, or a man/machine system? And what is passing
around the system—energy, products or information? The location of
command must then be identified. In the system formed by a typical
operating room, the surgeon was once in undisputed control, but with
the increase in complexity of modern operations, there is now more
likely to be a varying chain of command. At certain points, the anesthetist will regulate procedures, and at others the surgeon may become
subordinate to a person operating a heart/lung pump or monitoring
physiological pressures. The operation requires a team, rather than
a hierarchical system, and it is vital that this point be appreciated
by all who are involved. Similarly, in the piloting of an atomic submarine, command may swing between several officers, each monitoring
viewing screens. In this type of situation, the overall responsibility
must be carried by a generalist—a person who recognizes the limitations
of his knowledge and is prepared to delegate authority to those whose
expertise meets the demands of the moment.

The *boundaries* of the system should be described at this stage.
In the hospital context, does the system begin and end with admission
and discharge desks? Does it include the immediate neighborhood,
a whole city, a province or a state? Within the hospital, does the operating room system include the anesthetic room, the recovery room, and
provisions for transport back to the wards? What is the relationship
between a single, small system and larger systems? A car and its driver
may provide a reasonable boundary for a single designer, but if his
work is not to be wasted, it is necessary to take an occasional glance
at the larger system of urban highways. Is the size of vehicle appropriate
to anticipated speeds, road surfaces and traffic densities? Are the
exhaust emissions compatible with community standards? Is the
resistance to impact appropriate to local accident experience? The network of highways must be related in turn to overall transportation
policy (page 368). Failure to set sufficiently broad horizons of design
may yield cars that are readily destroyed by other road users, or a
highway system that causes intense urban congestion and even lethal
pollution of a downtown area.

Inevitably, certain *constraints* are imposed upon a system, and these
must be given due consideration in any modifications of design that

are suggested. Thus, in the matter of urban transportation, the taxpayers are prepared to allocate no more than a certain proportion of a city's budget to any solution that may be proposed. A relatively fixed number of people will attempt to enter the core of a city each day, and their tolerance of transportation by-products such as carbon monoxide and ozone is not readily increased.

After analysis is complete, *synthesis* begins. Possible alternative solutions are explored, usually for modification of the existing system, but occasionally for the design of a completely new system. Let us suppose it is planned to provide systematic health care for a provincial population of six million people. Possible approaches may be to enlarge, up-grade and expand an existing network of cottage hospitals, or alternatively to abandon such institutions and concentrate upon the development of a few superbly equipped and very large regional hospitals.

Arguments and premises used in synthesis must be carefully assessed for both validity and prejudice. An electrical engineer may be reluctant to suggest mechanical improvements to a system. A lecturer may be reluctant to consider such alternative methods of education as books and correspondence courses. And in the context of hospital planning, the traditional supporters of the cottage hospital may argue that the quality of highways in their part of the province does not permit transport of casualties over a long distance. If attention is directed simply to highways, their premise may be correct, but the less traditional approach of air evacuation by light aircraft or helicopter has not been given due weight. If the helicopter approach is to be adopted, horizons must be further enlarged. What will be the reaction of workers in nearby offices to the noise of frequent helicopter landings? And what impact will helicopter movements have upon the larger system of air traffic control around the city? All of these considerations must be incorporated into an adequate synthesis of the problem.

Models (page 351) of varying sophistication are helpful in synthesis. It is inevitably cheaper to discover an absurd solution in a simple model than in real life. Cost factors should be incorporated into the model, since the factor of absurdity is often financial. Aircraft have been developed at great trouble and expense, only to discover that their operating costs cannot be met by any conceivable fare schedule.

Where possible, the model should be simple. Much can be discovered merely by representing a system in cardboard. A greedy developer may wish to squeeze ten townhouses onto a plot of land originally occupied by two large homes, and he can profitably spend long hours trying to back a cardboard car into a model of a narrow and almost inaccessible garage. A hospital designer may play similar games, maneuvering cardboard beds, x-ray machines and other bulky items in and out of a model elevator. Sometimes, it is necessary to study how the human operator will react in the space left vacant by equipment. A mock-up, full-scale model is then created before embarking upon the expensive task of building permanent structures.

Block diagrams and schematics can be helpful. A very simple representation of the respiratory system is given in Figure 3. The controlled element is here the chest cavity, and the controller is the respiratory center. Information is passed from the chest to the controller via the sensory nerves, and appropriate information is then transmitted to the motor nerves, modifying the extent of chest movements. In designing automatically controlled resuscitators, a series of more sophisticated block diagrams and the corresponding electrical circuits are matched against the normal performance of the respiratory system (page 100).

In some instances, the system variables can be defined with sufficient precision to develop mathematical models (page 351). Linear programming techniques apply multiple regression equations to such problems as the optimizing of production and the matching of output and demand; for instance, a restaurant chain may wish information on the best, cheapest and most flexible items to incorporate in its menu. Many problems involve rather infrequent but randomly occurring events. Recourse may then be made to Monte Carlo methods (page 347), using some form of random number generator. Thus, when studying the performance of a modern highway network—and possible implications of its extension—the model must be capable of testing the effects of relatively rare contingencies, such as a tractor-trailer jack-knifing across three lanes of an expressway, or a sudden ice storm developing when the roads are too congested to permit the free access of salt trucks.

In addition to avoiding costly mistakes, modeling has the advantage that the handling of hazardous situations can be studied in a detached manner. Thus, a surgical team can be admitted to an operating theater and told to assume that their patient's heart has developed ventricular fibrillation. The various procedures used in resuscitation can be timed

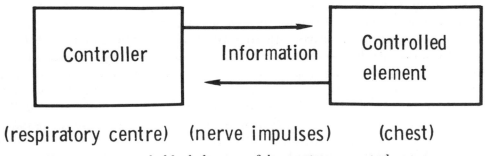

Figure 3. A simple block diagram of the respiratory control system.

relative to the cerebral tolerance of oxygen lack (four minutes) and then the same routine can be repeated with the lights suddenly extinguished at an early stage in resuscitation.

To be of real value, any model must be complete. This point was emphasized many times during the New York power failure of the late 1960's. During this emergency, some hospitals discovered to their chagrin that standby generators in the basement failed through flooding of the engine room or lack of lubrication of bearings, both pumping and lubrication being functions of a normal mains supply of electricity.

The objective of the modeler is to test all likely outcomes of the favored solution to his problem. Let us suppose that establishment of an acute coronary care unit has been suggested at a specific hospital. What will be the impact of this facility upon regional coronary mortality? Careful modeling may show it is of little value, particularly if the majority of fatal episodes occur during time occupied by:

1. the decision to call a physician,
2. the contacting of the physician,
3. the arrival of an ambulance,
4. transport to the hospital, and
5. admission through the outpatient department.

More effective approaches might be to publicize the early symptoms of myocardial infarction and/or to equip ambulance drivers with the knowledge and apparatus needed for cardiac resuscitation. Furthermore, the model is incomplete unless it tests the impact of the intensive care unit upon the general operations of the hospital. Are lives being lost in other wards because these are being starved of good nurses and necessary equipment?

A water management model may examine the consequences of removing phosphate from detergents, but will other soap additives have a greater detrimental effect upon ecology? And widening the boundaries of the model, what impact will the proposed changes have upon the sterility of clothing emerging from a communal laundry and the function of dishwashing machines used in the city restaurants?

Performance criteria are necessary for an objective evaluation of the various possible alternatives that are to be explored. In the case of the acute coronary care unit discussed above, the prime criterion would undoubtedly be the change in mortality from myocardial infarction, but an appropriate analysis would also include indices of useful recovery (such as the proportion of patients returning to normal employment) and of efficiency in other hospital wards (such as the number of infected wounds following elective surgery, and specific death rates in the general medical wards).

At this point, it is desirable to test the various hypotheses that have emerged by means of a *pilot run*. The system used for this purpose should be sufficiently flexible that it can be redesigned in the light of experience gained from the pilot operation. Unfortunately, many organizations carry out pilot runs without establishing adequate performance criteria. It is then almost impossible to learn from the trial. Teachers are guilty in this regard. Enormous changes in the patterns of school and university education have been made over the past ten years, but because performance criteria are lacking, there is no way to assess whether these costly changes have been advantageous or not.

A *cost/benefit analysis* should next be undertaken. Although the concept is unpopular with the general public, there are many areas where a price must be set upon human life and health. A proposed solution to a problem may be less than optimum in terms of the life expectancy of the individual, and yet more expedient in relationship to the overall goals of the community. Has the vast expenditure of research funds upon heart transplant operations been justified in terms of the rather small number of patients who have had their lives extended by one or two artificial years? Or would the same money have been better directed to some related area of research such as the prevention of heart attacks? Can one equate the vast economic burden of cardiovascular disease with what may be the even greater cost of the physical facilities and motivational drive needed for a community-wide program of preventive exercise?

On a more prosaic scale, one may equally apply cost/benefit criteria to urban expressway construction. Does a ten-minute reduction in homeward journey time justify a vast municipal expenditure upon new roads? In making such a judgment, has account been taken of the limited time for which the expressway will remain adequate? And has allowance been made for the social costs of disrupting established neighborhoods?

In the past, decisions on such questions have often been unduly influenced by groups with an inherent bias. Allocation of research funds has been at the discretion of medical specialists with a liking for exotic projects. Apportionment of community resources for physical recreation has been a battleground between ratepayers anxious to preserve property values and entrepreneurs anxious to build tall apartment blocks on any remaining open space, with neither side particularly interested in recreation. Expressway construction has been debated by potentially displaced homeowners and commercial interests anxious to see a city *grow;* often, the city roads commissioner—again by no means a disinterested party—has played an important role in resolving this last category of disputes.

There seems a need in today's complex problems for a generalist

who can bring knowledge from many scientific disciplines to bear
upon a decision and yet reach an independent judgment. Such a role
is increasingly being filled by the *systems engineer.* One group of
consultants in Toronto has reviewed and reported upon such disparate
problems as the organization of a metropolitan traffic computer system
(page 364), a grouping and reorganization of health science faculties
within the University of Toronto, and a consolidation and regrouping
of some 150 poorly spaced United Churches within the Toronto
presbyteries. The value of an impartial and objective assessment of
competing traffic systems, rival faculty *empires,* and churches that
would close "over the session clerk's dead body" can readily be
appreciated.

THE MAN/MACHINE INTERFACE

Detailed study of a system may reveal that the flow of information
is incomplete, or proceeding with some difficulty. To return to our
simple example of Figure 3, let us suppose that the normal basis of
chest movement is temporarily abolished by administration of the mus-

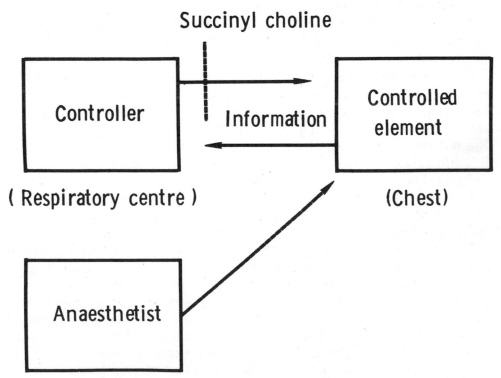

Figure 4. Simple block diagram illustrating the control of respiration during the
administration of succinyl choline.

cle relaxant succinyl choline. The motive force is now supplied by
an anesthetist, either directly (by manual compression of an anesthetic
bag), or indirectly (by the use of a respiratory pump). The situation
is as shown in Figure 4. The normal flow of information from the
controller (respiratory center) to the controlled element (chest) is interrupted. The anesthetist has little problem in transmitting information
(airflow) to the chest, but lacks any ready feedback from the controlled
element. Normally, the respiratory centers synthesize a variety of information such as arterial pO_2, pCO_2 and pH (page 101). It is technically
difficult for the human mind to make a suitably weighted combination
of this information, and accordingly the anesthetist may make an arbitrary decision as to his main purpose. Is he, for instance, charged with
maintenance of normal body oxygenation? If so, he will be guided
by such clinical signs as the color of his patient, and his task will
be made easier by the provision of selective information (continuous
monitoring of arterial oxygen pressure).

The diagram of Figure 4 may be extended to provide a specific
example of a *man/machine interface* (Fig. 5). Information regarding
blood pO_2 is detected by a pair of sensors (the eyes); in other systems,
a variety of forms of information (light, heat or sound) may be detected
by the several specialized receptors within the human body (page 116).

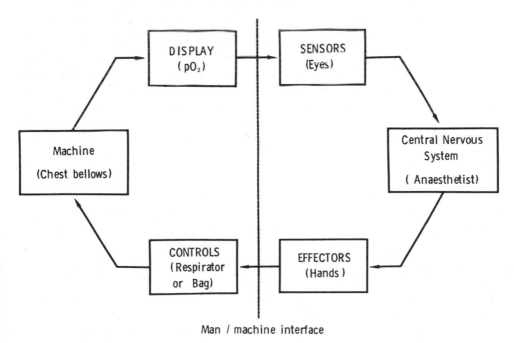

Man / machine interface

Figure 5. The man/machine interface, as exemplified by controlled respiration under
the care of an anesthetist.

The sensors translate the information into nerve currents, mainly on a basis of pulse frequency modulation, and data then reaches the *central nervous system*, where memory (page 84) and acquired skills (page 240) are used to weigh the input against (a) the intended setting, and (b) other relevant information. At a certain stage in an operation, the anesthetist may decide that greater relaxation and thus deeper anesthesia is needed. In effect, his *purpose* has been modified, and he may be prepared to tolerate a temporary departure of oxygen pressures from their optimum values. On the basis of such decisions, nerve pulses pass to the *effector organs*—in our particular example, the voluntary muscles controlling hand movements. The hands are then applied to the *controls* of the machine, possibly a simple anesthetic bag, possibly a complex arrangement of knobs varying inflation pressures, volumes and cycle lengths. The *machine* in our instance is the paralyzed chest bellows, and information is fed to this in the form of a gas volume or a gas pressure. Resultant changes in the state of the machine are passed to a *display*. Often, this may take the form of dials or an oscilloscope screen, although for the specific problem of reporting blood oxygen tension, a neon digital display is usually adopted.

The objective of the designer is to facilitate the transfer of information across the man/machine interface, in our example between the anesthetist and the patient. Detailed review may suggest that both blood gas homeostasis and adequate muscular relaxation are worthwhile objectives, but that other variables such as blood sugar are of secondary importance and can be considered as outside the boundaries of the system. The ideal approach might be to develop an analogue display that presented the anesthetist with a correctly weighted synthesis of selected variables such as blood gas composition, blood pressure, cardiac frequency and muscular relaxation. However, this would inevitably add to the complexity of the system, and before embarking upon such an approach, a careful cost/benefit analysis would be needed.

Irrespective of the complexity of the functions incorporated into the display, its outward characteristics must be carefully matched to the anticipated characteristics of the user. Thus, a system that relies upon color coding (as do some parts of an anesthetic machine) immediately leads to the possibility of error in a significant proportion of the male population (page 164). Similarly, the mechanical characteristics of the controls must be chosen with the user in mind. Are the taps of an anesthetic machine too stiff to respond to the female wrist? Are the necessary adjustments too fine to be carried out by an aging and somewhat tremulous staff member? And how does the overall system respond when faced by a critical incident such as an unexpected fall of blood pressure, or the need to maintain homeostasis in a patient with advanced lung disease?

PREDICTION OF FUTURE LOAD

An important function of the systems engineer is to predict the *future load* that will fall upon a system. This is sometimes difficult. One pertinent local example is the Toronto International Airport. Passenger arrival and departure facilities are arranged on two separate levels at the center of a circular tower. Some seventy departure gates are arranged about the periphery, and eight floors of indoor parking are provided immediately above the main building. It seems a perfect ergonomic arrangement, with minimal walking required of passengers burdened by heavy suitcases. Unfortunately, the rate of growth of air traffic in southern Ontario was sadly underestimated, and within five years of construction, the building was carrying twice its intended passenger load. Approaches to arrival and departure lounges were blocked by long lines of cars trying to enter the parking tower, elevators within the tower were insufficient, and equally long lines of cars fought to escape past the two parking payment booths. Within the terminal proper, hordes of relatives (presumably not included in the original calculations) besieged the limited restaurant facilities. Several stopgap remedies became urgently necessary. Within the parking tower, one floor was sacrificed to provide additional exit booths, and further parking areas were designated about a half mile distant from the departure lounge. An equally distant warehouse was arranged as a terminal for charter aircraft—the antithesis of design, noisy and overheated, with no taxi stand and an ancient bus to convey passengers to one of the original seventy departure gates. At the same time, construction of a more permanent second terminal was begun, this will require a redesign of the system of approach roads, and passengers changing planes will be confronted with a lengthy underground passage (although possibly helped on their journey by a moving sidewalk).

Similar problems of load prediction have afflicted other systems. Already, the Metro Toronto traffic computer (page 367) has reached the point where it lacks capacity to control further signal lights. Already, the University health sciences division is under pressure to accept a 50 percent increase in medical student enrollment, and to change from a two- to a three-semester operation, with inevitable additional requirements of staff and space. Already, the survey of redundant churches is outdated by both a decline in attendance and a proposed amalgamation of United and Anglican denominations.

Nevertheless, it would be wrong to conclude that the planning efforts of systems engineers have all been abject failures. Sometimes the techniques of *program evaluation* and *operations research* have yielded very valuable practical conclusions. Disastrous overloading of specific stages in an operation have been avoided by the review of *critical paths* and the development of *queuing theory*.

Queuing theory, like other facets of ergonomics, originated under
the stress of war, when it became important to ensure that several
hundred bomber aircraft and the necessary fighter escort arrived from
scattered airfields in an appropriately spaced sequence. A similar
approach was needed in research and development of new weapons.
Often, the period of R&D exceeded the period of potential use,
and it was thus vital to avoid unnecessary delays. The entire operation was charted, from initial laboratory research to production of the
final prototype, and critical stages were defined relative to a permitted
time scale of perhaps six years. More recently, the same type of approach has permitted the prediction of the date of manned space mis-
sions with remarkable accuracy. Increasing civilian application of queuing theory is seen in such problems as the control of cars entering a freeway (page 369), customers leaving a supermarket, and complex multi-
stage surgical procedures; thus, in a renal transplant operation, it is essential that both donor and recipient are prepared for the organ exchange at coincident times.

THE NON-SYSTEM

At almost every point in life—in the home, in industry and in the
armed services—one encounters the *non-system*, a machine which
incorporates a myriad of operational problems, with obvious failure
to consider interactions between the controller and the equipment
he must use.

Let us look at the average kitchen range. The switch is hidden behind
a number of dangerously hot saucepans, and is of the push-button
variety, giving no indication whether it is in the *on* or *off* position.
A sparkling display of lights suggests that certain heating rings are
in operation, but the lights are so far removed from the rings that
it is difficult to tell to which elements they refer. The rings heat rather
slowly, and it is thus difficult to correct a wrong decision by direct
observation. It is hardly surprising that occasionally saucepans are overheated or melted!

Another simple example of poor display is the standard aircraft
altimeter. Here, dual dials indicate 1000 feet and 10,000 feet increments
of altitude. From a mechanical viewpoint, there is rarely any problem.
Nevertheless, the display sometimes fails in its prime purpose of passing information on altitude to the controller of the aircraft. Reading
errors of 1000 feet are possible, and have apparently been responsible
for several crashes. Similar difficulties may arise in the reading of
both laboratory and domestic gas meters. It is all too common for a
technician to report the ventilatory cost of a first bicycle ergometer

loading (300 kg-m/min) as 31 liters per minute, and that of a heavier loading (450 kg-m/min) as 19 liters per minute. The careless operator has read the dials incorrectly, adding 10 liters to the first volume, and thus inhibiting the second volume measurement of 10 liters.

A more complex example of the non-system is found in the navigation room of older military vessels. During World War II, the speed of destroyers increased to more than 40 knots, yet navigation equipment still resembled that of a slowly moving tramp steamer. Steering was achieved by turning a large and very heavy wheel. The helmsman found it almost impossible to apply the necessary torque while frantically endeavoring to keep his balance on a slippery steel deck pitching in an Atlantic gale. The engine room was at some distance from the navigation deck, and transmission of information between the helmsman and engine-room staff was dependent on telegraph and telephone. Neither were particularly effective methods of communication at a noise level of around 110 dB (page 120). The lighting of the display was so arranged that irrelevant information (the manufacturer's name) was well illuminated, but the important facts (the scale readings) were in deep shadow. Finally, the performance of all crew members was adversely affected by the overall environmental conditions (noise, motion sickness, and exhausting heat of up to 120° F). It need hardly be stated that the designer had not spent much time afloat, and had totally failed to consider the destroyer and its crew as an overall system.

IMPROVEMENTS IN DISPLAY

Displays may be either qualitative or quantitative in type. Where a *qualitative* display meets the purpose of a system, it should be used in preference to quantitative information. Simple examples of qualitative indicators are the oil pressure and alternator warning lights fitted to many cars. Their purpose is satisfied if a safe minimum of oil pressure is maintained and the battery is being charged. However, some vehicle owners like more information—is oil pressure as well maintained as when the car was new? Does the battery require vigorous charging for long periods? Some impression can be formed from the brightness of the warning lights, but reliance upon intensity is unsatisfactory; the eye is designed as a comparator rather than as an absolute light meter.

The petrol (gas) gauge provides an example of a *semiquantitative* display. Older gauges were calibrated in gallons, but it is now more usual to provide only one-fourth, one-half, three-fourths and full scale markings. There is again no purpose in displaying more information than the minimum to which an operator must respond. On a long journey, he can gauge the distance corresponding to a quarter tank of gasoline, so that when the quarter mark is reached he knows the safe driving distance before refueling is necessary. More detailed graduation of the scale is distracting, and in most vehicles is unwarranted by the accuracy of the dipstick that is coupled to the gauge.

The typical traffic signal is an example of a *multiple qualitative* display. Here, the vehicle operator must respond to a variety of cues provided by sequence, combination, size, shape and intermittency of signals.

A qualitative signal has the disadvantage that it is rather easily confused with background information or noise. Thus a red traffic light on Broadway merges into the background of competing neon signs. An intermittent signal such as a red traffic light may also appear for the first time while the eye of a driver is distracted by reading perhaps thirteen or fourteen other directions, warnings and route indicators that clutter the average street corner.

Quantitative displays may provide a direct digital read-out, but more commonly an analogue is used—a moving pointer, a moving scale, a moving bar, or even a representation of a vehicle in motion.

Digital display is being used increasingly on scientific apparatus, sometimes with linkage to a paper-tape or computer card printout. It has the important advantage that an objective reading is obtained, independent of any observer bias. However, unless the equipment is very stable, the numbers may show an irritating oscillation. Further, a digital readout gives a less clear idea of the direction and rate of change of a variable than most analogue displays. Limitations of the human memory (page 84) restrict the total display to six digits, preferably arranged in two groups of three. It may be useful to incorporate a small *repeater* scale, showing the previous reading.

The *moving pointer* is perhaps the most common form of analogue. It has been used by the clockmaker for several centuries, but it would be rash to conclude from this that the features of a clock face provide a general model for analogue display; clocks are read correctly only because every citizen has had training in their use from early childhood. Errors that arise from a dual system of pointers have already been noted (page 9). The use of the entire 360° of a clock face allows a maximum presentation of information, but there is a danger of confusing readings at 0° and 360°; it is thus more usual to restrict the scale to the upper half or at most three fourths of the face. Details of the scale are governed by the information to be conveyed, but five to seven primary divisions with two to five subdivisions are the easiest to read. Scientists will sometimes attempt to interpret a scale to a tenth of a division, using a magnifying glass, but this is time consuming and unrealistic for most life situations—often, the best which can be achieved is an estimate of half a scale division. Too frequent secondary divisions lead to counting errors. If a substantial series of primary numbers must be displayed, arrangement around the outside of the dial avoids a cluttered effect; there is the added advantage that the numbers are not obscured by the pointer. On the other hand, an inside scale is inevitably shorter.

The *moving scale* is perhaps most familiar from its use on domestic bathroom scales. The choice is here dictated by the long series of numbers to be presented (commonly 0 to 280 pounds, with ten pound primary divisions and one-pound secondary divisions). To facilitate reading, a small magnifying lens is often built into the viewing window. A second useful application of the moving scale is for a logarithmic or other nonlinear display; there is less danger of confusion and interpolation is easier if the portion of the scale presented to view has approximately equal divisions.

The most universal *moving bar* indicator is the standard mercury or alcohol thermometer. Some car speedometers are arranged in this form, but no real advantage is gained thereby.

Certain *general considerations* apply to all of the analogue displays thus far considered. The only information shown in addition to the basic numerical scale should be a simple and unequivocal indication of function (e.g. mph). The need for removal of the manufacturer's name, and appropriate illumination of the scale have already been mentioned. If a glass front is necessary, problems of reflection and parallax should be overcome. The colors chosen for the gauge should provide maximum contrast without glare. Unless they are to be read mainly at night (page 158), a combination of mat black and white is preferred. The digits should be arranged in an upright position (a point not observed on some clock and speedometer faces). The size of the numerals and scale marks should be governed by the same considerations as proposed for type size (page 153). Edholm (1967) suggests that for a reading distance of one foot, the major marks should be 0.035 inches long and 0.005 inches wide, with 0.021 and 0.004 inches as the dimensions of subsidiary marks; the latter approach the threshold of angular discrimination (1 minute). Due attention must be paid to anticipated stereotypes, particularly a progression of scale readings from left to right and from below upwards. Lastly, adequate thought must be given to the overall layout of the display console, grouping related functions, and placing those that are most vital in the center of vision. Color coding and the use of other senses (hearing and touch, page 117) may help where much information has to be transferred across the man/machine interface.

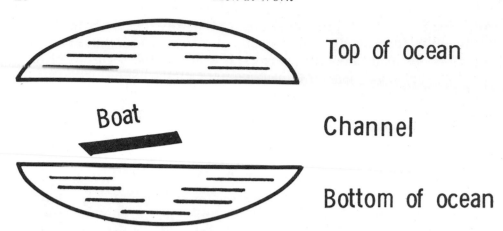

Figure 6. Schematic representation of an analogue display in the control room of an atomic submarine.

Complex analogue displays are used in atomic submarines (Fig. 6) and high performance aircraft. Information not immediately essential to the main purpose of the system is suppressed, although available if specifically requested. The maximum synthesis of information is of particular importance to the operator of the modern airplane. In a slow-moving biplane, a philosophy of placing dials wherever there happened to be a space in the display panel was not too serious a criticism of design, but the consequences of such an approach are disastrous when the speed of convergence of two aircraft (perhaps 3000 knots) approaches the capacity of the pilots to react.

Even where speed is not mandatory, the efficiency of human effort may be substantially improved by an alteration of display. Thus, the output of a biochemical auto-analyzer can be shown as a simple tabulation of variables such as blood sugar and hemoglobin level. However, with a little further automation, the machine can be programmed to print a normal distribution curve of blood sugar values appropriate to the patient's age and sex, and to indicate the position of the observed blood sugar reading upon this distribution curve. Selective information on the degree of abnormal function is then immediately available.

Analogue displays and computer printouts of data naturally have a high capital cost, and may also require expensive servicing. Their use must thus be justified by careful cost/benefit analysis. However, the principle of *screening out* redundant information can often be incorporated into a display at little or no extra cost. The oil pressure and alternator signals of a car have already been mentioned; a third example on the car dashboard is the engine temperature indicator. In most modern cars, this does not show the absolute temperature; a major part of the dial is allocated to the normal operating range. A driver can thus check very quickly the temperature of his machine relative to the anticipated norm.

IMPROVEMENTS IN CONTROL SYSTEMS

The *standardization* of both display panels and control levers could make a substantial contribution to accident prevention. The author has clear memories of renting an unfamiliar car at London airport, and driving for some forty miles before he discovered the location of the dip switch to his headlights. Others frequently confuse the windshield wiper and light switches, and may move the gear lever in the wrong direction because they have previously driven a different make of car. Negative transfer (page 248) is occurring. Annoyance may be caused to the motorist or to other drivers, and a sudden engagement of reverse gear could damage the engine. The manufacturers no doubt believe that a change in the shape and arrangement of display and control panels create the illusion of a *new* model, but in fact, they merely prevent the motorist from checking his visual impressions through the sensations of touch and muscle proprioception (page 96).

The human operator has certain *stereotypes* or preconceived notions as to how devices should operate. Some are almost universal, such as a vertical bar display reading from bottom to top. Others are regional or national. A light switch is moved upwards in the United States, and it is disconcerting to find the reverse in Europe. The letter C on a British tap signifies cold water, but a severe scalding can result from acceptance of this stereotype in France (C—chaud) or southern Europe (C—calor). Some stereotypes are peculiar to individual manufacturers. Thus, a motorist may break the wheel bolts on certain types of car before he discovers that the left-hand wheels are fitted with a left-handed thread.

The examples quoted thus far usually cause no more than annoyance. However, an error with a power switch can have fatal consequences. At one time, serious errors in the operation of controls were also frequent in aircraft, particularly under the stress of an emergency. Mysterious crashes would be traced to the operation of throttle instead of pitch controls, and to the lowering of the undercarriage prior to the braking flaps, instead of the reverse sequence. Fortunately, there is now much international agreement on the design of aircraft and ground-to-air control systems. All aircraft have their controls in similar positions, and a common touch and shape coding ensures that a particular control always has the same type of knob.

Another area where there has been progress is in the standardization of anesthetic equipment. The penalties for a wrong cylinder connection include both explosion and death; fortunately, errors have now been largely eliminated by a uniform color coding, and noninterchangeable cylinder regulators.

Many other forms of equipment still clamor for standardization. In deciding upon a universal stereotype, attention should be paid to the principle of *compatibility*. Within a control panel, all *on* switches should move in the same direction. If a control lever produces movement in a display or vehicle, the operator will take the necessary action more quickly and with less conscious intervention (page 96) if the control movement and the resultant response occur in the same direction. The amateur yachtsman adapts readily to the contrary movement of the tiller in normal sailing, but the conflict between control and response can contribute to errors in an emergency.

The impact of body form upon the design of controls is discussed further on page 305. In general, simple levers are provided where rapid movements are required, although considerable force can be applied to the tiller of a large yacht. Rotary controls, such as a steering wheel, are helpful in making accurate movements. Powerful movements (such as the unassisted braking of a car) are best carried out by the legs, while fine movements (such as the tuning of a radio receiver) are best carried out by the fingers.

AUTOMATED CONTROL SYSTEMS

The possibility of human error is much reduced by the use of automated control systems; further, the range of human endeavor is extended, and operations such as car driving can be brought within the scope of seriously handicapped patients (page 291).

Automatic controls now have a fairly long industrial history. One of the earliest automated machines was the windmill. Here, a small auxiliary set of blades turned the sails into the wind. As wind speed increased, the angle of the sails was adjusted to maintain a constant speed of rotation, and further controls varied the speed of the millstones to match the supply of grain.

When steam engines began to supplant wind and water power, a similar series of regulators were needed to replace the routine functions of the engine man. Watt's ball governor (Fig. 7) was an important invention in this area. Previously, careful throttle adjustment had been necessary to maintain speed as the load increased (for instance, in sawmill operations). However, Watt's governor sensed the machine output from the speed of rotation of two heavy balls. Due to their inertia, an increase of engine speed caused them to fly outwards and downwards against the tension exerted by a heavy spring. This move-

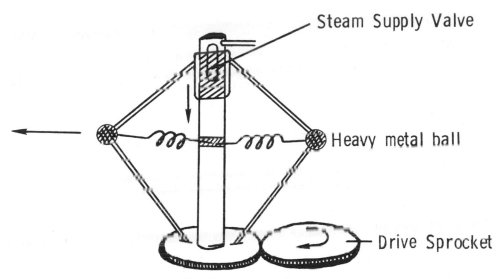

Figure 7. Ball-valve governor for a steam engine.

ment in turn displaced a sleeve valve, varying the steam supply to the engine.

The behavior of most servomechanisms can be described by appropriate differential equations. Maxwell first developed such mathematical statements for the ball-valve governor as early as 1860. Other gifted engineers of this period were Brunel and Trevithick. Brunel is perhaps best known for his railway bridges and tunnels in southwestern England (Clark, 1969), but in the present context we may note that in the year 1868 he designed the *Great Eastern*. This was an advanced form of double-hulled iron steamship equipped with a servomechanism to adjust the helm to water speed. Similar devices are still sometimes used on model sailing yachts—as the wind force increases, a larger tiller deflection is produced. Trevithick was the son of a tin-mine manager in Cornwall, and a pioneer builder of high pressure steam engines. As a young man, much of his time was spent in ensuring that the beam pumps kept the mines free of water. The pistons were driven in one direction by steam pressure, and lifted water during their reverse travel. Accordingly, a man spent his day operating steam and water valves with each piston cycle until Trevithick appreciated that simple levers could serve the same function.

Physiologists soon began to apply these principles to the interpretation of body function (page 91). Hering discussed the *Selbsteuerung* of the respiratory center (1868) and in the same decade Claude Bernard (1878) wrote upon the *Constancy of the Milieu Intérieur*. More recently, Cannon (1929) began to talk of the *Wisdom of the Body,* and a knowledge of servomechanisms has found increasing application in mechanical and electrical models of biological systems (page 103).

All of the authors cited thus far have concerned themselves with specific loop systems. Wiener (1948) made a significant advance by drawing attention to the general principle of the closed loop, thus opening the way to the design of such inventions as computers, high-fidelity record players, aircraft, missiles and even Keynesian models of the economic system.

THE CLOSED-LOOP SYSTEM

The essence of automation is to take the man out of a man/machine system. Let us suppose that a man has the misfortune to live in an area of marginal television reception. He wishes to watch a program emanating from station Z, and knows by experience he can get some semblance of a picture if his outside antenna is rotated by about 90° from the position previously adopted to watch station X. He decides to *close the loop* in the cheapest way possible, by sending his son into the garden with a *command* to rotate the antenna through 90°. The boy shouts when he thinks this has been accomplished, and the father now assumes the role of *comparator*, looking at both picture quality and the antenna position as seen through the window. He decides a further 15° of rotation are needed, and calls out of the door to this effect. So, by successive approximations, the antenna is finally adjusted with acceptable accuracy (perhaps to ± 1° of the intended deflection):

Command	Operation	Result	Error
+90°	+75°	+ 75°	− 15°
+15°	+10°	+ 85°	− 5°
+ 5°	+15°	+100°	+10°
− 10°	− 11°	+ 89°	− 1° (accepted).

Many operations still rely upon man to provide the final feedback link—for instance the sighting of artillery, the operation of a winch, the taxiing of an aircraft, and even the teaching of most university students.

Control is normally dependent upon *negative feedback* (page 91). Subtraction of the feedback from the command signal occurs in a *comparator;* for some purposes, the cheapest comparator is still provided by the human brain. The accurate measurement of time is important for the calculation of longitude, and Harrison (1760) devoted much energy to the design of a ship's *chronometer* that would keep the correct time under varying conditions of environmental temperature.

More recently, clocks have been devised with a feedback from the
domestic electrical frequency. However, now that time signals are
transmitted by radio and television, it is quite possible to navigate
a ship with a ten-dollar wristwatch, the wearer *closing the loop* very
simply by correcting his instrument against standard time.

Some comparators are essentially *regulators*, aiming the system
towards a predetermined target. Simple examples are the burner control
upon a domestic furnace, and the thermoregulator within the human
body. Other comparators are *servosystems*, seeking to match output
to a constantly moving target. Thus, a radar antenna is automatically
turned towards an object such as an enemy aircraft; feedback is proportional to signal brightness, and the antenna continues to rotate until
the signal is maximal. The system is engaged in *hill-climbing* (Fig.
8), seeking to maximize the value of y with respect to x. Such a machine
works very well if there is but a single *hill*. Unfortunately, in practice
there are often a large number of *hills* and *troughs* arranged about
a multidimensional space. In the example of Figure 8, it is possible
to prevent the radar equipment from wasting time upon the false signal
B if the human operator initially points the antenna in the approximate
direction of the prime target (A).

The need for a man to make such preliminary adjustments can be
avoided if functions of memory and comparison are added to the loop.
The signal brightness is noted at various rotations, several bright
points are compared with one another, and the antenna is directed
to re-explore the most promising. This type of search procedure is
currently used in routing transcontinental telephone conversations.
However, if the system is very complex, the time needed for an exhaus-

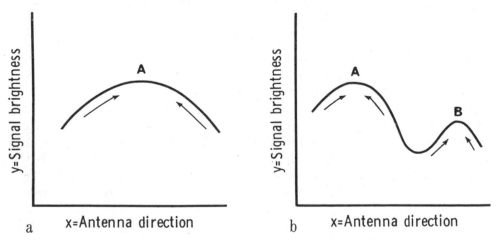

Figure 8. Relationship between signal brightness and antenna direction (a) single
hill, (b) two *hills*.

tive search of all functions can be prohibitive, and it is still necessary to ask a human technician to choose the initial settings. This situation is well exemplified by the role of the industrial chemist in initiating a radical change of production at an automated chemical plant.

The rate of *hill-climbing* can be speeded by making the equipment responsive to the first derivative of movement (for instance, antenna velocity) or even the second derivative (antenna acceleration).

The searching of a *hill* naturally presupposes agreement upon the purpose of regulation and upon methods of measuring optimal function. In the complex systems of the human body, the purpose of the system may change from time to time, leading to a varying command signal. While the domestic furnace continues to seek a hallway temperature of 75° F unless it is adjusted by some external agency, the body may alter its temperature command from 98.6° to 103° or even 104° F in response to changes occurring elsewhere within the system—intense physical activity in the legs, or microbial invasion of the chest. Traditional therapy for fever (antipyretics) merely makes an external resetting of the thermoregulator, and takes no account of the prime cause of the abnormal command signal. In fact, the altered command may be playing a useful role in counteracting infection. Many similar examples of variable command are found within the body. Thus, tendon reflexes normally protect the muscles and tendons from excessive tension, but the protective reflex concerned may be over-ridden to prevent the body from falling. Hypnotists are sometimes able to counteract normal inhibitory feedback and induce surprising increments of strength. The same phenomenon may occur with the arousal (page 215) of a sudden emergency. A mother who finds her child trapped under a car may momentarily develop the strength necessary to lift the vehicle and release the infant.

Even in fully automated equipment, it is necessary to incorporate provision for altered command and manual override. The latter may become necessary on account of machine failure, or some circumstance unforeseen when designing the system (as in the moon landings).

Complete closure of the loop may reduce the load upon the human operator to the point where vigilance is impaired (page 217). Thus, long stretches of the Pennsylvania Turnpike are completely straight, and at periods of low traffic density, there is a real danger that a driver may fall asleep. Expressways of more recent design, such as the New York Throughway, deliberately incorporate a few bends to counteract this tendency.

The *transfer function* for any given loop is given by the ratio of output e_o to input e_i (Fig. 9).

$$\text{Transfer function} = \frac{e_o}{e_i}$$

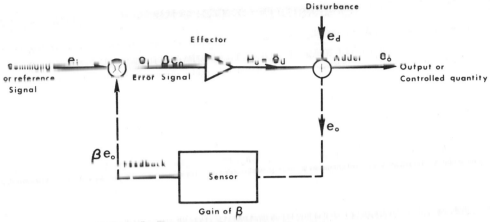

Figure 9. Example of a simple, closed-loop system.

The difference between this transfer function and the *gain* G of the effector amplifier deserves emphasis. The latter is equal to

$$G = \frac{\text{Effector output}}{\text{Effector input}} = \frac{e_o - e_d}{e_i - \beta e_o}$$

where e_d is the disturbing voltage, and β is the gain of the feedback loop. Notice that feedback has the effect of lowering the gain of the system; for this reason, the concept was initially disliked by radio engineers striving to amplify distressingly faint signals.

Solving for the output (e_o), we find

$$e_o = e_i \left(\frac{G}{1 + \beta G} \right) + e_d \left(\frac{1}{1 + \beta G} \right)$$

and in the absence of an external disturbance, this simplifies to

$$e_o = e_i \left(\frac{G}{1 + \beta G} \right)$$

There are dangers in applying transfer theory to human control loops, since the transfer functions of the body vary with the circumstances. The time required to hit a brake pedal in response to a traffic signal depends not only upon the mechanical characteristics of the system, but also upon the level of arousal in the driver. Furthermore, it is difficult to measure the gain G of most biological regulators. Consider the control of blood glucose. An initial blood sugar of 80 mg/100 ml is suddenly increased by 20 mg to 100 mg/100 ml. Quite rapidly, the body adjusts this to a lower figure of 85 mg/100 ml. In other words, e_d has been reduced by a factor of one fourth and βG is 3. However, in order to measure the gain G of the system, it would be necessary to disconnect the normal feedback loop, and this is not possible within the limitations of present experimental technology.

STABILITY OF THE LOOP

The stability of a control loop varies inversely with its speed of
operation. The need for rapid adjustment must be balanced against
the dangers of oscillation. This point is well brought out by human
arm movements. The arm forms a compound pendulum, and if the
muscles are relaxed it will oscillate with a natural frequency of several
cycles per second. When accurate movements are to be performed,
this tendency is overcome by periodic contraction of both the prime
movers and antagonistic muscles, giving a heavily damped *controlled*
movement (page 173). However, in other circumstances (for instance a
cricket or a tennis stroke) the antagonists are completely relaxed, and
an uncontrolled ballistic stroke is initiated by a sudden contraction
of the prime movers.

Extreme *instability* develops if feedback changes from negative to
positive sign. In the circuit of Figure 9, e_o is now added to e_i, exaggerat-
ing the tendency to change. Examples of positive feedback within the
body include formation of the embryonic blastula, blood clotting, the
generation of a nerve impulse(page 92) and coitus. A simple instance
of an unstable man/machine system is provided by a motorist who
drives too fast along a winding lane. By the time he applies corrective
movements to the steering wheel, he is already more than 90° out
of phase with the lane, and attempts at steering worsen the trajectory
of the car (Fig. 10). One remedy is to slow down. The time required
for steering adjustments remains essentially unchanged, but because
of the slower vehicle speed, steering is now perhaps 30° out of phase,
and has a useful effect upon the course of the car. The other *trick*
adopted by the experienced driver is to anticipate the settings of throttle
and steering wheel that will be required. *Phase advance* overcomes
the problems inherent in a slowly responding control loop; it is found
in the human body, for instance in the dual proprioceptive system
of the limbs, where responses of muscle tone occur to not only the
prime signal but also its first derivative.

The possibility that a system may become unstable should never
be overlooked. The North American power network is normally exceed-
ingly stable, but because of incomplete modeling, the possibility of
instability was not appreciated until a sudden and widespread failure
of the system occurred in the late 1960's. This bizarre episode was
apparently triggered by increasing surges of power around the Great
Lakes, and led to various subsidiary problems through flooding of
engine rooms, failure of bearing lubrication, and absence of emergency
power sources at such vital locations as Kennedy International Airport.

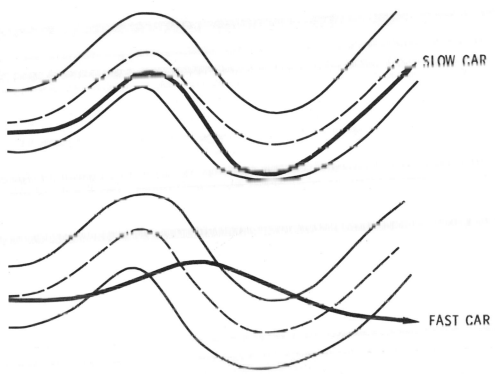

Figure 10. An illustration of the transition from negative to positive feedback. In the upper panel, a motorist drives slowly along a winding lane, and makes useful steering movements. In the lower panel, his movements are more than 90° out of phase, and attempts at steering worsen the trajectory of the car.

MAINTENANCE

A special case of instability is a breakdown in the equipment itself. This is a problem in many parts of the world. In North America, the difficulty arises from the high cost of maintenance. Often, repairs to a car or a television set exceed the cost of replacing it. Here, the ergonomist can help (1) by insisting upon a modular design with readily accessible checkpoints for each module, and (2) by devising more efficient and logical search procedures for the repair technician (page 353). In the underdeveloped countries, tractors and pumps shipped under foreign aid programs sometimes fail at an early date due to lack of maintenance. It is easy to blame the African or the Asiatic, but this is perhaps unfair in view of his general lack of mechanical background. If a machine fails from incomplete or incorrect maintenance, the basic fault is not with the user, but with the designer. Considerable progress has been made in designing cars that require minimum lubrication, and hopefully the time is not far distant when all machines will require only very infrequent maintenance.

PART I

BIOLOGICAL
FUNDAMENTALS

CHAPTER 3

THE PHYSIOLOGY OF WORK

IN THIS SECTION, WE SHALL examine rather briefly the physiological basis of human movement and will apply this knowledge to specific industrial tasks. Practical applications to machine operations and the design of working areas will follow a review of anthropometric considerations to be discussed elsewhere (page 305).

RATE OF ENERGY RELEASE

Man is inevitably subject to the principle of the conservation of energy (page 168). Thus, he can only work upon some external system if he depletes his personal energy stores. Work can continue in such a manner for minutes or even hours, but ultimately the laborer becomes fatigued unless food is provided to replenish these stores.

The Proximate Energy Source

Muscle contraction is initiated through the chemical combination of two long protein filaments—actin and myosin. As these join to form actomyosin, cross-bridges between the filaments shorten, producing a typical muscle movement.

Energy is required for this reaction, the proximate source being a store of adenosine triphosphate (ATP) within the active muscles. When the ATP is split to adenosine diphosphate (ADP) and a phosphate radical, there is a substantial release of free energy, variously set at 7 to 12 kilocalories per mole of ATP.

The precise caloric yield of a single ATP molecule is of more interest to the biochemist than to the ergonomist. The latter may be content to note that:

1. the reaction is exothermal,
2. the quantity of energy liberated is large relative to many biochemical reactions, and
3. a substantial proportion (about 40%) of the released energy can be applied to the combination of actin and myosin.

Unfortunately, body stores of ATP are very limited. Muscle biopsy suggests that resting muscle contains about 5 mM per kg of wet weight. There are some 28 kg of muscle in an average man, but it is rare to find more than 20 kg involved in heavy work. The usable store

41

of ATP is thus some 100 mM, or 1.1 kilocalorie, energy sufficient for perhaps 0.5 seconds of maximum effort.

A second local intramuscular store of energy is creatine phosphate (CP). When CP is split to creatine and phosphate radicals, there is again a substantial release of free energy (about 11 kilocalories per mole), and this energy can be applied to the resynthesis of ATP. Biopsy shows an initial concentration of some 15 mM of CP per kg of wet muscle. Assuming perfect coupling of the ATP and CP reactions, the total phosphagen store (ATP + CP) would thus provide no more than 4.4 kilocalories of energy.

Usage of phosphagen and exhaustion of labile oxygen stores together create an oxygen debt, described as the *alactate debt* to distinguish it from a somewhat larger debt associated with accumulation of lactic acid. The alactate debt must be repaid after a bout of intense activity. Repayment is an exponential process, with a halftime of about 22 seconds.

If a worker makes a maximum effort, the alactate debt could theoretically be deployed within two seconds. However, it is more usual for phosphagen usage to extend over at least 8 seconds. The *capacity* and *power* of the reaction are conveniently expressed in units that will subsequently be applied to more sustained work—the equivalent transport of oxygen. In such terms, the maximum alactate debt is about 22 ml/kg of body weight, and if accumulated uniformly over 8 seconds, the power of the reaction is 165 ml/kg min.

Under normal industrial circumstances, more sustained effort is required. If oxygen is available, the phosphagen is resynthesized by the breakdown of glycogen and glucose to CO_2 and water, while if the immediate oxygen supply is insufficient, additional energy is derived from the anaerobic breakdown of hexose molecules to pyruvate, with subsequent conversion of the latter to lactate. Standard equations of thermodynamics can be applied to both aerobic and anaerobic processes. Thus, the change in heat content of the system $\triangle H$ is given by

$$\triangle H = \triangle E + P \triangle V$$

where $\triangle E$ is the change of internal energy, and $P \triangle V$ is the accompanying external work. The change in free energy of the system, $\triangle G$, is given by

$$\triangle G = \triangle H - T \triangle S$$

where T is the absolute temperature, and S is the randomness or entropy of the system. Combining the two equations, the evolved heat is given by

$$\triangle G + T \triangle S - W$$

where W is the work performed.

Muscular activity is a complex process, involving changes in both heat content (enthalpy, $\triangle$ H) and configuration (entropy, T $\triangle$ S). Inevitably, certain simplifying assumptions are made in calculating mechanical efficiency. In particular, the energy content of the phosphagen molecules must be estimated for *in vivo* reactions, and it is necessary to assume that free energy $\triangle$ G is similar to enthalpy $\triangle$ H at 11 kilocalories per mole.

Under aerobic conditions (Fig. 11), 39 moles of ATP are conserved for each mole of glycogen-derived hexose that is metabolized. One mole of hexose phosphate yields 672 kilocalories, so that the efficiency of ATP resynthesis is given by 39 × 11/672, or 63 percent. The overall mechanical efficiency of aerobic work is about 25 percent, so that the various steps involved in the transfer of free energy from ATP to actomyosin must have an efficiency of about 40 percent (Banister, 1971; di Prampero, 1971). Rather higher efficiencies have been noted during the forward reaction (Banister, 1971), and it may be that the efficiency of coupling between ATP and actomyosin drops with cyclic activity. Unfortunately, it is difficult to measure all components of the external work performed; in most activities, energy is dissipated in acceleration and deceleration of the limbs, and estimates of mechanical efficiency thus lack precision. Under anaerobic conditions, efficiency is much

(a) aerobic

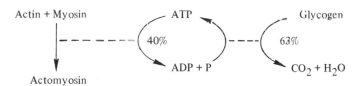

(b) anaerobic

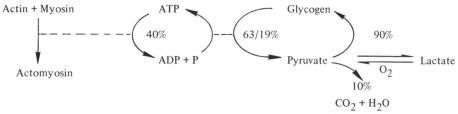

Figure 11. The efficiency of energy transformation in the body.

lower; if glycogen is the source of fuel, only 3 moles of ATP are generated in place of 39, while if glucose is used, only 2 moles are generated in place of 38. Fortunately, much of the lactate that accumulates can be resynthesized to glycogen once oxygen is again available (Fig. 11). Margaria and his associates set the proportion of resynthesis as high as 90 percent (di Prampero, 1971), giving an overall efficiency of anaerobic glucose usage of $\frac{40 \times 63 \times 10}{100 \times 19 \times 1}$, or about 13 percent.

Several practical lessons may be drawn from this brief biochemical excursion:

1. Because the efficiency of anaerobic work is low, twice as much food will be needed to replenish energy stores if a given task is performed anaerobically.
2. Since the reserves of anaerobic fuel—glucose and glycogen—are limited and resynthesis is slow, fatigue may arise from exhaustion of these reserves.
3. Much free energy is dissipated as heat; at least 75 percent of the energy consumed in aerobic activity, and 87 percent of that for anaerobic work must be passed to the environment if body temperature is not to rise.

The Role of Lactic Acid

In the absence of oxygen, resynthesis of phosphagen becomes dependent upon the breakdown of hexose to pyruvate, with conversion of the latter to lactate. Relatively large quantities of lactate accumulate in the active muscles, and a substantial proportion also diffuses into the bloodstream. Some of the latter may be oxidized in other parts of the body while the primary activity is in progress (Keul *et al.*, 1972), but the major portion persists into the recovery period as the lactate component of the oxygen debt. Part is then reconverted to glycogen within the liver, using energy derived from oxidation of the remainder to CO_2 and water (Fig. 11).

The magnitude of the lactate debt is determined in part by glycogen stores. If muscle glycogen has been depleted by a day of relatively heavy aerobic work, or repeated bursts of anaerobic effort, the final sprint for the homeward-bound bus is unlikely to produce a very large increase of blood lactate. Motivation is also a determinant of peak lactate levels, since anaerobic work leads to a distressing general breathlessness coupled with local sensations of fatigue, weakness and pain within the active muscles. If a young and well-rested man undertakes maximum exertion for about a minute, he will reach a final blood lactate of 100 to 120 mg/100 ml,* and the concentration within active muscles will be three or four times as great. Blood lactate readings are usually lower in an older worker, dropping to a maximum of perhaps 60 mg/100 ml at the time of retirement; however, it is uncertain how far this discrepancy reflects reduced ·glycogen storage in an older person, and how far it is due to poorer motivation.

The blood lactate readings that follow maximum effort in a well-rested person of a given age group provide a useful clue to motivation, but since equilibrium between the active muscle and the bloodstream is reached rather slowly, the *capacity* and *power* of the lactate reaction are better estimated in other ways. If a man runs up a long flight of stairs at top speed, it is possible to measure the anaerobic work that has been performed; alternatively, the excess oxygen consumption can be examined during the first fifteen minutes of the subsequent recovery period. A *curve-stripping* approach is used to distinguish the alactate from the lactate debt (Fig. 12). The energy yield of the latter is about 15 kilocalories, deployed during the first 40 seconds of maximum effort. Expressed in terms of equivalent oxygen usage, the lactate *capacity* of a young worker is about 3 liters (45 ml/kg), with a power of 68 ml/kg min (di Prampero, 1971). Repayment occurs after cessation of effort; the added oxygen consumption wanes with a halftime of 10 to 15 minutes, but full replenishment of glycogen reserves may take a day or longer.

Aerobic Power

Since both alactate and lactate sources of anaerobic energy can be depleted within 40 seconds, most types of industrial activity depend upon a steady supply of oxygen from the atmosphere to the working tissues. If the effort has a duration of one minute, about half of the maximum potential energy release is aerobic. With 5 minutes of activity, 80 percent is oxygen dependent, and over one hour of work 98 percent of metabolism is aerobic in type.

Human power is thus determined very largely by maximum oxygen intake. Figures vary with age, sex and physical fitness. The young male office worker will probably have an aerobic power of 3 liters/min (42 ml/kg min), while the young manual laborer may be nearer to 4 liters/min (48 ml/kg min). However, traditional differences between the office worker and the laborer are disappearing with automation,

---

* Because of the time required for lactic acid to diffuse into the bloodstream and reach other parts of the body, both arterial lactate and breathlessness may be greatest 1 to 2 minutes after ceasing activity.

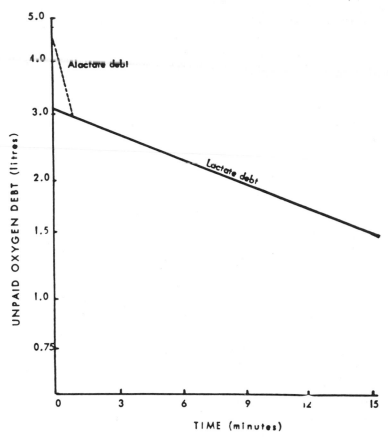

Figure 12. Curve-stripping approach used in determining lactate and alactate oxygen debts. The least squares method is used to fit a semilogarithmic line to excess oxygen consumption between the second and fifteenth minutes of recovery. Extrapolation to zero time permits the calculation of the lactate debt as the area under this line. A second line is now fitted to oxygen consumption in excess of that required for repayment of the lactate debt over the period 0 to 2 minutes; this permits calculation of the alactate debt. (From Roy J. Shephard, *Alive Man!* Springfield, Illinois, Thomas, 1972.)

and the aerobic power of both groups now depends more upon recreational than industrial activity. Aerobic power diminishes by about 10 percent per decade of adult life, so that at retirement values average 25 to 30 ml/kg min. Women have a poorer aerobic power than men, partly because their bodies contain more fat, and partly because cultural conventions set lower standards of physical activity for the female. The young stenographer or university girl has a maximum oxygen intake of 35 to 40 ml/kg min, and this diminishes to 22 to 26 ml/kg min at retirement.

One liter of oxygen yields 4.7 to 5.1 kilocalories of energy, depending upon diet. Taking a rounded figure of 5 kilocalories per liter, the average young man with an aerobic power of 3 liters/min can thus develop a work rate of 15 kilocalories per minute for several minutes. At the age of 65, this potential has dropped to 8 to 10 kilocalories/min in the men, and 6 to 7 kilocalories/min in the women.

The determination of aerobic power and the associated electrocardiographic examination form an important part of an industrial medical assessment—not only in terms of predicting the ability to undertake arduous physical work, but also from the viewpoint of diagnosing insipient coronary disease.

The standard laboratory approach when testing a young man is to conduct treadmill exercise at increasing slopes and/or speeds until an increment of workload no longer produces an increment of oxygen consumption (Fig. 13). The individual is then said to have reached his directly measured maximum oxygen intake.

The treadmill approach has been used by some employers with large and well-equipped ergonomics laboratories, but is not generally

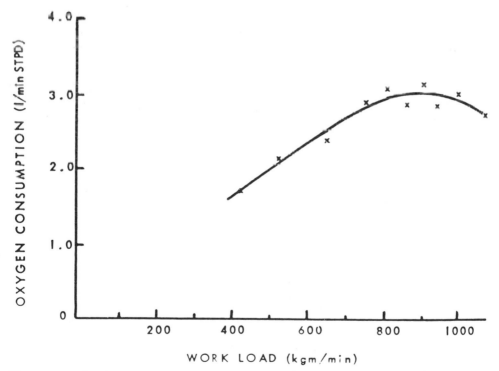

Figure 13. The direct measurement of maximum oxygen intake. A patient runs on a treadmill at increasing slopes until a plateau of oxygen consumption is reached. (From Roy J. Shephard, *Alive Man!* Springfield, Illinois, Thomas, 1972.)

popular. Apart from the need for expensive and bulky equipment, there are real problems in motivating older men and women to maximum effort. Further, there are undoubtedly small but finite risks of muscular injury and cardiac arrest when testing older people. Such risks need clearer definition (Rochmis and Blackburn, 1971), but *a priori* it seems more likely that harm will result from maximum effort on a treadmill than from more moderate exercise on a step or a bicycle ergometer.

It is thus usual to predict maximum oxygen intake from the relationship between pulse rate and oxygen consumption or pulse rate and work load during sub-maximum effort. A line fitted to four or more paired measurements at increasing loads (Maritz *et al.*, 1961) can be extrapolated to the anticipated maximum heart rate of the individual (Fig. 14), or similar calculations can be carried out by means of nomograms (Åstrand, 1960; Margaria *et al.*, 1965). Under the conditions of a careful laboratory experiment, submaximum tests can estimate aerobic power with an accuracy of about 10 percent (Shephard *et al.*, 1968b), but it is doubtful whether the same accuracy can be maintained in routine tests. Anxiety such as may accompany an annual medical examination increases the heart rate in submaximum effort and thus decreases the predicted maximum oxygen intake. If the room is hot, a similar effect may occur. The worker may be tired from the day's duties, prolonged standing, a sleepless night, or a previous evening of dissipation; all of these factors, together with recent meals, smoking and minor infections distort the pulse response to submaximum effort and thus the predicted maximum oxygen intake. Further, while the relationship between oxygen consumption and pulse rate is relatively

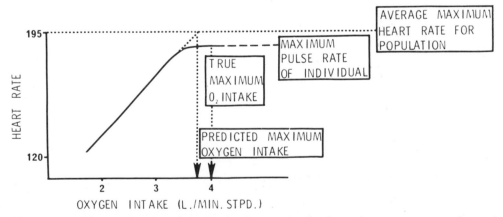

Figure 14. The prediction of maximum oxygen intake from the response to submaximum effort. Four paired observations are made at increasing loads, and a line fitted to the data is extrapolated to the individual's theoretical maximum heart rate. (From Roy J. Shephard, *Endurance Fitness.* Toronto, University of Toronto Press, 1969.)

linear in a young person, this is less true of the older worker who may be affected by various forms of chronic disease.

There is also some uncertainty regarding maximum heart rate. This is apparently diminished slightly at the altitude of cities such as Johannesburg and Mexico City, and may also be somewhat lower in workers who have a high level of physical fitness. On the average, the maximum heart rate diminishes from 195/min at age 25 to about 160/min at 65; however, there is a considerable scatter of rates among young workers, and even more variation in the superimposed rate of aging. A variety of chronic diseases may depress the heart rate at any given age, and in some workers symptoms such as angina or extreme breathlessness may prevent the attainment of the theoretical maximum figure.

Finally, the laboratory worker is often interested in the characteristics of a *population*. A randomly distributed error of 10 percent per person can then be reduced to an uncertainty of 1 percent by testing 100 subjects. However, the industrial physiologist usually wishes to know the working capacity of the individual employee, and therefore he cannot improve upon the accuracy of his procedure by an increase of his sample size.

For these various reasons, some authors have queried the usefulness of extrapolation to a predicted maximum oxygen intake. One possible alternative is to report by close interpolation the amount of work that can be performed at a pulse rate of 170 beats per minute, the P.W.C.$_{170}$ (Wahlund, 1948). A somewhat similar proposal is to report the pulse rate at an oxygen consumption of 1.5 liters/min (Cotes, 1966). Both of these tests suffer the disadvantage that they are too easy for the young worker and yet are difficult or even beyond the competence of an older person. The most recent, and probably the most reasonable proposal is to record all data such as workload and ECG configuration at a target pulse rate, corresponding to 75 percent of aerobic power in an average person of the age specified. With a little rounding, this reduces to a very simple series of numbers:

Age	Target Pulse Rate	Ascents of 18-inch Step (Men)
25	160/min	21/min
35	150/min	19/min
45	140/min	17/min
55	130/min	14/min

Physiologists have spent much time in debating the optimum mode of exercise for a submaximum performance test (Andersen *et al.*, 1971). The simplest approach is undoubtedly a step test. If the rate of ascent is paced by a metronome, and the subject is persuaded to stand erect upon the steps, the work performed can be calculated quite precisely from the product of step height, rate of ascent and body weight. A good quality electrocardiogram can be recorded from standard chest leads and there is no problem in collecting expired gas for the measurement of oxygen consumption if this is desired. The main limitation of the step test is the difficulty in making ancillary measurements such as blood pressure.

A bicycle ergometer is relatively expensive ($300 to $3000). It is often claimed to be unique in that it provides a known work load, thereby obviating the need to measure oxygen consumption. However, such claims are not well founded. To obtain accurate absolute figures for the work performed, it is necessary to couple the pedals periodically to an expensive torque calibrator. Even in a relative sense, the efficiency of bicycle ergometry is not much more consistent than that in a carefully performed step test. The arms are immobilized, facilitating the measurement of blood pressure, but against this advantage a disproportionate load is thrown upon one muscle group (the quadriceps). Further, in the step test, it is possible to set a single 75 percent loading for all men of a given age, since work is performed mainly against body weight; on the other hand, 75 percent bicycle ergometer loadings must be individually calculated. For these reasons, there is much to commend the step test for routine use.

Let us suppose we are examining a 25-year-old worker. The simplest type of assessment would do no more than note the pulse rate and electrocardiographic changes after five minutes of stepping at 21 ascents per minute. If the man is of average fitness, he will reach the target pulse of 160/min. If he is in superior physical condition, the final pulse rate will be no more than perhaps 150/min, while if he is unfit it may reach 170/min.

The next stage of sophistication calls for testing at four intensities of effort, starting at perhaps 8 ascents/min for three minutes and progressing through two further three-minute periods at 12 and 16 ascents per minute; the final loading can then be adjusted to 20, 21 or 22 ascents per minute in order to achieve the target pulse rate. Data can then be reported as the rate of stepping at the target pulse, or (if staff are used to thinking in such terms) as the maximum oxygen intake, predicted by one of the procedures already discussed.

In situations where a highly paid job such as test pilot hangs upon the result of an examination, the appropriate method of testing has yet to be found. A rapid heart rate due to anxiety seems inevitable. Two possibilities worth exploring are (1) the measurement of resting pulse rate while the worker is asleep, and (2) the estimation of cardiac stroke volume at the target pulse rate. The exercise stroke volume is an important determinant of oxygen transport, and at a loading of

75 percent of aerobic power is unlikely to be influenced by emotional factors.

Subsequent Limitations

When work must be sustained for several hours, the maximum potential rate of energy expenditure may become curtailed by a failing oxygen transport system or an exhaustion of food reserves.

The principal determinant of oxygen transport is the maximum car diac output; this could be progressively diverted to inactive tissues, and might also fall if fluid reserves were depleted. The regulation of body temperature thus poses an early threat to the system. We have noted already that because human effort is mechanically ineffi- cient, between 75 and 87 percent of expended energy is normally dissipated as heat. In the first few minutes of effort, accommodation is possible through a rise of body temperature, but the maximum poten- tial heat storage (300 kilocalories) provides for no more than fifteen to twenty minutes of effort. An increase of skin blood flow is thus an important concomitant of sustained activity. Different authors (An- dorsen et al., 1971; Shephard, 1968) have set the skin flow in near maximum effort at 5 to 20 percent of cardiac output, and in order to cope with this increasing flow there must be a corresponding diver- sion of blood away from the active muscles. At the same time, other factors are reducing the maximum stroke volume and thus the maximum output of the heart. The rising body temperature leads to a wide dilata- tion of the capacity vessels of the limbs. The hydrostatic pressure within the muscle capillaries rises, and local metabolic changes increase capil- lary permeability. There is thus a loss of fluid from the circulation into the active tissues; the weight of some muscles increases by as much as 20 percent, and this makes major inroads upon a circulating blood volume of 5 liters. Unless the climate is cool, the blood volume is further depleted by sweating; this can reach a peak of 1 to 2 liters/hour.

Both fluid loss and the accompanying excretion of mineral ions must be made good if the worker is not to suffer progressive exhaustion over the working day.

Food Reserves

To this point, we have considered food reserves as relatively limitless. In the case of fat, this is perhaps justified: the average worker with 20 percent body fat (14 kg) carries an energy store of almost 100,000 kilocalories, enough to meet his needs for a month or more. Fat can provide energy for the resynthesis of phosphagen, but as much as 75 percent of the fuel for vigorous rhythmic effort is drawn from carbohydrates. The brain is unable to metabolize fat, and if muscle blood flow is restricted by effort, some of the active muscles also may lack the oxygen needed to metabolize fat.

Carbohydrate stores are quite small, comprising some 400 gm of glycogen within the muscles, a further 100 gm within the liver, and perhaps 3 gm of usable glucose in the bloodstream—a total reserve of about 2000 kilocalories. If three quarters of the muscle mass is active, the usable reserve drops to 1600 kilocalories, and if a man works for 100 minutes at 20 kilocalories/min, drawing three quarters of his energy from carbohydrate, 94 percent of the available store will be exhausted. It is unlikely that depletion proceeds uniformly; some muscles perform anaerobic work, and these muscles will be fatigued in less than 100 minutes. Once the glycogen is used up, aerobic work depends upon the mobilization and transport of fat. This is not in itself seriously limiting, but difficulty arises because almost every physical task has an anaerobic component. This can now be performed only at the expense of an already low blood sugar.

Restoration of glycogen reserves proceeds most readily if the worker is provided with sugar. More complex carbohydrates and protein can also be converted to glycogen, but only the glycerol component of fat can be used in this way. Irrespective of diet, replenishment of intramuscular glycogen may take a day or longer, and it is thus helpful to spare the glycogen reserves of the heavy worker by periodic sweetened drinks. Solutions are preferable to sweets (candy) since the fluid and mineral balance can be restored at the same time. Some athletes like very strong glucose solutions (e.g. 40 gm/100 ml). Physiologists have recommended lower concentrations (< 10%) to avoid delays in the emptying of the stomach. However, glucose is absorbed through the gastric mucosa, and providing the stomach is not unduly distended, a delay in emptying is not necessarily a serious criticism of a concentrated drink. The maximum rate of transport of glucose between the liver and the working muscle is about 1 gm/min, so that there is little advantage in burdening the body with more than 100 gm of glucose per hour. If excessive peaks of blood sugar are created, much of the glucose will be excreted in the urine; the optimum arrangement is thus a series of small but frequent sweetened drinks.

INDUSTRIAL APPLICATIONS

The Anaerobic Component

A worker is rarely required to exert his maximum anaerobic power, except in sprinting for the exit stairway at the end of a day's shift. Nevertheless, it is fairly common to find an anaerobic component in daily activity. The transition from aerobic to anaerobic effort is not clear-cut, and in some forms of rhythmic work lactate will accumulate at loads calling for an expenditure of no more than 50 to 60 percent of aerobic power (Shephard *et al.*, 1968a). The probable explanation is that whereas aerobic conditions are well maintained in most muscle groups, a few muscles are contracting at a high percentage of their maximum voluntary force, thereby restricting or completely interrupting the local blood flow.

If a task is continued, the blood lactate may gradually drop back from their initial peak toward resting levels. Some authors have considered this a reflection of a delayed circulatory adjustment to exercise; however, the timing (decrease over 10 to 15 minutes) is out of keeping with such an explanation. The work performed may diminish as the muscles warm up, but the main basis of the phenomenon is probably a progressive rise of systemic blood pressure facilitating perfusion of the most active muscles.

If more vigorous effort is required, the lactate does not fall. On the contrary, it continues to increase until limiting blood concentrations are reached (Kay and Shephard, 1969). The discomfort of the worker is shown by a steady rise of pulse rate, ventilation and gas exchange ratio (CO_2 output/O_2 intake). Obviously, work of this severity cannot be maintained for any long period. French industrial physiologists have categorized workers in terms of the *puissance maximale supportée* (P.M.S.)—the heaviest load that can be sustained for twenty minutes without such changes in pulse, ventilation and gas exchange ratio (Sadoul *et al.*, 1966). If this approach is used, several quite lengthy ergometer tests are needed on every worker, and unfortunately the final answer is task specific; the P.M.S. for the use of a hand-supported drill, for example, is very different from that for bicycle ergometry.

The *pulse deficit* (Davies, 1970) is a somewhat similar measure. As originally proposed, it presumed that the basic circulatory adaptation to work was complete within one minute, and that any subsequent increment of pulse rate over the first five minutes indicated an accumulating oxygen debt. Details of procedure have been criticized (Shephard, 1970) on the basis that the circulatory adaptations to effort are often incomplete after five minutes; nevertheless, the changing pulse rate at a given work load provides a good indication of overall stress, including not only lactate accumulation, but also thermal load, loss of mechanical efficiency, and cumulative psychological pressures.

If a task involves uniform and rhythmic contraction of large muscles, for instance a soldier marching with a heavy pack upon his back, it may be possible to develop 80 percent of aerobic power before there is a significant accumulation of lactate. If a heavy load is thrown upon one muscle group, for instance the quadriceps of a man pedaling a delivery tricycle, quite large amounts of lactate may appear at only 50 to 60 percent of aerobic power. Lactate is particularly likely to accumulate if isometric work is performed, either in supporting body weight or in lifting heavy external objects. Practical examples include the awkward posture of a miner working in a narrow coal seam, a carpenter fixing a shelf that is almost beyond his reach, and the efforts of a laborer lifting the handles of a well-laden wheelbarrow. When heavy work is performed by relatively small muscle groups such as those concerned with handgrip, fatiguing concentrations of lactate develop locally within the muscle, but because the bulk of tissue involved is small relative to the total volume of body water, the increase in blood lactate is barely measurable (< 10 mg/100 ml).

Anaerobic limitation of human performance can be overcome in several ways. Postural problems can often be corrected. If the miner cannot stand in the coal seam, it may be possible for him to lie down as he works. Modifications of chair and bench heights (page 190), handle lengths of paint brushes and screwdrivers, and other adaptations of design reduce the need to maintain awkward and costly body positions. If fatigue is restricted to one muscle group, it may be possible to redesign equipment so that the load is distributed more uniformly. Alternatively, the worker can be taught to use a wider range of muscles. In some instances (such as writer's cramp) it may be possible to carry out the same task without developing such intense muscular effort. An improvement of personal fitness yields two important practical gains; to return to our example of the delivery tricycle, the cost of pedaling at a given speed may drop from a highly anaerobic 70 percent of maximum oxygen intake to a much more acceptable 60 percent of aerobic power. At the same time, an increase of quadriceps strength throws a lesser relative load upon this muscle group, facilitating local blood flow.

Despite these various maneuvers, a task may still exceed aerobic tolerance. Recourse is then made to *rest pauses*. These are effective in allowing not only oxidation of accumulated lactate but also a reduction of body temperature. The relative lengths of work and rest periods can be adjusted to maintain an acceptable constancy of such indices as pulse rate, ventilation and gas exchange ratio. One common arrangement is to allow ten minutes of rest in each hour. In an assembly line operation, this provides a useful opportunity for mechanical adjustments. However, if work is of an individual nature, it is preferable to arrange more frequent pauses. Irma Åstrand and her associates (1960) have shown that a very high rate of working can be sustained with little build-up of lactate if periods of activity last for less than one minute. It may seem uneconomic to halt activity this frequently, but with a little thought the ergonomist can often build a work pause into the overall task design. Thus, a man stacking heavy boxes on a trolley can pause to label each box before it is lifted into position.

The Aerobic Component

If a machine is to be operated continuously for long periods, it is undesirable to run it at more than 50 percent of its rated power. A similar restriction of loading is necessary when interpreting the aerobic power of the worker. A maximum oxygen intake test is restricted to a few minutes. As the length of a task is increased, so the permissible intensity of activity diminishes. Bonjer (1968) has recommended setting the limit at 63 percent of aerobic power for one hour, 53 percent for two hours, 47 percent for four hours, and 33 percent for eight hours; intermediate times can be calculated from the formula:

$$\text{Allowable load (\% aerobic power)} = 32.3 \, (\log 5700 - \log t)$$

where t is the duration of the activity in minutes. Other authors have proposed 35 to 50 percent of aerobic power as an acceptable ceiling for a normal eight-hour working day. The figure adopted is necessarily influenced by the spread of aerobic power within the working population, and the possibility that the individual employee can function at less than the average pace. In the case of team or conveyor belt operations, the 35 to 50 percent load must be chosen for the least well-endowed member of the labor force. The conditions of employment affect the choice of loading. Aerobic power may be diminished by high altitude, prolonged standing or heat. Account must also be taken of the size of active muscle groups, awkwardness of posture, the intensity of superimposed peaks of effort, and the energy used in traveling to and from work. Most miners now travel to and from the coalface by electric train, but at one time they walked through several miles of narrow, steep and rough underground passages at the beginning and the end of each day. Again, a man who cycles ten miles or stands for an hour in a crowded bus will have a lesser reserve of physical energy on reaching work than will the car driver.

The most widely accepted eight-hour ceiling is 40 percent of aerobic power. Where employees have complete freedom to choose their rate of working, this standard is commonly adopted (Hughes and Goldman, 1970). It is undoubtedly a conservative figure, since under the adverse climatic conditions of South African mines, the Bantu manage to work at 50 to 60 percent of maximum oxygen intake. Nevertheless, the majority of such *native* workers are on a one-year contract, and it is by no means certain how their bodies would react to forty years of daily effort at a 50 to 60 percent loading. Furthermore, a conservative ceiling is likely to provide a contented work force, with a minimum of cumulative fatigue, injuries and avoidable absenteeism. Given current developments in automation and power equipment, a 40 percent loading places few limitations upon what an employer can achieve.

The Long-term Component

Problems of thermal balance are discussed elsewhere (page 66). The need to balance food intake and energy expenditure was noted by the Germans during World War II, when it was found necessary to increase the food supply to prisoners working in the Ruhr coal mines in order to sustain a reasonable productivity. Undernutrition still influences output in some underdeveloped countries (page 296). In Western nations, a palatable diet remains important to morale, and where a worker is entirely dependent upon food provided by the employer (as on a cargo boat or in an isolated northern community) an increase in the quality and variety of canteen fare may make a significant contribution to productivity.

The relative merits of fat and carbohydrate for the man performing hard physical work are still debated. Certainly, the Eskimo performs very hard physical work while consuming a high fat diet. The proportion of fat used during activity increases with a habitual fat diet, and this tends to spare intramuscular stores of glycogen. On the other hand, glycogen stores are increased by a few days on a high carbohydrate diet. Thus, if a worker knows he must engage in unusually demanding activity, he may prepare himself by a substantial period on a high fat diet, switching to a high carbohydrate diet a couple of days before undertaking the special project.

If hard physical work is to be performed soon after eating, it is well to avoid a heavy and greasy meal, as the combination of vigorous effort and a full stomach can induce vomiting; many work canteens seem to ignore this precept, and serve very greasy food.

MUSCLE STRENGTH AND THE WORKER

Isotonic and Isometric Activity

Two main types of muscle activity are distinguished by the physiologist. In an *isotonic* contraction, the muscle is allowed to shorten at constant tension, with the performance of external work. In an *isometric* contraction, the muscle is held to a constant external length, and work is performed internally as the muscle components slide over one another. The division into isotonic and isometric activity becomes somewhat artificial in the real world of industry; however, some activities such as walking or the rhythmic pedaling of a bicycle are largely isotonic, while others such as the carrying of a heavy box are largely isometric.

Rhythmic, isotonic work is limited mainly by the considerations discussed in the previous section. Isometric contractions begin to occlude muscle blood vessels if the contraction exceeds 15 percent of maximum voluntary force, and complete occlusion occurs at approximately 70 percent of maximum force (Lind and McNicol, 1967). Intense isometric efforts are thus fatiguing and (because of the anaerobic work) mechanically inefficient. They also lead to a large rise of blood pressure, with attendant risks of precipitating a coronary attack (myocardial infarction) or rupture of the arterial wall (aneurysm), particularly in the older worker.

Strength Measurements

Extreme strength is rarely required in modern factories. However, it is helpful for the ergonomist to know the average force that a healthy person can exert in different body positions, and details of an individual's strength may be needed when designing vehicles or arranging employment for patients with pathological forms of muscle weakness.

Isotonic strength is difficult to measure, partly because it must be estimated through a series of successive approximations, and partly because the force developed varies with the speed of muscle shortening and thus the internal work that is performed.

The force that can be developed varies greatly with the fixation of the worker; the helmsman on a slippery and pitching deck (page 25) cannot exert anything approaching the effort recorded in a laboratory where adequate counterpressure is available. The angulation of the joint greatly influences effective strength (Rohmert, 1971; see Fig. 15), as does the site of attachment of a load; where possible, heavy articles should be designed to allow carriage close to the body, thereby minimizing leverage.

The strength of female workers is about two thirds of that found in men. Both sexes show a marked decline of strength during the last two decades of employment.

Strength and Design

Where a heavy force must be developed by the workers, the length of the lever arm and the required torque is so adjusted that the task

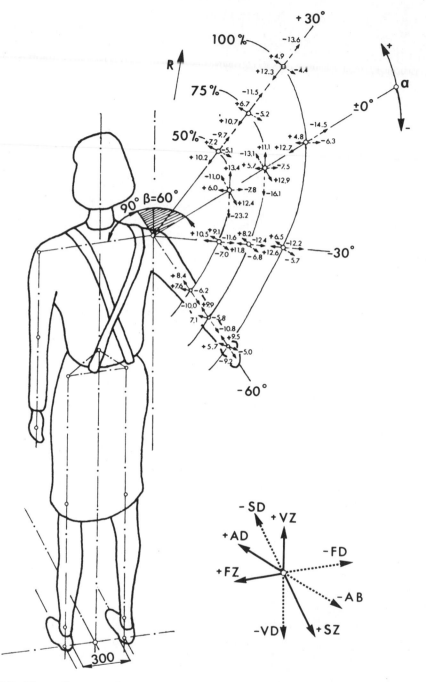

Figure 15. The influence of joint angulation on isometric force. (Reprinted with permission from W. Rohmert, In *Frontiers of Fitness*, ed. by R.J. Shephard. Springfield, Illinois, Thomas, 1971.)

falls within the competence of 95 or 98 percent of anticipated users.

The arm is capable of exerting a greater force when pushing or pulling in a forward direction than when moving in a lateral direction. More force can be developed in the horizontal than in the vertical plane, although if a vertical movement is desired, this is best performed with the arm at the side of the body (McCormick, 1957). The force of a horizontal thrust can be doubled if counterpressure is provided for the feet, a point readily appreciated when trying to row a small boat.

The hand may be used in gripping tools or in operating the ratchet on a hand brake. An initial force of 40 to 50 kg can be developed over much of the working life; however, this maximum force can only be sustained for twenty to thirty seconds. The endurance of an isometric contraction shows a hyperbolic relationship to force, expressed as a percentage of maximum contraction (Fig. 16). The turning strength of the hand is commonly needed to operate a refrigerator door. The level of the lever should coincide with elbow height, since maximum torque can be developed when the elbow is held at 90°. The mid-angulation of the lever should be horizontal, since the force of pronation

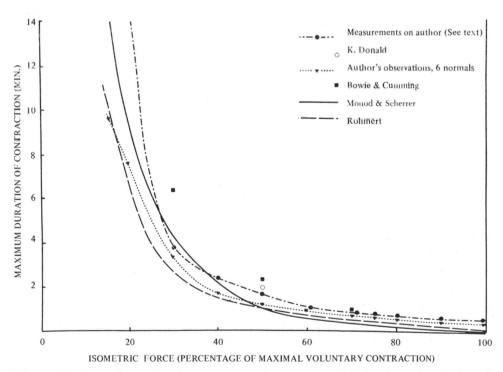

Figure 16. The relationship between isometric force and endurance. Curves based on observations of Monod and Scherrer (1957), Rohmert (1960), Bowie and Cumming (1971), Donald and Lind (1967), and the author's unpublished data.

is related to the extent of initial supination, while the force of supination is related to the initial pronation.

The back musculature is involved in the lifting of heavy weights. Incorrect techniques of lifting and weak musculature lead to a serious toll of back injuries, both prolapse of the intervertebral discs and non-specific strains of the sacroiliac joints (pages 183).

The legs may exert maximum isometric effort while applying mechanically operated car brakes or turning the rudder bar of a small aircraft. In both cases the force is exerted from a sitting position. If both feet are used, a force of some 250 kg can be sustained for thirty to forty seconds. The optimum level for the development of a braking action is 15 cm below the seat; if the pedals are set 30 cm below the seat, the force that can be exerted is much smaller.

SPEED AND ACCURACY OF MOVEMENT

Reaction Time

A *simple reaction time* is the time required for a single response to a simple stimulus. Reactions to touch or sound are generally faster than reactions to a visual cue, and may occur within 120 to 140 msec. The observer must make some movement in response to the cue. In the experimental situation a light key is pressed, and movement time makes a negligible contribution to the total reaction period. However, in a more usual ergonomic situation (for instance, when applying a car brake in response to a red light), the foot must move from the accelerator to the brake pedal, and develop a substantial force in the latter situation. The true reaction time can be identified by fitting a sensor that detects the initial lifting of the accelerator pedal. This occurs within 170 to 180 msec. However, the total time to application of the brake is much longer (350 to 450 msec). The speed of the movement component is decreased when heavy resistance is encountered, and power brakes thus make a useful contribution to road safety by speeding the overall reaction time.

The main disadvantage of power controls is a loss of feedback with the lesser resistance to movement of the muscles and joints. Fortunately, the feedback of information on the distance moved seems more important than that relating to the pressure exerted, and devices such as power steering and power brakes thus have a considerable range of travel.

In heavy traffic, the reaction time may be much slower than that measured in the laboratory. The driver must watch road signs, lane markings, turning vehicles and pedestrians in addition to signal lights, and the response is a *complex reaction time* with many cues. Typical experiments show that with three stimuli in place of one, the auditory reaction time is increased from 220 to 360 msec, and the visual reaction time is lengthened from 260 to 400 msec. Reactions may be speeded by a stronger stimulus, hence the trend to larger and more powerful traffic signals. The response is also faster if anticipated (one virtue of adding an amber traffic signal).

Positioning Movements

With positioning movements, a control is carried from its initial setting to a predetermined final position. Practical examples include the return action on a typewriter carriage, the emergency braking of a car, and the acceleration of a vehicle to a speed limit such as 35 miles per hour. In the last example, the position of the foot is controlled through an intermediary gauge (the speedometer dial).

The time required for positioning is not directly proportional to the distance moved. Thus, let us suppose that a typist has the misfortune to type a scientific table without the aid of a tabulating system. It will take her only a little more time to slide the carriage through twenty-five spaces than through five, since her rate of movement is much faster over the middle part of the traverse. As the required position is approached, the carriage is slowed by contraction of the antagonist muscles, and unless the typist is very experienced at preparing tables in this way, time may be occupied with correction of an overshoot. Naturally, the speed of movement is much increased if a stop is provided so that overshoot cannot occur.

The speed of arm movements shows minor differences with direction. *Single-handed movements* are performed somewhat more rapidly in an anteroposterior than in a lateral direction. If control is visual, accuracy is greatest when the target is a short distance (17 to 18 cm) away from the worker and set at an angle of about 60°.

Two-handed movements are performed most rapidly when each of the controls is set 30° off center. The accuracy of visual control is greatest when the target lies straight ahead, and it deteriorates rapidly if it is more than 30° off center.

Kinesthetic Control

Most common tasks are controlled by kinesthetic (position) sense rather than by direct observation of the limb; this is an important expression of learning (page 241). If the task calls for a short range of movement, the desired end position is commonly passed. In contrast, a long movement often falls short of the target. However, there is quite a range of hand movements (10 to 50 cm) where accuracy is reasonably consistent. The right hand is normally more accurate than the left, and both hands are more accurate below than above the horizontal plane. Movements towards the body are performed more consistently than those in the opposite direction, one reason for the match-striking pattern of the habitual smoker. Movements from above downwards tend to overshoot, because position sense is distorted by the weight of the limb. This problem of kinesthesis is exaggerated when gravitational forces are abnormal (for instance, during space or underwater activity).

Repetitive Movements

The operation of a Morse key is a simple example of a repetitive movement. Speed is more important than accuracy. The maximum rate of tapping is set at 10 to 14 taps per second by the natural frequency of the fingers, but the preferred rate is slower (1.5 to 5 taps per second). A right-handed person can use the right hand faster than the left, and the index finger functions slightly faster than the fourth finger. However, tapping speed is not the usual limiting factor in Morse telegraph operation. If a response is required from the receiver (such as noting down a letter), a complex reaction time of at least 0.5 seconds must be allowed between stimuli.

Tracking Movements

Tracking movements may be continuous or discontinuous. Practical examples include the steering of a car, the following of an aircraft by a searchlight, and the feeding of a bolt of cloth into a sewing machine. In each case, the movement initiates a direct feedback through the altered visual stimulus. As with discrete positioning, there is a tendency to overshoot small and to undershoot larger adjustments, while movement time bears no direct relationship to the distance traveled.

Serial Movements

Serial movements are characterized by relatively independent actions that must be performed in the correct sequence. Examples include typing and the playing of a piano or an organ. The overall speed for this type of activity is greatly increased by practice. The main impact of training is upon manipulation time, and the gross travel time shows little improvement. Presumably, the latter is an inherent characteristic of the individual's motor skill. The practical implication is that performance can be improved most easily by reducing the manipulative component of a task. Thus, the operation of a manual telephone switchboard can be speeded by the substitution of switches for the more usual switchboard pins. The speed of an individual action is influenced by the direction of travel (a downward movement proceeding faster than a horizontal one). Preceding and subsequent actions also have a significant influence; thus a pianist will find it easier to play an adjacent note than one that is removed by several octaves.

Isometric Reactions

Occasionally, it may be necessary to hold a control in a fixed position. This involves a sustained contraction of both agonists and antagonists, like any isometric contraction, it is a fatiguing process. Where possible, controls should be arranged to lock in position, or (if rapid adjustment is needed in an emergency) power assistance should be provided so that the force exerted by the limb is less than 15 percent of a maximum voluntary contraction.

Tremor may be a problem in positioning a control, particularly for the older worker, Oscillations can be minimized by incorporating damping elements into the control system, and providing support for the body (seating) or the active limb (an arm rest).

DESIGN APPLICATIONS

Knobs

Knobs may be designed for finger operation, as in radio and television tuners, or (if more torque is needed) for wrist movement. Small knobs are manipulated more slowly, but also more accurately than those of larger size. The travel time for any given muscle group is increased if more torque is required, but the speed of final adjustment remains unchanged. If substantial torque is to be developed through a small knob, a gearing arrangement may be used. This permits accurate adjustment, reducing manipulation time at the expense of an increase in travel time. If very low gearing is required, a crank may be turned faster than a knob.

The accuracy of knob adjustment is greater for the right hand than for the left. It is also improved by a substantial angle of turn, and by adoption of conventional target positions (such as 0, 90, 180, 270 degrees). As with the typewriter carriage, movement to stereotyped end positions can be speeded by a partial or complete locking device (as in a four-position switch on an electric cooker). In some instances, the operator may perform the coarse adjustment, and leave the final manipulation to a servosystem; this principle is used in some television tuners.

Cranks

If the speed of rotation is decreased, the force that can be exerted upon a knob or handwheel increases; nevertheless, even with extremely slow movements, the potential torque does not exceed 25 kgcm, and for heavier loads a crank must be used. If a large force is to be exerted, this is best initiated from 45° below or 45° above the horizontal line, the direction of rotation being immaterial. Crossbars offer no advantage over a single-ended crank.

FITNESS OF THE WORKER

Whether cardiorespiratory power or muscular strength is needed in a given industry, there is no question that the worker who is physically fit will have an advantage over the man who is out of condition. This is particularly true of the older employee, since the natural decrement of aerobic power and strength hampers performance of many tasks readily undertaken by the younger employee; a 10 to 20 percent gain of function through a physical training program is enough to bring many activities that were becoming fatiguing within the compass of an older person. Physical fitness is particularly important under conditions of climatic stress, since there is an interaction between heat acclimatization and physical training. Good physical health can also counteract minor absenteeism; there is a general elevation of mood and minor infections no longer reduce function to the point where absence from work seems mandatory. Greater strength and flexibility with improved reaction times diminish the risks of injury on the job. Team spirit may also be fostered by participation in group activities, with their opportunities for cooperation, development of status, and relief of pent-up aggressions; however, careful management is necessary if the theoretical advantages of a physical activity program are to be realized in practice (page 236).

The techniques used in persuading employees to achieve the desired gains of physical fitness are varied. Exercise breaks may be provided in company time, and in the armed services there may even be a sports afternoon. The employer may construct a sports ground or provide free membership in an athletic club. Alternatively, the employee may be threatened with dismissal if appropriate standards of personal health are not maintained. Nevertheless, the problem of motivating the employee to seek positive health remains a vexing question.

CHAPTER 4

MAN AND HIS ENVIRONMENT

THE CHARACTERISTICS OF THE WORKING environment have a profound impact upon interactions between man and any equipment that he may be using. Indeed, the topic is so vast that this brief section will do no more than indicate certain types of problem that may arise. Consideration will be given to the psychology and the physiology of environment.

THE PSYCHOLOGY OF ENVIRONMENT

The performance of today's worker is influenced more by psychosocial than by physical or physiological factors. Few current industrial tasks stress physiological systems fully, and indeed gross physical insults—such as extremes of heat and cold—have been largely banished, thanks to modern technology. This leaves pride of place to questions of arousal and motivation.

Arousal

We shall discuss later the characteristic inverted U-shaped relationship between arousal and task performance (page 215). The optimum level of arousal depends upon the personality of the individual, the difficulty of the task he must perform, and the environment in which he must work. A skillful employer will manipulate the industrial surroundings to realize this optimum, increasing the arousal of one group of his staff by such devices as brightly colored furnishings (page 216), brisk music, and an increase of work pace, yet decreasing the arousal of a second group by providing them with soft chairs, soft lighting, and sound-absorbing carpets, drapes and ceiling.

The solo worker is often under-aroused, particularly if he must sustain vigilance during the night hours. Techniques of maintaining his arousal and thus his performance are discussed elsewhere (page 215). Under-arousal may also be associated with boring and repetitive production-line work. The benefit of bright lights, bright music and a cooler working environment can here be supplemented by an increase in relative task difficulty. This may be brought about by (1) an increase of task complexity, (2) an increase in the speed of production, or (3) the employment of less skillful personnel.

There are still some working situations where the physical environment provides excessive stimulation, through extreme heat, cold or noise (page 67), and this can induce and sustain a state of over-arousal. More commonly, the stress is psychological rather than physical. Information may be presented faster than it can be handled by the brain (Poulton, 1970; page 85). Pressure may be applied to executives and sales workers to achieve excessive and unrealistic objectives. Other threats to the security of employment may be seen in automation, advancing age or chronic disease. Stress may be created by lack of sleep, perhaps a product of disturbed circadian rhythms secondary to shift work or intercontinental travel (page 268). Even an unfamiliar foreign city can create a vicious circle of increasing arousal, sleeplessness and poor working performance in a harassed executive.

Many employees, faced by a state of chronic over-arousal, react by self-medication with alcohol or barbiturates. Such treatment is at best palliative, and rapidly leads to addiction if the problem is not tackled at its source. The good employer should thus be watchful for signs of over-arousal.

Motivation

The psychosocial environment includes both the cultural background of the individual worker (page 233) and the group attitudes of factory staff. The motivation of the individual and the general *climate of opinion* have a marked influence upon willingness to accept and learn new techniques, to operate machines at maximum potential and to avoid unnecessary damage to expensive equipment.

Motivation, like arousal, can be modified by adverse physical conditions—an excess of heat, cold or noise can reduce the enthusiasm of a work force to the point where productivity falls, morale is at a low ebb, and equipment is used carelessly or inefficiently. To a lesser extent, the reverse is true of unusually good working conditions.

THE PHYSIOLOGY OF ENVIRONMENT

Rather than conform to a traditional physical classification, environmental problems will be discussed in terms of altered physiology—oxygen transport, muscular activity, sensory feedback and other disturbances of normal function.

Oxygen Transport

We have noted elsewhere (pages 55) that fatigue is likely if a worker operates at more than 40 to 50 percent of maximum oxygen intake over an eight-hour shift (Burger, 1964; Åstrand, 1967). Exposure to extremes of heat, high altitude and high pressure make such a situation more likely to arise.

A combination of vigorous physical activity and excess *heat* may be found in deep mines, in some blast furnaces, bakeries and the engine rooms of older ships (Leithead and Lind, 1964; Shephard, 1971; Kerslake, 1972). Within a few minutes of exposure, the potential oxygen transport is reduced by a diversion of blood flow from the active muscles to the blood vessels of the skin, in an attempt to dissipate accumulating body heat. Under adverse conditions, the skin vessels may receive as much as 25 percent of cardiac output (Shephard, 1968). Sweating also begins within a few minutes, and the cumulative loss of fluid (up to 2 liters per hour) is supplemented by edema formation in the peripheral tissues, further restricting cardiac output and thus oxygen transport. At the same time, there may be a small increase in the oxygen cost of activity, partly because the fatigued worker is more clumsy, and partly because the rising body temperature increases the oxygen needs of physically inactive tissues. With repeated exposure, acclimatization occurs. This is partly a matter of learning; the worker discovers tricks that avoid the hottest parts of the factory. Sweat is also produced earlier and in greater quantities, so that the need for a large skin blood flow is reduced.

The approach to provision of a thermally satisfactory working environment is typically empirical. Tests are made to check whether a desired combination of work load and heat burden can be sustained without a progressive rise of pulse rate over the course of the day (Brouha, 1960; Horvath *et al.*, 1961; Brent, 1968). If readings show a progressive deterioration, then the environment must be improved or rest pauses extended until stability is attained.

The comfort of the working environment can often be improved by increased air movement; a fan commonly lowers the *effective* temperature by 2° C, enough to make the difference between inefficient and efficient work (Kerslake, 1972). The main disadvantage to a desk worker is that his papers are frequently disarranged. Another simple and inexpensive remedy is a dehumidifier; a reduction of relative humidity can make a large difference to *effective* temperature. Complete air-conditioning systems are being fitted increasingly in both factories and offices, as it becomes recognized that the capital costs of such installations can be recouped through increased productivity. In some parts of the world, factories are being built without windows or are even buried underground to avoid radiant heat from the sun. Where the industrial process itself produces much heat, it may be uneconomic to cool the entire factory; alternatively, the worker can be provided with an air-conditioned control cab, an air-ventilated suit or a specially cooled rest area.

With acute exposure to *high altitudes*, oxygen transport is impeded by the decrease in density of the inspired air. Impairment of highly skilled psychomotor tasks such as night flying can be detected at altitudes as low as 4000 ft (McFarland, 1953; Gillies, 1965). A reduction of maximum oxygen intake first appears at 5000 to 6000 ft, and develops progressively as the altitude is further increased (Goddard, 1967; Margaria, 1967; Jokl and Jokl, 1968). Industrial operations such as mining are possible at altitudes of up to 19,500 ft, but supplementary oxygen is desirable for skilled work above 10,000 feet, even if the physical effort involved is quite light.

In physiological terms, the man who is working at altitude finds it impossible to increase his breathing to the point where the sea level ventilation is restored. In consequence, there is a drop in the partial pressure of oxygen in alveolar gas, an impairment of oxygen diffusion across the lung membranes, and a drop in the oxygen content of arterial blood. Assuming that the completeness of oxygen extraction in the tissue capillaries remains essentially unchanged, less oxygen is thus transported for each liter of blood that is pumped around the circulation. At altitudes in excess of 10,000 feet, the maximum cardiac output also diminishes, due to a drop in both maximum heart rate and maximum stroke volume. As maximum oxygen intake falls, the heavy worker is inevitably brought closer to a fatiguing load (40% to 50% of aerobic power), with slower recovery following effort. During the first few days at altitudes of 10,000 to 18,000 feet, motivation is decreased and the sense of fatigue is augmented by mountain sickness—a syndrome characterized by headache, nausea, vomiting, disturbed appetite, irregular breathing, irritability, lassitude and sleeplessness. Although the symptoms are rather nonspecific, they apparently have a physiological basis in a rising pressure of cerebrospinal fluid, a lack of oxygen in the brain, and an excessive washout of carbon dioxide from the body.

With more prolonged altitude residence, acclimatization occurs (Shephard, 1973). An adjustment of body buffer systems resets the respiratory chemostat, thereby permitting a greater ventilation and a partial restoration of alveolar gas pressures. The oxygen carrying capacity of unit volume of blood is augmented by an increase of hemoglobin concentration, and adaptations of enzyme systems occur at the tissue level. Nevertheless, oxygen transport is not fully restored; the heart rate is still abnormally high in submaximum effort, and the maximum oxygen intake remains below its sea level figure. Even well-acclimatized miners refuse to live above 18,000 feet, or to work at more than 19,500 feet.

Problems of gas transport are also encountered when working under very *high pressures*, as in certain diving operations (Miles, 1962; Lambertsen, 1967; Bennett and Elliott, 1969). At a depth of 300 feet (approximately 10 atmospheres), the density of respired gas is such that the work of breathing is greatly increased, and a large part of the available oxygen supply is diverted to the muscles of respiration. At the same time, ventilation is impeded, and there is a substantial build-up of carbon dioxide in the tissues. The simplest remedy is to reduce the density of the respired mixture, replacing nitrogen by helium.

Muscular Activity

The efficiency of muscle activity is influenced by the local tissue temperature. This is well recognized by the athlete who warms up prior to a contest. Mechanisms of benefit include an increase of peripheral blood flow, a reduction of tissue viscosity and an increase in the activity of tissue enzymes.

However, the typical athlete is concerned with a relatively brief feat of superb performance; with the more regular rhythm of activity anticipated in industry the muscles concerned reach a steady temperature within about five minutes. Thereafter, the problem is heat dissipation rather than warm-up. Even in a *cold* climate, if adequate clothing is worn, the worker is only likely to encounter loss of muscle power, stiffness, and a risk of muscle tears during the first few minutes of activity.

Prolonged exposure to *heat* can lead to either progressive muscular weakness (heat neurasthenia) or painful muscle cramps (Wyndham, 1973). Both problems are associated with an excessive loss of mineral ions such as sodium and potassium in sweat, and possible exhaustion of mineral ion regulating mechanisms in the adrenal cortex. The affected worker may become irritable, with poor performance in activities calling for teamwork. If hot working conditions are unavoidable, symptoms can be obviated by taking additional salt at meals, supplementing this if necessary by drinks containing mineral ions.

Sensory Feedback

An appropriate sensory feedback is an essential constituent of any man/machine system (page 21). Unfortunately, many types of environment distort normal sensation.

Problems peculiar to the visual and oculogyral systems are discussed elsewhere (pages 144).

Fine motor tasks are performed awkwardly in the *cold* (Gillies, 1965), partly because of impaired feedback from pressure receptors in the numbed fingers, and partly because clumsy gloves are worn. Reflex control of fine movements is hampered by incomplete relaxation of antagonistic muscles and increased tissue viscosity. With severe chilling, central coordination of movements in the brain may also suffer. The problem has practical significance in Canada; each winter a num-

ber of workers become so cold that they fall to their death while con-
structing tall buildings. Protective measures include the provision of
windbreaks and good windproof clothing. The latter must allow some
adjustment of insulation so that its wind-resisting properties are
not destroyed by an accumulation of sweat when vigorous work is
undertaken.

Clumsiness may also arise in the *heat*. Here, difficulty in performing
fine work arises because the skin surface is covered with sweat, and
small objects cannot be gripped securely.

Communication is essential in a team operation. Nevertheless, it
can be quite difficult to arrange, both in the rarefied atmospheres of
high altitude and also in the compressed foreign gases used in *underwater exploration* (Bennett and Elliott, 1969). Speech frequencies are
distorted not only by the unusual gas conditions but by resonance of
breathing equipment (Davies, 1962), and voices may become virtually
unrecognizable (page 123). Attempts have been made to devise electronic
equipment that will restore normal sea level frequencies of speech,
but so far, these have not been too successful. Reliance must thus
be placed on simple, unequivocal commands, often given mechanically
(for example by pulling on a rope).

Other Disturbances

Each of the hostile environments we have discussed can influence
not only the performance but also the health of the worker, and often
the impact of the environment upon health is modified by physical
activity.

An excessive *heat* burden can lead to fatal hyperpyrexia (Wyndham,
1973), and the likelihood of such a contingency is influenced by the
rate of metabolic heat production. Excessive *chilling* may lead to fatal hypothermia (Gillies, 1965); physical activity protects against this
by increasing the output of body heat, but such a gain is largely offset if increased quantities of cold air are forced over or through the clothing by vigorous limb movements (Kerslake, 1972). Acute pulmonary
edema is a potentially fatal complication of exposure to *high altitudes*
(Jokl and Jokl, 1968); the pathology is still somewhat in dispute, but
one factor seems a sudden return to heavy physical work in a mountain-
ous region with inadequate acclimatization. The *underwater habitat*
places many restrictions upon the worker, relating to the toxicity of
oxygen, inert gases, and CO_2, and problems of gas compression and expansion (Miles, 1962; Lambertsen, 1967; Bennett and Elliott, 1969);
physical activity hastens the appearance of both oxygen and CO_2 intoxi-
cation.

Lastly, we may note the mechanical interaction between man and
his tools. It follows from Newton's laws of motion (page 168) that when

equipment such as an electric drill is set in motion, there is a tendency for the operator to move with the machine. In a normal environment, it is easy for the worker to brace himself against such forces, but it is very difficult when working underwater or in outer space (page 147).

Considerable thought is thus being devoted to the invention of tools that impart a minimum torque to the operator.

CHAPTER 5

FATIGUE IN INDUSTRY

FATIGUE IS ONE OF THE MOST COMMON complaints of the worker. It is also a prime target of the ergonomist, as he seeks to improve the interaction between man and machine. Nevertheless, it remains an elusive phenomenon, and for this reason is very difficult to correct. It may result from a combination of several forms of stress, mental, physical and physiological. The physiological type of fatigue is perhaps the best understood, and we shall devote most of the present section to its consideration.

PSYCHOLOGICAL ASPECTS

Psychological fatigue or mental tiredness is a very real complaint, particularly if a man is employed in repetitive work on a production line (Burger, 1964). It may arise acutely, but is more often chronic; it then has an emotional or situational rather than a physiological basis. It can often be qualified in such terms as a diminution in the quality or quantity of industrial output, with an increasing loss of usable time, absenteeism and a high labor turnover (Linden, 1969). Nevertheless, it remains essentially a subjective phenomenon, with no obvious physiological or biochemical accompaniments. Occasionally, there may be evidence of overwork, in itself often a sign of some more general psychological problem. But more usually, the employee will complain that his capacity exceeds that of the task he is required to perform. He becomes acutely conscious of the uniformity and monotony of his work, loses interest in his job and develops a poor and irregular rate of production.

The basic situation is one of under-arousal, with progressive loss of vigilance (Poulton, 1970; page 217); however, the picture may be complicated by adverse psychological reactions to either the monotony of the work itself or the associated environment. Heat exhaustion, for example, may be marked by anxiety, irritability and even belligerency (Wyndham, 1973).

Psychological fatigue can usually be distinguished from physiological tiredness because the worker demands a change rather than a rest. Productivity can be improved by the various devices suggested for

72

increase of arousal (page 215). There is much to commend a ten-minute mid-morning calisthenics break for this purpose.

PHYSICAL ASPECTS

The physical problems associated with adverse environmental conditions are discussed on pages 66 to 71. A hostile environment may increase the liability to physiological fatigue, either directly (by impairing oxygen transport or weakening muscular power) or indirectly (by increasing clumsiness or adding the encumbrance of protective equipment). Some constraints (such as poor lighting or excessive noise) may lead to fatigue by increasing the psychomotor demands of a task. Other unpleasant features of the environment may merely disturb the attitudes of the worker, making him less resistant to psychological fatigue (Hashimoto *et al.*, 1971).

PHYSIOLOGICAL ASPECTS

There is little question that physiological fatigue has an objective basis—indeed, it can be demonstrated in an isolated muscle preparation. It is often defined as a *diminution of working capacity*, but is more conveniently shown and measured as a failure of homeostasis at a fixed rate of working—a rising pulse rate, ventilation, gas exchange ratio, core temperature and blood lactate. Despite these objective signs, there may be a wide disparity between such evidence of fatigue and subjective feelings of tiredness, emphasizing the substantial margin between the physiologically possible and the psychologically acceptable.

The physiological causes of fatigue are varied, including problems of general and local circulatory homeostasis, a lack of biochemical requirements (glucose, glycogen and mineral ions), and an exhaustion of central regulatory mechanisms.

General Circulatory Homeostasis

The ability to maintain general circulatory homeostasis is indicated by the maximum oxygen intake (page 45). While an effort equivalent to 100 percent of maximum oxygen intake can be sustained for a few minutes, it is not possible to operate at more than 60 to 70 percent of aerobic power over a one-hour period, and fatigue is likely if more than 40 to 50 percent of aerobic power is demanded over an eight hour working day (Burger, 1964; Åstrand, 1967; Bonjer, 1968; page 55); this last figure can conveniently be translated into a pulse ceiling of 120 to 130/min for a young person, and 110/min for an older worker.

Why is it not possible to continue working indefinitely at the maximum oxygen intake? This is partly an expression of circulatory fatigue. As exercise continues, an increased proportion of the total blood volume is found in the peripheral circulation, reflecting not only the increased demands of heat dissipation but (probably) an exhaustion of sympathetic regulatory mechanisms. Sweating and a progressive loss of fluid into the active tissues also contribute to the reduction of central blood volume; there is therefore a progressive decline in the stroke volume of the heart, with a tachycardia in submaximum effort, and a proportionate diminution of maximum oxygen transport. A second important consideration is that effort at more than 50 percent of aerobic power inevitably involves an anaerobic component, and if the intensity of activity surpasses this threshold there is a progressive and fatiguing build-up of anaerobic metabolites over the working day (Kay and Shephard, 1969).

The 40 to 50 percent limit is essentially situational, and may need to be lowered if working conditions are unfavorable. Thus a hot climate leads to an enhanced rate of fluid depletion and a more rapid onset of circulatory fatigue. Likewise a poor posture or the use of small muscle groups enhances the rate of lactic acid accumulation.

Fortunately, most industrial demands fail to reach 40 to 50 percent of the worker's aerobic power (Brown, 1964a; page 55). However, any circumstance that lowers the maximum oxygen intake of the individual (such as old age, illness, malnutrition or even a general lack of physical fitness) makes this form of fatigue more likely.

Local Circulatory Homeostasis

Local circulatory homeostasis implies an ability to perfuse the actively contracting muscles. This may be difficult to ensure during vigorous dynamic activity, and becomes impossible during strong sustained isometric contractions as when supporting the body or lifting heavy weights. Laboratory experiments show that interruption of the muscle circulation commences when a muscle is contracting at 15 percent of its maximum voluntary force and becomes complete at 70 to 80 percent of maximum voluntary force (Lind and McNicol, 1967; page 53). In the absence of blood flow, fatigue is rapid. The genesis of mysterious fatigue factors is now largely discounted, but in the absence of perfusion, work must be performed anaerobically, with an inevitable accumulation of lactic acid. Some authors have complained that there is not a close relationship between increases in blood lactate and the onset of fatigue. However, the probable explanation is that their measurements are made in the wrong place; the significant factor is the intramuscular accumulation of lactic acid. When the intramuscular pH has dropped to about 7.0, this inhibits further chemical activity, and fatigue is felt.

Interruption of the local circulation is more likely to occur in the worker who has weak muscles than in a person who is strong. It is also more likely when the task is carried out by a small group of muscles. The intrinsic and extrinsic muscles of the eye are particularly prone to fatigue (Brown, 1964b). Excessive attempts at accommodation of the lens may arise from poor lighting, bad handwriting or printing, and a need to handle very small objects. Difficulties are exacerbated by uncorrected errors of visual refraction or a poor balance of the extrinsic eye muscles. Glare may likewise lead to fatigue of the sphincter pupillae muscles.

Any posture that calls for the support of either body weight or a heavy load can lead to this class of fatigue. Posture is an important consideration in lifting (page 185), carrying, and other forms of work performed from the standing position. The body weight itself can impede the circulation to some regions such as the buttocks while sitting. A good design of chair (page 321) allows the adoption of several possible sitting positions without loss of efficiency on the part of the worker.

Biochemical Requirements

Under aerobic conditions, the fuel for activity may be either fat or carbohydrate; the relative usage of the two energy sources varies somewhat with diet, training and the intensity of effort, but as much as 75 percent of vigorous work can be based upon the metabolism of carbohydrate (page 51). Neither the immediate glucose content of the blood (around 6 gm, 24 kilocalories) nor the potential mobilization of sugar from liver glycogen (1 to 2 gm/min, 4 to 8 kilocalories/min) are sufficient to meet maximum demands, and the body is thus heavily dependent upon local reserves of glycogen within the muscle fibres (1.5 gm/100 gm of tissue, total about 400 gm).

If the worker is engaged in near maximal aerobic activity, glycogen reserves are exhausted in 90 to 100 min (Hultman, 1971). Thereafter, he is dependent upon the transport of fat to the active cells. This is sufficient to sustain at least 50 percent of maximum aerobic power, but unfortunately cannot provide fuel for anaerobic activity. Thus, there develops a sensation of intense weakness and fatigue whenever postural or isometric effort must be made. If anaerobic activity cannot be avoided, it is undertaken at the expense of a further depletion of an already low blood sugar. Blood glucose readings of 50 mg/100 ml or less are possible with severe and protracted effort. The brain is dependent upon carbohydrate for its metabolism, so that the falling blood sugar may add problems of central nervous system fatigue to the peripheral weakness of the muscle fibers. Measurements of oxygen cost and integrated electromyograms both show more clumsy movements during fatigue, although it is not clear whether this indicates a failure of central coordination or an attempt to disperse the task over less fatigued muscle groups. Certainly, the speed of motor reactions can be increased by the administration of small doses of glucose or sugar at this stage, and a given muscle may still respond to direct stimulation when it is no longer responsive to impulses from the central nervous system.

Other Factors Contributing to Fatigue

Simonson (1971) has reviewed the many other factors that can contribute to fatigue. Almost every step from the genesis of the nerve impulse to the final restoration of the *status quo* can be implicated.

The disease of myasthenia gravis is associated with marked weakness and fatigue. Here, the problem seems a failure of transmission at the neuromuscular junction. Excessive stimulation of a given nerve leads to some depletion of transmitter substances in normal individuals (page 89), but it is unlikely that this contributes significantly to industrial fatigue.

A second medical condition associated with profound weakness and fatigue is Addison's disease—a deficiency in the secretions of the adrenal cortex. Cortisol is necessary to maintenance of blood sugar and liver glycogen stores, and aldosterone promotes sodium retention and potassium excretion. It is thus tempting to hypothesize that the stress of prolonged work leads to an exhaustion of the adrenal cortex, with consequent weakness and fatigue. In experimental animals, administration of pitressin and aldosterone increases transcellular gradients of potassium and sodium, and reduces fatigue. In men who are forced to work hard in very hot conditions, there may also be an exhaustion of the adrenal cortex, but there is little evidence that this occurs when strenuous exercise is carried out under more temperate conditions. Some animal studies have shown depletion of neurosecretory material in the pituitary hypophysis, but the main hormonal response to sustained activity is a depletion of the adrenal medullary secretions, adrenaline and noradrenaline.

Lastly, it is possible that exhausting effort may induce some very slowly reversible tissue injury. Disregarding gross pathologies such as tendon ruptures and fatigue fractures of bone, there remain a number of possible subcellular changes. Evidence of altered membrane permeability can be found in modifications of the ionic balance of the plasma, the escape of various intracellular enzymes into the bloodstream, and the appearance of albumin in the urine. Tissue biopsy shows parallel changes in the ultrastructure of muscle, heart and nerve (Banister, 1971), including swelling of both mitochondria and active muscle fibers. It has been suggested that these changes are a necessary concomitant of an adaptive response to heavy exercise. Nevertheless, the dividing line that separates such appearances from irreversible tissue injury is a fine one, and it would be surprising if ultrastructural changes could develop without concomitant alterations of function (fatigue).

PART II

NEUROMUSCULAR CONTROL OF WORK

CHAPTER 6

HUMAN CYBERNETICS AND
THE DESIGN OF THE NERVOUS SYSTEM

IN A SUBSEQUENT CHAPTER, we shall see how mechanical and electrical machines can be arranged to make simple decisions, and to carry out elementary mathematical operations in a logical sequence (pages 345). However, let us first examine how the central nervous system of the human operator deals with similar problems. Our discussion will not be restricted to the central controller, but will encompass also feedback loops that regulate processes in the peripheral parts of the body. Just as digital and analogue computers are used to monitor and control the output of a distant industrial plant through the deployment of suitable sensors and the feedback of duly processed information to critical, rate-limiting steps within the factory, so the human brain controls the scale of many body functions through appropriate cybernetic loops. Sometimes, the loops are extremely complex, but clues as to their nature can be obtained by modeling the suspected arrangement with an analogue or a digital computer, the observed behavior of the model being compared with anticipated responses to steady state and transient variations in the external environment.

ORGANIZATION AND BEHAVIOR

The evolutionary development of sophisticated control mechanisms has important implications for the behavior of living organisms. A single cell is commonly defined as a mass of protoplasm containing a nucleus. It forms an easily comprehended anatomical and physiological unit, and in the lowest forms of life, such as the unicellular amoeba, the mechanisms for reception of information, central coordination and the making of an appropriate response are all present within the single cell.

As multicellular organisms evolved, it would have been possible for groups of cells to live in close association, while independently receiving and responding to signals from the outside world. However, a differentiation of function proved more practical and efficient. The function of a differentiated organism comprises the summed responses of constituent cells, although the concept of *life* implies rather more than a simple summation. Thus, the heart used in a transplant operation is technically alive, but the donor is hopefully *dead* before his segment of the procedure is initiated; in a physiological sense, the control loops linking the heart with other body areas have undergone irreversible failure.

The differentiation of cells concerned with the input of information (receptors) from others concerned with the initiation of a response (effectors) represented a substantial evolutionary advance. But the possible range of responses was still very limited. Another important landmark was the differentiation of nerve cells to transfer information from receptors to effectors. The prototype of this arrangement is still found in creatures such as the jellyfish, where the connecting nerve cells form a diffuse network. If a jellyfish receives a signal on any part of its receptor surface, there is a uniform, diffuse and rather poorly directed response.

Further differentiation permitted an allocation of function within the neural cells. In the earthworm, for instance, we find groups of nerve cells organized into ganglia. Responses then tend to be localized to that area of the body associated with the stimulated ganglion. A simple *reflex arc* may be described, comprising a receptor, connecting cell and effector. Normally, this initiates a stable, localized and irreversible response; however, there is some possibility that the local reaction may be reinforced or inhibited by information transmitted from ganglia in adjacent body segments.

The next stage in evolutionary history seems to have been the development of association neurones. In a creature with this organizational pattern, one receptor is linked to several possible effectors, allowing the alternatives of a local or a general response. A fairly direct relationship can be shown between the number of association cells and the place of an organism in the evolutionary scale. Certainly, man has many more association neurones than any other species.

The general response mediated via association neurones is still crude. It is seen in patients who are unfortunate enough to have sustained a permanent injury of the spinal cord. Slight stimulation of the skin in the lower half of the body produces flexor spasm of the legs and abdominal wall, evacuation of the bladder, profuse sweating, erection of the penis and even seminal emission. Obviously, a general response of this type can be acutely embarrassing to both a patient and a pretty young nurse. The evolutionary remedy in an uninjured man has been a further allocation of function. The head ganglion has enlarged to become the brain. Here, decisions are made. If the nurse expresses an interest in showing her etchings, an *and* decision may be reached. But more commonly, considerations of ethics and/or expediency require an *or* or a *not* decision. The main function of the head ganglion seems to inhibit undesirable association pathways. This emphasis is apparent even at quite primitive levels of organization. Thus, if certain forms of annelid are cut in half, the back portion squirms in a purposeless fashion because it is denied the normal flow of inhibitory impulses from the head ganglion.

METAPHYSICS AND BEHAVIOR

Progressive elaboration of the head ganglion has given rise to the complex brain of present-day man. Stimuli now not only evoke an outward response, but also give rise to conscious sensations. On many occasions, the response of an effector organ is apparently *voluntary* in nature, and it may be extremely difficult to pinpoint an initiating stimulus. *Thought* seems even more open-ended, often being devoid of either stimulus or response. This raises an interesting metaphysical problem fundamental to human ergonomics. Can all behavior be described in physical and mechanistic terms, or does man have an element of volition, choice and free will?

From a historical perspective, scientists have argued strongly in favor of a mechanistic interpretation of life. This, surely, provided the drive behind Borelli's studies of muscles as machines, and Voit's experiments on the conservation of energy within the body. Early twentieth century scientists such as Bayliss and Huxley—perhaps as a reaction to opposition from the religious establishment—pressed their mechanistic philosophy with a fierce missionary fervor that bordered upon the tedious.

Most reasonable people today, whether laymen or scientists, accept a substantially mechanistic view of human behavior. Observed activity can ultimately be described in terms of an appropriate sequence of sensory information, central correlation and initiation of an effector response. But this does not make man a mere automaton, with no control over his behavior. We are rarely dealing with the easy cause and effect sequence typified by the knee-jerk that follows a blow upon the patellar tendon. Let us allow ourselves to be drawn back to the pretty nurse. If she talks about showing us her etchings, we are not in the realm of simple respondent behavior (page 207). There is no inevitable, involuntary reflex that drives us relentlessly towards the door of her apartment. The simple beckoning of immediate auditory and visual cues is counterbalanced by a vast array of *and, or* and *not* signals from the higher areas of the brain. Our apparently *voluntary* choice is determined by past experience, or as the psychologist would term it, by operant conditioning (page 207).

The power of any digital computer depends largely upon the size of its memory (page 331), and the unique feature of the human brain relative to commercial computers is the size and scope of its memory bank. Man has a vast capacity to set up within his cerebral cortex reverberating neuronal circuits. These store important items of information, and can be used at some later time to initiate or modify the response of an effector organ. All of the significant events of a lifetime mold the subsequent behavior of the individual. Thus, a willingness to work hard may result from childhood exposure to the protestant work ethic, and alterations of child-rearing practices may have more impact upon economic productivity than an extensive study of immediate motivation in the adult (page 232). Our concept of *voluntary* behavior also has significant implications for such problems as the maintenance of *law and order*. Positive reinforcement (page 251) of acceptable behavior from birth is more effective in molding a social conscience than the use of negatively reinforcing deterrents in the adult period. We can well agree with the advice of Paul on the development of personal ethics through a deliberate long-term program of operant conditioning:

> "Whatsoever things are true, whatsoever things are honest, whatsoever things are just, whatsoever things are pure, whatsoever things are lovely, whatsoever things are of good report; if there be any virtue, and if there be any praise, think on these things."
>
> Phil. 4:8

MEMORY AND THE HUMAN COMPUTER

Let us now look a little more closely at some specific design features of the central controller, the human brain. It is helpful to use an analogy for this purpose, noting as we do so that our concept is colored by the contemporary cultural heritage. Just as Borelli found it meaningful to liken the heart to a winepress, so we are attracted to the parallel between the functional layout of the brain and that of a modern digital computer (Fig. 17). The model illustrated was developed by Broadbent (1958) of the Applied Psychology Unit at Cambridge University.

In the place of punch cards or tape, the *input* of new information comes from specialized receptor organs (the eyes, ears, nose, skin and so on). The rate of operation of the *accumulator* (page 330) is finite, and unfortunately it has to contend with a highly variable flow of information from the receptor organs. Thus, there is a *short-term memory store* somewhat analogous to the buffered input of third-generation digital computers. Access from the short-term store to the accumulator is governed by an input selector gate. As in many university computer facilities, levels of priority are assigned to different items of information on a semirational basis. In general, those concerned with survival are processed in preference to those that deal with abstract questions of philosophy. But occasionally, a dubious judgment is made; thus, a

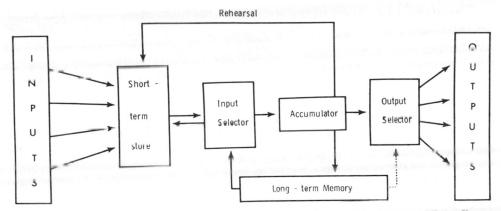

Figure 17. Schematic diagram of the human computer, based on studies of Broadbent (1958).

driver may process information on the skirt length of a girl boarding a bus, rather than on the tail signal of a vehicle immediately in front.

Here we see an important basis for the human factor in accidents (page 355). The accumulator—as in most computers—can only carry out one operation at a time, and the input gate must select the required information within the time span allocated for a particular operation. This point was demonstrated by Welford (1952). His study antedated the miniskirt by some years, and he set his subjects two rather simple tasks (moving a short lever either upwards or downwards in response to the sound of a bell or of a hammer impact). If the second signal was received prior to execution of the first response, the reaction time for the second signal was always lengthened by *at least* the overlap between the second signal and the initial response. In terms of Broadbent's model, the second signal could not be transferred from the short-term store to the accumulator until the first signal had been safely registered in the output selector. Variations in the level of arousal may be envisaged as altering performance by causing an inappropriate setting of the input selector; this may occur if there are too many extraneous signals (over-arousal) or insufficient signals to which a response must be made (under-arousal). In either case, the limited *conductance* of the information processing channel becomes jammed with irrelevant material.

On account of the switching processes used in the human computer (page 89), the speed of operation is frustratingly slow relative to modern electronic equipment. If a simple *and* response is required (such as pressing a lever when a bell is heard) the reaction time may be 0.2 seconds, and if the complexity of computation is increased, then the reaction time is lengthened as the logarithm of the number of alternative responses that must be distinguished (Merkel, 1885).

The size and stability of the short-term memory store is rather critical in view of the slow speed of data processing. Unfortunately, this aspect of the human brain compares poorly with computer back-up stores such as magnetic tape. If two digits are stored for only two seconds, there is a 56 percent chance the information will be forgotten, and with three seconds of storage, the incidence of error rises to 70 percent. One simple remedy is to *rehearse* the information (Kay and Poulton, 1951). Let us suppose we are told a telephone number, and have no pencil within reach. We pass the digits as rapidly as possible through our accumulator and back to the short-term memory. Now the limitations of the accumulator become apparent. It functions so slowly that no more than six or seven digits can be rehearsed before some start to fade in the short-term store. Current ergonomic implications include the design of telephone directories, car license plates and postal codes.

Present expansion of both telephone usage and direct dialing systems is making a seven digit code increasingly inadequate. It might be possible to reduce dialing errors, and even to use an eight digit system if the first few digits could be used very rapidly, before fading had occurred. Unfortunately, the conventional dial telephone operates at the very slow rate of one digit per second, and there is thus much interest in high-speed push-button dialing systems. The relative merits of letter and number codes are still hotly debated (Conrad, 1960 and 1967). In earlier days, letter combinations often had some associative value (for example, the telephone code MARble arch or the car license plate CAernarvonshire). But with the need for an increasing number of possible combinations, letters have now lost their associative value. Random letters are harder to learn than digits, although more information is conveyed per symbol (Cardozo and Leopold, 1963). Current British car license and postal codes both use a combination of letters and digits. In a seven-symbol sequence, the most vulnerable positions are the fifth and sixth symbols, and the digits should thus be placed in these two positions. A somewhat longer string of symbols can be retained if they are heard (7 to 8) than if they are read (5 to 6). It may be that rehearsal is normally an auditory-type procedure, and that visually presented material must be recoded before it can be rehearsed.

Short-term memory is vital to the shorthand typist and to the simultaneous translator. Poulton (1955) found that simultaneous input (speech) and output (writing or translation) led to a significant increase in errors. The most efficient technique in transmitting such information seems a short but fast burst of speech, followed by a pause for transcription or translation. The optimum length of phrase is set at six words by the capacity of the short-term memory, and the speaker should thus attempt to complete an appropriate grammatical form within this span. Lengthy displacements of the verb cause severe problems when translating English into German, or vice versa. If suitable six-word phrases are separated by appropriate pauses, as many as 180 words can be presented per minute, with a repetition error of no more than 6 to 7 percent.

Rehearsal not only conserves information in the short-term store, but also helps to establish it in the *long-term memory*. Material in the long-term store will persist for hours, days or even years. The long term capacity is surprisingly large. Neurologists have estimated that the numbers of facts in a textbook such as *Gray's Anatomy* exceeds the number of nerve cells in the cerebral cortex. Nevertheless, many generations of medical students have memorized the entire work and after an interval of a year or more have regurgitated its contents to the evident satisfaction of their examiners. The secret to this apparent achievement of the impossible lies in the association of ideas and the grouping of related facts. The capacity of the human brain to digest information—to associate and group concepts—enormously extends its potential, and this is an important reason why the decision-making capability of the human operator is unlikely to be supplanted by the largest and most sophisticated of computers.

Much of the information within the long-term memory is essentially factual. But there is also a significant storage of technical skills. Let us suppose that a secretary is asked to type a letter. The auditory input passes through the short-term memory, and assuming that the staff member making the request has sufficient seniority, the input selector immediately channels this information through the accumulator to that part of the long-term memory where motor skills have been stored. Then, signals pass through the output selector to the appropriate digital muscles. At the same time, information is fed back through the input selector to the short-term memory, keeping the typist aware of what her fingers are doing. The relative distribution of signals between input and output selectors depends much upon skill and experience. As a given task is fully learnt, much less conscious thought is given to it—indeed, it is said to become *automatic*, and there is a corresponding increase in the speed of performance for a given level of accuracy. Furthermore, because less information is directed back through the input selector, *spare capacity* develops, with an ability to perform other tasks simultaneously. Thy typist, for example, can gossip cheerfully while producing a tolerable letter. A second good illustration of spare capacity is provided by an experienced car driver (Brown and Poulton, 1961). When the novice is first confronted by the bewildering array of gauges, pedals and levers that control the average car, his task seems impossible even with undivided attention. Yet as experience grows, almost every movement becomes automatic. A bus driver can sell tickets, make change and generally attend to the needs of his passengers while the vehicle is in motion. However, passengers would do well to ponder the finite capacity of the human computer. The wrath of the average bus driver is provoked when simple problems are presented in very dense traffic, and when excessively complex demands are made on an empty road. Studies of experienced police drivers (Brown and Poulton, 1961) have led to rather similar conclusions. Subjects were required to detect any new numbers introduced into an eight digit sequence presented by a tape recorder at four-second intervals. Under all road conditions, there were more errors when driving than when at rest, but the percentage of errors was further increased when the route passed through congested shopping streets. Perhaps because the vehicles were operated by highly trained police drivers, the subsidiary task caused no deterioration of driving skill (as assessed by vehicle speed and the number of control movements that were made).

ELECTRONICS OF THE HUMAN COMPUTER

A man-made digital computer consists in essence of a vast array of on/off switches (page 332). In the human central nervous system, responses are rather more complex, and there is thus a greater potential for adaptation to varied requirements. Switching behavior will be illustrated by the properties of a simple reflex arc, although similar principles apply elsewhere within the central nervous system.

The arc (Fig. 18) comprises a receptor input via sensory nerve fibers, one or more connecting *internuncial neurones*, and an output to the

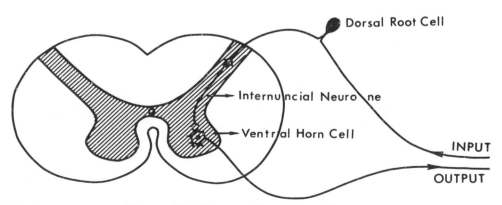

Figure 18. Diagram of a simple reflex arc.

effector muscles via motor fibers originating in the ventral horn cells. Variations in response arise through the mechanisms illustrated in Figure 19. There are no simple on/off switches. Instead, a chemical substance transmits a signal across the *synapse*, the small gap separating the terminals of one nerve fiber from the cell body of the next. There is some argument as to whether the quantity of transmitter liberated is a continuous or a discontinuous variable, but there is general agreement that a critical quantity of transmitter substance must accumulate before a discharge is initiated in the adjacent cell. Let us suppose that two internuncial fibers each impinge upon a single ventral horn cell (Fig. 19a). If a single fiber is stimulated, this liberates 0.75 units of transmitter substance, a quantity insufficient to reach the *switching* threshold. However, if both fibers are stimulated simultaneously, and each liberates 0.75 units of transmitter, the threshold of one unit is surpassed and the cell fires. *Summation* is said to have occurred. This is normally a spatial summation; temporal summation is also theoretically possible, but is less likely in practice, since the transmitter substance is rapidly destroyed at the surface of the ventral horn cell. In the second illustration (Fig. 19b), four internuncial nerve fibers impinge on three ventral horn cells. If the upper two are stimulated, each liberates one unit of transmitter substance, and two ventral horn cells

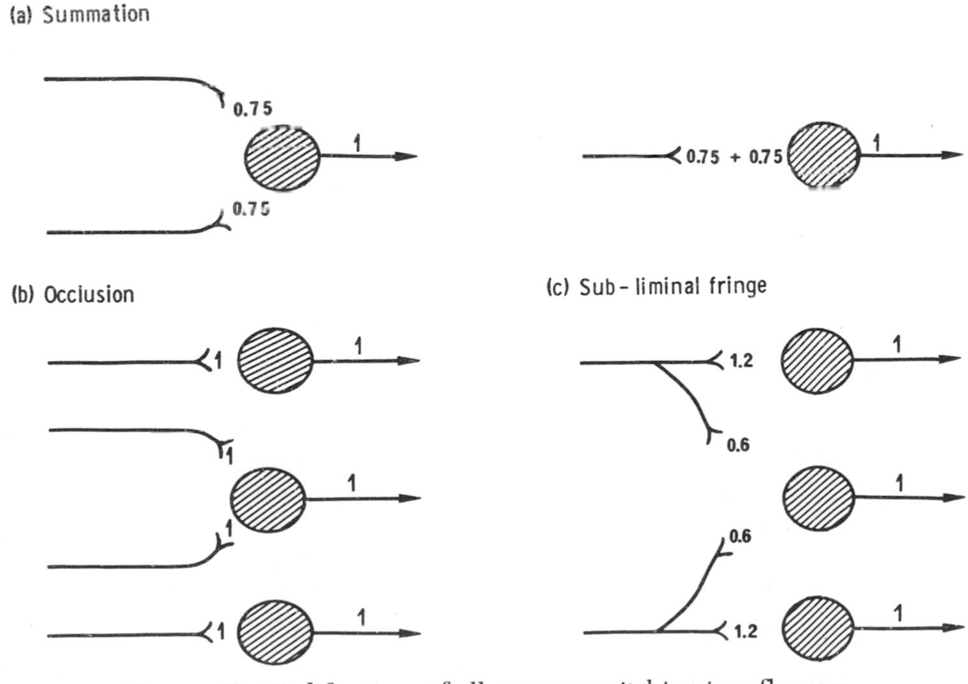

(a) Summation

(b) Occlusion

(c) Sub-liminal fringe

Figure 19. Modifications of *all-or-none* switching in reflex arcs.

will discharge. If the lower two fibers are stimulated, again two ventral horn cells will fire. However, if all four fibers are stimulated simultaneously, only three ventral horn cells can respond. *Occlusion* is said to have occurred. The third sketch (Fig. 10c) illustrates the concept of a *subliminal fringe*. Here, two internuncial fibers each liberate an effective stimulus at the ventral horn cell with which they are most closely related. However, they also liberate smaller and individually ineffective quantities of transmitter in the region of a third ventral horn cell. Thus, if both fibers are stimulated simultaneously, a response may be elicited from all three cells. Finally, there is a possibility of *inhibition*. In some instances, the chemical that is liberated at the synapse, far from stimulating the ventral horn cell, actually decreases the likelihood that it will respond to the normal transmitter substance. As with stimulation, summation, occlusion and subliminal fringe effects can be demonstrated for inhibitory fibers. The importance of the inhibition of inappropriate reflexes has already been stressed (page 83). The development of appropriate inhibitory feedback loops has done much to increase both the precision and the versatility of responses in higher species.

There is further potential for a flexible rather than an *all-or-none* response at the level of the effector organ. A suitable gradation of activity is often achieved by varying the number of *motor units** involved in a given movement. The effector organ includes both active muscle and series elastic elements, and because the elastic elements relax rather slowly, the response to two closely successive contractions exceeds that to a single muscle twitch. With a series of suitably spaced contractions, a *tetanus* is developed with three to four times the force of a single twitch. Individual motor units normally develop tetanic contractions out of phase with one another, and in this way a smooth contraction of the muscle as a whole is ensured. The power output of a muscle is markedly influenced by both its speed of shortening and the magnitude of any external resistance to its contraction. With an infinitely light load, shortening can be very rapid, but most of the energy of muscle contraction is dissipated against the internal viscous resistance of the effector organ. With an infinite resistance, no shortening occurs. Muscle force is now large, but again no external work is developed. The conversion of chemical energy into useful work reaches an optimum when the muscles are contracting rhythmically at about 30 percent of their maximum force. A number of machines such as the bicycle operate at this level of loading.

The slow speed of operation of the human brain is readily understood

---

* A motor unit comprises the several muscle fibers supplied by a single ventral horn cell.

in terms of its switching mechanism. The rate of diffusion of the chemical transmitter through a fluid medium is inevitably rather slow, and a finite delay is imposed at each synapse. This is in the order of 0.1 to 0.2 msec. If due allowance is made for further delays in the neural transmission lines connecting the central nervous system with receptor and effector organs, an estimate can be formed of the number of synapses involved in any given reflex.

Unlike most electromechanical switches, the human nervous system does not show sharp *on* and *off* transients. Transmitter substance may accumulate in the subliminal fringe, so that the number of ventral horn cells involved in a given activity increases with its repetition. *Recruitment* is then said to have occurred. When the stimulus is withdrawn, there is often an *after-discharge*, due partly to persistence of transmitter and partly to passage of the signal through long and reverberating internuncial pathways.

An appreciable interval is required between successive operations of the switching mechanism. During this *latent period*, there is a restoration of the chemical *status quo*. If a given switch is used repeatedly, *fatigue develops* (page 72). The switch then fails to respond until a prolonged recovery interval has been allowed. The onset of fatigue seems associated with a much less ready reversal of metabolic change at the synapse. Such factors as impairment of local blood supply, oxygen lack and depression of neural cells by toxic agents all hasten the appearance of fatigue.

FEEDBACK LOOPS IN THE BODY

The Concept of Cybernetics

Cybernetics is an important component of the discipline of ergonomics. The term "cybernetics" was first introduced by Norbert Weiner in 1948. The concept carries a connotation of control through the feedback of information, and ranges in a comparative way over the theory of nerve networks, electronic computing machines, servo devices for the control of machinery, and other systems for the processing of information (see also pages 32–35). We shall here look briefly at a number of networks for the regulation of body function, develop some comparable analogue circuits, and examine the possibility of using these circuits to replace the human controlling system.

A simple electrical loop is illustrated in Figure 20. The feedback line regulates the output of a D.C. amplifier, A. The output voltage E_o is opposite in sign to the input voltage E_i, and is scaled by the ratio of feedback resistance (R_f) to the input impedance of A (R_1):

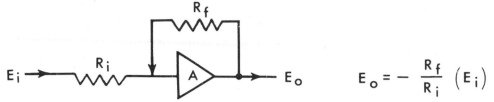

Figure 20. A simple example of an electrical feedback system.

$$E_o = - \frac{(R_f)}{(R_i)} E_i$$

Details of the input impedance term are discussed on page 35. The majority of body systems are more complex than this example. In general, they are characterized by negative feedback—in other words, the initiation of a change inhibits its continuance. This leads to stability of the loop (page 36). However, there are a few examples of positive feedback within the body, and we shall first consider one such mechanism—the transmission of a nerve impulse.

The Neural Generator—A Positive Feedback System

We have noted that the neural switchboard does not have a clear-cut on/off mechanism. Nevertheless, positive feedback ensures relatively sharp transients of activity, and an all-or-none response within a given fiber.

Normally, there is a high concentration of sodium ions external to the neural membrane, and a high concentration of potassium ions within the neural cytoplasm. The corresponding electrochemical potential E is given by the formula

$$E = \frac{RT}{nF} \log_e \left(\frac{K^+_{in}}{K^+_{ex}} \right)$$

where R is the gas constant, T is the absolute temperature, F is Faraday's constant (the number of coulombs of electricity required to liberate one chemical equivalent of any substance), n is the valency of the ions, and K^+_{in} and K^+_{ex} are the ionic concentrations on either side of the cell membrane. Sodium ions are excluded from the cell by metabolic effort—the *sodium pump* mechanism. However, if the metabolic activity of the membrane is inhibited by certain chemicals, or the size of the *pores* is increased by a distortion of the cell membrane, an unstable state is created. If the disturbance exceeds a certain threshold value, positive feedback develops; the potential difference across the cell membrane decreases progressively until a discharge occurs (Fig.

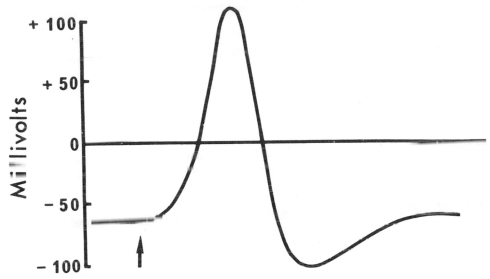

Figure 21. Electrochemical potential at the neural membrane. The voltage of the interior of the cell is expressed relative to extracellular fluid. At the point indicated by the arrow, a strong stimulus initiates positive feedback, with generation of a nerve impulse.

21). The disturbance propagates itself through the cell body and along the nerve fiber as a discreet impulse, and is followed by a *refractory interval* before a further response can occur. During the recovery phase, the mechanism that increased membrane permeability is inactivated, and the normal ionic balance is progressively restored, as sodium ions are pumped out of the cell body. Initially, the refractory period is absolute—no response can occur, irrespective of stimulus strength—but in the later phases of recovery, there is a relative refractory period, when a further discharge can be initiated by a high intensity of stimulation.

The strength of stimulus needed to generate a nerve impulse depends upon both its duration and its rate of presentation. The ergonomist is particularly interested in the behavior of sensory end organs, but the majority of experimental physiologists have worked with large nerve fibers, applying electrical stimuli with sharp *on* and *off* transients. Under such conditions, it is possible to define (Fig. 22) a *rheobase* (the minimum stimulus required to initiate a discharge over an infinite period) and the *chronaxie* (the time required to initiate a discharge when the stimulating current is of twice the rheobasic strength).

In real life, many stimuli develop gradually, and the *shape* of the on-transient then has an important bearing upon whether the discharge threshold is reached. Many sensory receptors *adapt* rather quickly

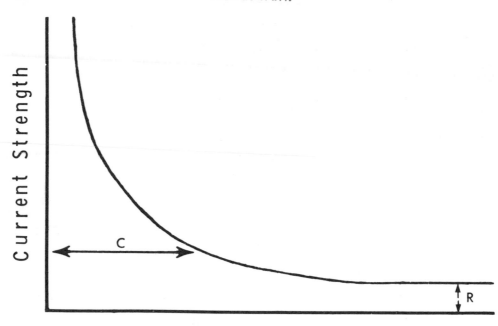

Figure 22. The relationship between effective stimulus length and current strength. R = rheobase (threshold for stimuli of infinite duration), C = chronaxie (effective stimulus length for current of twice rheobasic strength).

to a new stimulus level, and can only detect sudden changes in the external environment. Lighthouse designers have known for many years that a flashing beam provides more effective warning than a steady light. Currently, an increasing number of road traffic signals are also being changed from a steady to a flashing mode (page 365). Pain receptors, in contrast, have a very slow rate of adaptation. A chair design that causes a painful ischemia of the sciatic nerve (page 321) will probably become more rather than less uncomfortable with a prolonged period of sitting.

We have noted already that the positive feedback arrangement within the nervous system tends to give an *all-or-none response*. Once a propagated disturbance has been generated, its magnitude is independent of the initiating stimulus. This is convenient for the switching of electrical circuits within the brain, but presents a problem when information is needed on the quality rather than the mere presence or absence of a stimulus. The quality of peripheral skin stimulation, for instance, must be inferred from the depth of responding receptors (deep or superficial), the area of response, and its duration. If the stimulus is strong, individual receptors may fire more than once. All incoming sensory information is referred to *association areas* in the cerebral cortex. Here, further judgments as to the quality of a stimulus are made. Consider a pilot manipulating a knob on an aircraft control panel. He may dimly perceive the knob with the peripheral part of his visual field (page 151), and note that it seems the correct size, shape and color. Shape and texture will be checked from the area and depth of pressure receptors stimulated by digital contact, and further sensory information will be derived from temperature receptors, muscle and joint proprioceptors (as torque is applied to the knob), and auditory feedback (both local clicks and more general changes of engine noise). If the sum of this information matches impressions stored in the association areas of the sensory cortex, then the pilot will continue to manipulate the knob until the desired setting is obtained.

Muscle Tone—A System with Positive and Negative Feedback

The system regulating the tone of the skeletal muscles provides an interesting example of both positive and negative feedback. Specific stretch receptors are located within *intrafusal* muscle fibers. Since these fibers are arranged in parallel with the muscle proper, an increase in their tension is relieved by contraction of the muscle. Under such circumstances, negative feedback occurs (Fig. 23). The loop passes from the intrafusal receptors to the ventral horn cells supplying the muscle proper, facilitating discharge; this corrects the rise of tension in the intrafusal fibers and abolishes the discharge of the intrafusal receptors. A good example of this sequence is provided by the knee jerk. The quadriceps muscle is placed under tension by an external force (a brisk tap on the patellar tendon); the intrafusal receptors initiate an equally brisk contraction of the muscle proper and this in turn suppresses their discharge.

Let us now consider the situation when the ventral horn cells supplying the quadriceps discharge. A voluntary contraction is initiated in the muscle proper. Why is it that negative feedback from the intrafusal fibers does not promptly inhibit muscular activity? The answer is provided by a second nerve pathway. As the muscle contraction is initiated via the pyramidal system, impulses also pass via the reticular formation and an extrapyramidal system, initiating parallel shortening of the intrafusal fibers. In terms of the electrical model (Fig. 20), the setting of resistor R_f is adjusted to reduce negative feedback. This pattern of operation is useful, since it permits the body to detect an unexpected resistance. Let us suppose that a man is sawing through a plank of wood; normally, the discharge of intrafusal and extrafusal fibers is well coordinated. However, if the saw touches a nail, the intrafusal

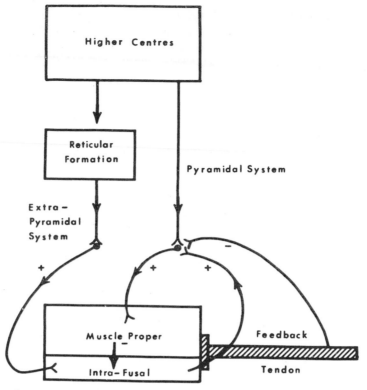

Figure 23. The regulation of muscle tone. Note that the intrafusal receptors correspond with the spindle organs of Figure 22.

and extrafusal contractions immediately get out of phase with each other, and the person senses that length-tension relationships within the muscle are inappropriate. Detection of *normal* resistance to a given movement plays an important role in the learning of many industrial tasks.

A third possible method of varying loop activity (Fig. 23) is to initiate a shortening of the intrafusal fibers via the extrapyramidal system. As the intrafusal tension rises, the ventral horn cells supplying the muscle proper are activated, relieving tension within the system. In essence, the setting of the resistance R_f (Fig. 20) has again been modified, this time with the objective of increasing feedback. Many of the *automatic* movements characteristic of a skilled operator are brought about in this way.

One objection to the basic system of Figure 23 is that the tendons are unprotected. If the tension within a tendon rises, muscle contraction initiated via the intrafusal fibers will increase it still further. This positive feedback arrangement seems calculated to produce tendon injury. A subsidiary feedback thus originates from tension receptors within the tendon; the threshold of this loop is high, but once stimulated it inhibits contraction of the corresponding muscle group.

Bioengineers have speculated whether additional properties are conferred by the dual system of muscle and tendon receptors. It seems possible that intrafusal organs respond mainly to the rate of change of tension, while tendon organs report *static* levels of tension (Buller, 1966). If this is indeed the case, then the relative proportions of dynamic and static feedback can be regulated through a change in the tension of the intrafusal fibers, as mediated via the extrapyramidal system. Engineers may recognize a parallel between this system and the variable phase advance used to minimize oscillations in a mechanical or electrical servo system (West, 1953). An arrangement to minimize unwanted oscillations is particularly important in the human body, since many limb movements are carried out at or near the natural frequency of the part, and the lengthy nerve pathways and slow synaptic transmission predispose to instability.

Regulation of Posture and Arousal

Problems involved in the control of posture and arousal provide a logical extension to our thoughts on the basic reflex arcs regulating muscle tonus. The simple system of Figure 23 is now coupled to a more complex network, having many routes for the input of sensory information, and many feedback loops (Fig. 24). Sources of postural information include receptors sensing static and dynamic head posture (the labyrith of the inner ear), stretch receptors in the neck and other body muscles, visual impulses, and receptors sensing the relative pressures on different parts of the soles of the feet.

The head is normally restored to its resting position as a response to impulses from the labyrinthine receptors. This creates a differential stretching of the neck muscles, which is in turn sensed and corrected by muscle stretch reflexes until an equilibrium posture is restored. Because his center of gravity is high, a standing man is potentially in an unstable equilibrium. However, the feedback loops seem remarkably economical in maintaining balance, and electromyograms show little electrical activity in the trunk and thigh muscles while standing.

The central control of posture occurs mainly within the cerebellum. This compares information from the eyes, ears and pressure receptors with the static and dynamic tensions of the postural muscles, and initiates any necessary adjustments in the feedback loops (page 96) controlling muscular activity. Most regulation of posture is automatic, requiring no conscious thought. Nevertheless, parallel pathways travel to the sensory cortex via the thalamus, permitting conscious perception

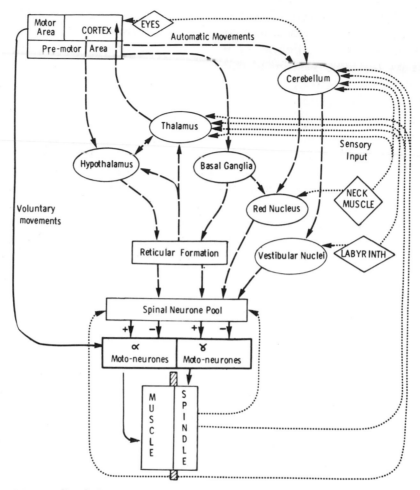

Figure 24. Feedback loops involved in the control of body posture. (From R.J. Shephard, *Alive Man!* Springfield, Illinois, Thomas, 1972).

of posture (see. also page 87). Voluntary alterations of posture are achieved via the pyramidal pathways, but the automatic adjustments that occur during the performance of normal daily tasks are initiated via fibers that link the pre-motor area of the cerebral cortex with the cerebellum.

The various sensory receptors and motor units of the body are spatially represented in inverse order over the surface of the contralateral cerebral cortex. The area of representation varies with the importance of the movement to be controlled, a large part of the available surface being allocated to the regulation of hand and fine finger movements.

Other regions of the cerebral cortex receive stimuli directly from the eyes and ears. The association areas (page 87) serve to interpret and meld information in a meaningful way. Occasionally, congenital or acquired lesions may disturb this synthetic process, creating a variety of perceptual-motor handicaps. Learning of an industrial skill may be difficult for such individuals, but if the defect is recognized and suitably modified training programs are instituted, many patients with perceptual-motor handicaps can become useful members of the labor force.

Another important area of the neural network illustrated in Figure 24 is the reticular formation of the brain. Activation of the reticular formation leads to an increase of postural tone, and depression to its reduction. The reticular formation is active when a person is aroused. Psychological aspects of arousal are considered elsewhere (page 215), but we may note that the postural control loop passing through the reticular formation has an input from the cerebral cortex. Let us suppose that a lady is bullied while she is attempting to learn to drive a car; an excessive cerebral input activates her reticular formation, making her "scared stiff." In effect, the sensitivity of her intrafusal receptors has been increased to the point where muscle tone is making any form of skilled activity very difficult. Counteraction of excessive muscle tone is a prerequisite to a successful driving lesson.

We may note also that if the sensory input to the loops shown in Figure 24 is inadequate, performance may deteriorate through loss of arousal (page 215). This is perhaps most obvious for tasks with a substantial mental component, but a minimum muscle tone is also needed for optimum physical performance. The night driver is particularly vulnerable to loss of arousal. The darkened cab of the vehicle, a comfortable seat, a pleasant neutral temperature, silence and the absence of other vehicles contribute to relaxation, inadequate arousal, and ultimately of sleep. The remedy is to increase the flow of sensory information; auditory stimulation is increased by listening to a radio or talking to a companion, the eyes are deliberately active in searching the road ahead, and the cab temperature is kept somewhat cool. If the highway is well-designed, periodic bends in the road provide an input to the eyes and the muscle proprioceptors. On long journeys, these various arousing maneuvers must be supplemented by regular breaks that involve vigorous physical activity. Many long-distance drivers also make recourse to drugs such as caffeine and the amphetamines. These produce a pharmacological arousal, but should not be taken on a regular basis; in many countries, the use of amphetamines is now illegal.

INDETERMINATE FEEDBACK NETWORKS

The Analogue Approach

Immediate knowledge of a body system is sometimes insufficient to draw a complex feedback network such as Figure 24. An alternative approach is to make an analogue of the system, and to add elements until the responses of the model mirror those found in real life. The ergonomics of analogue experimentation may be summarized as shown in Figure 25. In essence, the human mind serves as a comparator; it assesses the error of the analogue response relative to that of the system it represents, and feeds back new ideas to the analogue until an acceptable error of representation is achieved. This approach has been widely used in the study of respiratory and cardiovascular control mechanisms, and some of those with an interest in modeling have gone further, using their model to control body function when the normal mechanisms are suppressed by anesthesia.

Current Knowledge of Respiratory Control

The prime controlled variables are the composition of the cerebrospinal fluid and the blood. Information on these variables is fed to the respiratory comparator, and depending on its sensitivity, alterations in the activity of the respiratory muscles are induced via the reticular formation (page 99).

The comparator consists essentially of two groups of closely inter-related neurones on the ventral surface of the medulla (Fig. 26). One cell group is connected mainly with the muscles of inspiration, and the other with the muscles of expiration. Both groups of cells have

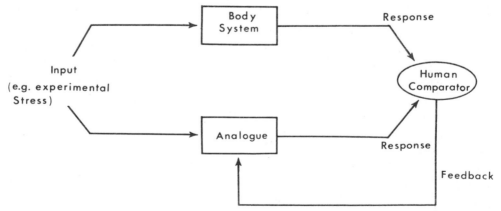

Figure 25. Diagram illustrating the development of an analogue of a body system (Jackson, 1960).

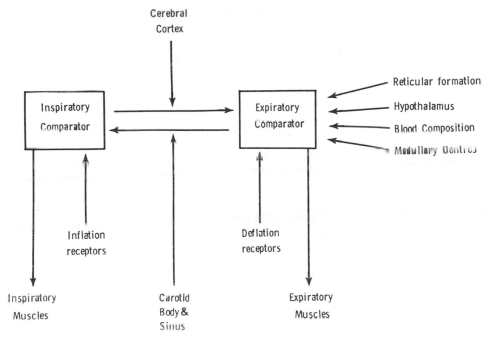

Figure 26. Information flow in the respiratory comparator.

an inherent rhythmic discharge even when isolated from the rest of the body. In cybernetic terminology, there is an output from the respiratory controller even in the absence of an error signal.

The prime signal seems a local change of tissue pH, and the sensitivity of the controller to a given change of CO_2 pressure depends upon the acid/base buffering status of medullary tissues (Brooks *et al*, 1965; Torrance, 1966). The immediate response is to local changes in the composition of blood perfusing the inspiratory and expiratory comparators. CO_2 is a stimulant, and oxygen lack a depressant. But such information is powerfully supplemented by the discharge of central and peripheral chemoreceptors. The central chemoreceptors report the composition of the cerebrospinal fluid in the lateral part of the fourth ventricle, while peripheral chemoreceptors (sensitive to both oxygen lack and a rising CO_2 pressure) are located in the carotid bodies. Control theorists have speculated that the peripheral mechanisms may provide a rapid coarse control of ventilation, and that fine control is mediated via the central chemoreceptors (Duffin, 1971).

Impulses from many other sources impinge on the medullary comparators; for simplicity, all may be viewed as altering the sensitivity of response to the prime controlled variables. Thus

　　1. *the higher centers* of the brain may increase or decrease controller sensitivity. Examples include the drive to respiration from

the motor cortex during physical activity and the voluntary inhibition of breathing when a worker senses exposure to a toxic gas.

2. the *reticular formation* modulates ventilation in parallel with
   the level of arousal. Thus, a fighter pilot who is over-aroused
   when performing complex maneuvers during combat will over-
   breathe—sometimes not only exhausting his oxygen supply but
   also producing symptoms of CO_2 lack in the body.

3. the *hypothalamus* modulates sensitivity with changes in the temperature of the body core. This particular control loop is well
   developed in animals that pant in the heat.

4. other inputs are derived from stretch receptors in the lungs, muscle and joint receptors (Fig. 24), cutaneous receptors, and centers
   regulating the cardiovascular system.

Specific Characteristics of the Control System

1. *Sensitivity.* Many man-made servosystems require a large error
signal for successful operation. However, the respiratory controller
appears to be extremely sensitive. Thus, arterial and tissue CO_2 pressures are maintained to within 1 to 2 percent of initial resting values
despite a fifteen- to twenty-fold increase of CO_2 production during
vigorous physical activity.

2. *Pre-set controls.* The respiratory comparators are pre-set to
respond well in one particular environment (for most people, normal
sea-level conditions). If a worker is suddenly moved to a different
environment, he may suffer several days or even weeks of disability
while necessary adjustments of feedback and controller sensitivity are
made. Thus, at high altitudes (page 66), the need for a larger respiratory
minute volume must be reconciled with a shortage of the prime controlled variable (carbon dioxide).

Recourse may be made to more primitive, coarse feedback loops,
via the carotid bodies. These can sustain ventilation with varying success while the body undertakes various adaptive measures. At altitude,
a renal excretion of bicarbonate increases controller sensitivity, so that
an adequate ventilation can be sustained at a lower CO_2 pressure;
an increase of blood hemoglobin level also permits greater oxygen
carriage per unit volume of blood, and thus reduces the need for hyperventilation.

3. *Lengthy transmission lines.* The exchange of the controlled variables (CO_2 and oxygen) occurs within the lungs, but the sensors reporting the partial pressures of these gases are located in the carotid bodies
and the medulla. Depending upon the cardiac output, there is thus
an inevitable phase lag in the adjustment of ventilation to changes
in CO_2 and oxygen pressures.

Limitations of Analogues

Because of the complexity of body systems, most models incorporate a number of simplifying assumptions (Grodins *et al.*, 1954). Limitations will be illustrated with specific reference to respiratory control.

1. *Linearity.* Mathematical representation of any system is helped by assuming that the elements respond in a linear manner. Unfortunately, this condition is rarely satisfied within the body. In the case of respiration, even the relationship between the controlled variable (blood composition) and the controlling variable (ventilation) is nonlinear.

2. *Single controlled variable.* A single controlled variable is usually assumed. For the respiratory system, this may be the tissue CO_2 pressure, the mixed venous CO_2 pressure, or the arterial CO_2 pressure, the first assumption being nearest to the truth.

3. *Nonphasic controlling variable.* Many body functions such as ventilation are phasic in nature, but for convenience are modeled as a continuous function of time, ignoring such problems as differences in the volumes of inspired and expired gas.

4. *Wastage of controlling variable.* A variable fraction of the respired volume is *wasted* in ventilation of respiratory dead space. Wastage is a nonlinear function of tidal volume. Nevertheless, most models assume a fixed percentage of wastage.

5. *Instantaneous equilibration.* In many models, processes of equilibration are assumed to be instantaneous. Respiratory models usually assume immediate equilibration of blood and gas phases, yet the blood requires a finite time to traverse the pulmonary capillaries, and its composition is changing in a nonlinear manner throughout this time.

6. *Spatial homogeneity.* Most modelers are content to consider the lungs as a single unit, whereas it is well recognized that ventilation/perfusion ratios differ in various parts of the lungs; alveolar and pulmonary capillary gas pressures are in fact functions of both space and time.

In a similar way, the tissues are assumed to form a homogenous pool, whereas they consist of many parallel elements, each with a characteristic pattern of metabolism and blood flow, and each having a characteristic length of transmission line connecting it with the gas exchanging surface of the lung.

7. *Problems of phase boundaries.* Solubility functions are needed to describe the movement of gases between air, blood and tissues. The functions for CO_2 are frequently assumed to be linear and equal for both blood and tissues; the influence of blood oxygenation upon the solubility factor is also commonly ignored.

8. *Other problems.* Many other problems arise in specific models. Thus, respiratory analogues commonly ignore both the pulsatile nature of blood flow, and also potential interactions between blood gas composition and the conductance of the transmission lines (cardiac output).

Possible Techniques of Modeling

The first stage of modeling is to draw a simple diagram of the system. Grodin's concept of respiratory control is illustrated in Figure 27. Assuming that an investigator has the necessary mathematical exper-

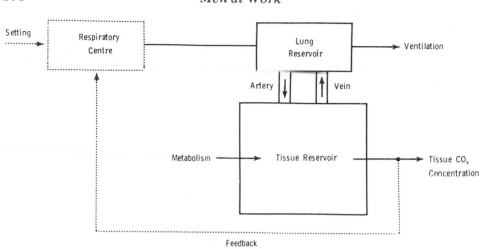

Figure 27. Simple model of the respiratory center (Grodins et al., 1954).

tise, it is possible to formulate algebraic or differential equations governing operation of such a system. However, the complexity of physiological controls makes the necessary equations very cumbersome. Many human biologists thus elect the simpler approach of using an electrical or mechanical analogue.

The basic Grodins model (Fig. 27) can be represented by quite a simple electrical circuit (Fig. 28). The conductances ($1/R_1$ and $1/R_2$) correspond to ventilation and blood transport, the capacitances (C_1 and C_2) represent lung and tissue volumes, the initial charges applied to these capacitances (Q_1 and Q_2) represent the initial quantities of CO_2 in the lungs and tissues, the current I represents the steady inflow of CO_2 from body metabolism, and the voltage E represents the concentration of CO_2 in inspired gas.

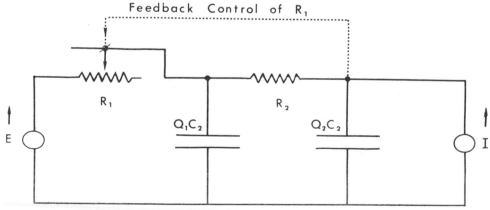

Figure 28. The electrical equivalent of the Grodins model, shown in Figure 27.

Some authors have used direct analogues of this type, with networks of conductances, capacitances and inductances (Fig. 39). Unfortunately, the tolerance limits of most electrical components are too broad to permit accurate analysis, and it is more usual to devise a corresponding circuit for a conventional analogue (Fig. 29) or digital computer (Campbell and Matthews, 1968; Duvelleroy, 1970).

Accuracy of Model

Having formulated a program for analogue or digital computer, it is possible to test the accuracy with which it represents the normal body response to a transient or steady-state disturbance. In the respiratory model, the disturbance could arise from an increase of inspired CO_2 concentration (an increase of E, Fig. 28), or from an increase of tissue CO_2 production (an increase of I, Fig. 28). The latter perhaps has more relevance to the problems of man at work, but most investigators have preferred to work with changes of inspired CO_2 concentration (presumably because it is then easy to induce sharp *on* and *off* transients).

The steady-state response of the Grodins model to an increase of inspired CO_2 is quite reminiscent of real life (Fig. 30); little increase

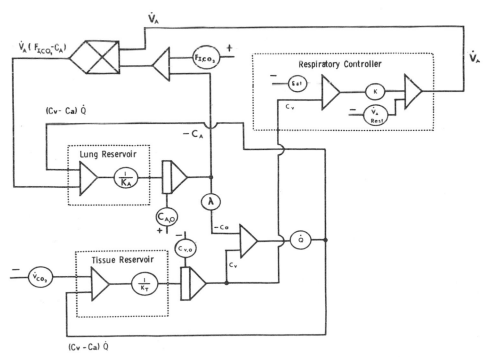

Figure 29. Grodins model of the respiratory system arranged for analogue computer simulation; compare with basic diagram of Figure 27.

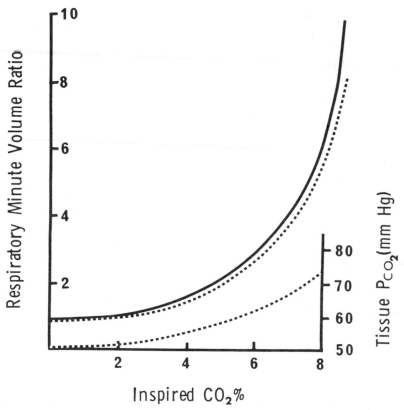

Figure 30. Steady-state response of Grodins model to inspired CO_2. Broken line indicates approximate results to be anticipated in a healthy young man. Dotted line indicates tissue P_{CO_2} (Grodins *et al.*, 1965).

of ventilation occurs with less than 2% CO_2, but thereafter there is a rapid and progressive increase of ventilation. At any given setting of the controller, a specific error signal is needed to sustain a given increase of ventilation; although Grodins describes this as an increase of "tissue P_{CO_2}," he in fact uses the CO_2 content of mixed venous blood. The size of the error signal diminishes as the controller sensitivity is increased.

A more exacting test of the model is provided by its response to a transient CO_2 signal. Normal patterns for a healthy man are illustrated in Figure 31. The on-transient of ventilation is relatively slow, and the off-transient is more rapid. Alveolar gas composition shows some *overshoot* during both *on* and *off* transients. As the concentration of inspired CO_2 is increased, the adaptation of ventilation proceeds more slowly, and the overshoot of alveolar gas composition becomes less obvious (Padget, 1927–8). If *reasonable* values for the various constants are inserted in the Grodins model, it behaves in approximately the ro

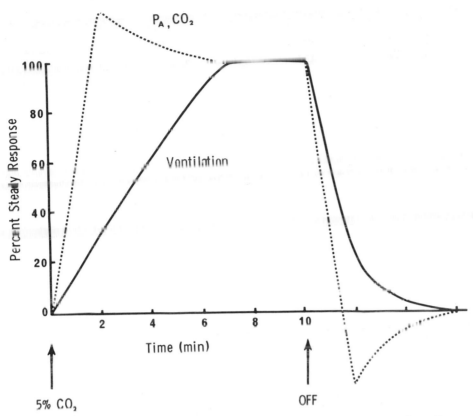

Figure 31. Response of a normal subject to inhalation of 5.4% CO_2 (based on an experiment of Padget).

quired manner (Fig. 32). However, the speed of the transient and the extent of overshoot are not matched with great accuracy (Lambertsen *et al.*, 1965; Defares, 1963). On the other hand, the reported speed of adaptation in real life varies quite widely, depending upon the care that has been taken to check that a new steady state has been reached.

Improving the Model

Steps may now be taken to improve the model as suggested in Figure 25. Defares (1963) pointed out that it was questionable to assume a uniform distribution of cardiac output to the tissue pool. The inhalation of CO_2 also seemed likely to induce a major increase of cerebral blood flow. He thus proposed a more elaborate model (Fig. 33) embodying three conductances ($1/R_1^-$, $1/R_2^-$ and $1/R_3^-$) proportional to ventilation, general blood flow and blood flow to the controlling centers. Although theoretically more accurate, there was unfortunately no means of deter-

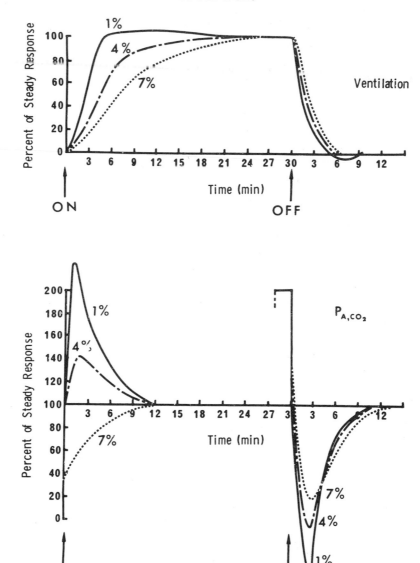

Figure 32. Response of Grodins model to varying percentages of CO_2. Assumptions: tissue CO_2 production 263 ml/min, tissue fluid volume 40 liters, cardiac output 6 liters/min, sensitivity of respiratory center 2 liters/min/mmHg, intercept for respiratory center approx. 95 liters/min, lung volume 3 liters, slope of CO_2 dissociation curve 4 ml/liter/mmHg, intercept 320 ml/liter.

mining several of the constants needed for his model, including the effective tissue pool (C_3), local metabolism (I_3) and local blood flow ($1/R_3$) to the controller. Despite these difficulties, Defares claimed that his "third stage" model gave a better representation of transient events than did the Grodins system.

Feedback Control of R_1

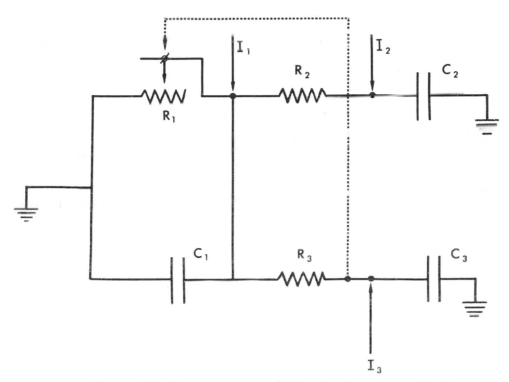

Figure 33. Defares model of respiratory control. Note that in contrast with Figure 28, the cardiac output has been divided between two conductances $\frac{1}{R_2}$, supplying the general tissue reservoir, and $\frac{1}{R_3}$, supplying the controller; each has an independent input of CO_2 production (I_2 and I_3). Further, for technical reasons, the voltage E has been replaced by a current ($I_1 = E/R_1$).

Yet more elaborate models (Milhorn, 1965a and 1966) have added oxygen sensors. The complexity of the system is then such that the corresponding differential equations must be processed by a digital computer.

PRACTICAL VALUE OF CYBERNETIC ANALYSIS

The average ergonomist is interested in practical objectives. What then is the practical value of a cybernetic analysis such as that described above? Apart from the general insight it gives into techniques of regulating complex systems, specific by-products may be listed as follows:

Identification of Controlled Variable

Before a system can be devised to replace normal respiratory controls, it is essential to define the controlled variable. It is now generally accepted that the prime variable is tissue pH, but studies such as those of Grodins played a significant role in establishing this fact. In particular, the phase lag between changes of alveolar gas composition and ventilatory response pointed to regulation at some distance from the lungs, with time required for both diffusion of CO_2 to the regulating site and saturation of the tissues to a new higher CO_2 pressure.

Studies of Controller Sensitivity

The sensitivity of the respiratory controller is normally estimated by means of a laborious and discontinuous series of measurements of alveolar ventilation and blood gas composition. However, accepting the indications of the models (that alveolar gas composition can represent the controlled variable, and that alveolar ventilation is the controlling variable), it is possible (Belleville and Seed, 1959a and 1959b) to make very rapid plots of controller sensitivity (Fig. 34).

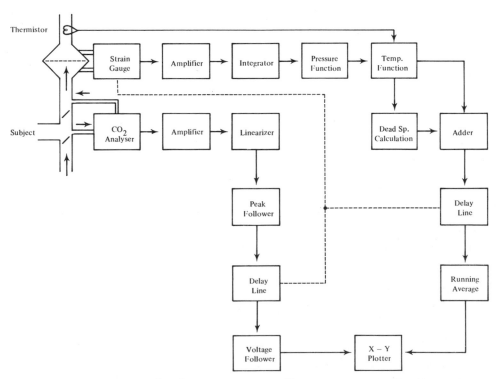

Figure 34   Arrangement for plotting sensitivity of respiratory controller (Bellville and Seed, 1959a).

Ventilation is sensed by a conventional screen flowmeter and strain gauge. An amplifier first converts the signal to a *single-ended* (unipolar) form, and a voltage detector then senses the beginning and the end of expiration; this information is fed to delay lines. An integrator converts the flow to a volume signal, and suitable pressure and temperature functions are introduced to correct this volume to standard conditions. The dead space is calculated as a predetermined fraction of tidal volume, and is subtracted prior to calculating a running average of expiratory ventilation.

Carbon dioxide concentrations are sensed at the mouth by means of an infrared gas analyzer fitted with a suitable data linearizing circuit. The end-tidal CO_2 concentration is taken as the controlled variable, and this value is applied to an x-y recorder via a circuit including a peak follower, delay line and voltage follower.

Authors such as Belleville and Seed have used x-y plots of alveolar ventilation and end-tidal CO_2 concentration to study changes of controller sensitivity during treatment with depressant drugs such as codeine and morphine.

Automatic Control of Respiration

During anesthesia, normal mechanisms for the control of respiration are suppressed. The traditional approach to patient management during anesthesia (Fig. 35) has been to establish an alternative control loop through the anesthetist (page 21). He uses his special senses to detect the condition of the patient—his color, pulse rate, blood pressure and so on, and compares this input with a setting of *expected normal values* stored in his cerebral cortex; negative feedback is then provided by appropriate modification in the rate of compression of an anesthetic bag.

From an ergonomic point of view, the arrangement is poor. The task of the anesthetist is tedious, and loss of vigilance could endanger the patient. One remedy is to devise a machine to take over the routine management of ventilation, leaving the anesthetist to handle the more difficult problems of induction and instrument failure. In the machine, electrical sensors detect controlled variables, and feed this information to a comparator; the error signal is then used to modify the rate of operation of a mechanical ventilator.

It is obviously but a small step from the circuit of Figure 34 to a system for the automatic control of breathing (Fig. 36). The carbon dioxide pressure is *shaped* as described above, and is then compared with a predetermined setting. The output of the comparator is scaled to indicate the additional alveolar ventilation needed to restore equilib-

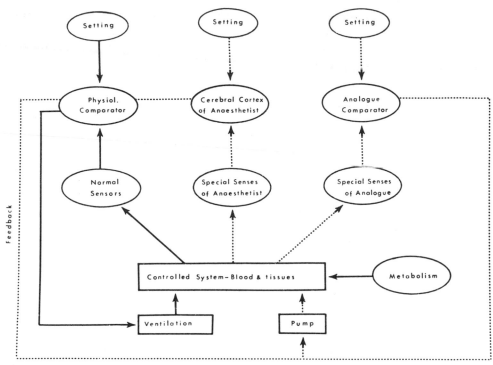

Figure 35. A comparison of normal ventilatory control with that achieved by (a) an anesthetist and (b) an analogue device.

rium. This reading is added to the observed alveolar ventilation, and the combined output is then fed to the respiratory pump.

On-line Calculations of Oxygen Consumption

The traditional approach of the laboratory worker has been to record the readings of his various instruments one by one and to complete the calculations later, in the leisure of his office.

This pattern of laboratory work has several disadvantages. It is slow,

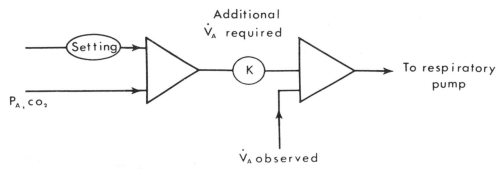

Figure 36. A simple system for the automatic control of alveolar ventilation.

and there is a real possibility of error or observer bias in the reading of the various instruments. An unsatisfactory experiment often remains undetected until the results are calculated, and there is little possibility of modifying experimental design in the light of patient reactions to an initial test.

Many of these difficulties can be overcome through on-line data processing (Lambertsen and Gelfand, 1966; Auchincloss *et al.*, 1968). In the case of oxygen consumption, it is necessary to monitor the expired gas volume and oxygen and carbon dioxide concentrations, using techniques of the type shown in Figure 34. The information is then processed through an analogue computer as illustrated in Figure 37. Practical problems of on-line techniques include:

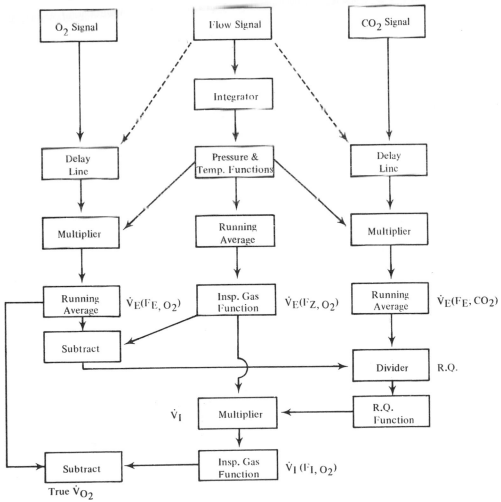

Figure 37. Circuit for the on-line calculation of oxygen consumption. Some of the steps involved are shown in greater detail in Figure 34.

1. *Variations in instrument response times.* Before processing data, it is vital that the signals for oxygen, CO_2 and ventilation have a common time base. Unfortunately, many oxygen sensors have rather slow response characteristics, and available high speed instruments (fuel cell, discharge tube and mass spectrometer) all have serious technical limitations.

2. *Accuracy of analogue computation.* A progressive attenuation of accuracy occurs as data is fed through the various amplifiers of an analogue computer (page 336). This difficulty can be largely overcome by an early conversion of data to digital format.

Investigation of Physiological Problems

Having defined the control circuit, it is possible to study abnormalities of operation that may occur in disease or during exposure to unusual environments. In the context of respiratory control, there

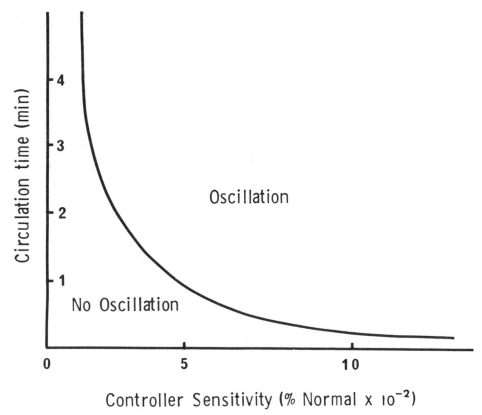

Figure 38. Combinations of circulation time and controller sensitivity at which oscillation developed in an analogue of the respiratory control system (Milhorn and Guyton, 1965).

is the puzzling symptom of Cheyne-Stokes breathing—a waxing and waning of ventilation seen commonly in patients close to death. The somewhat analogous picture of *intermittent breathing* is also encoun tered in workers on first moving to high altitudes.

Early hypotheses to account for these phenomena included an inherent waxing and waning of controller sensitivity due to a poor medullary blood flow, and an instability of control loops. Milhorn and Guyton (1965) were able to use an extension of the Grodins model to show that the second hypothesis was correct. Instability in any control system arises from either a lengthening of control loops or a diminution of damping. Milhorn and Guyton found a clear-cut relationship between the effective length of the control loop (circulation time), sensitivity of the controller and the presence or absence of oscillation (Fig. 38). With normal sensitivity, oscillation did not occur until the circulation time was lengthened from ten seconds to two or three minutes; this type of situation might occur with the sluggish circulation and large intracardiac blood volume of terminal heart failure. Oscillation could occur also with a normal circulation time but a thirteen-fold increase of controller sensitivity. In the early days at high altitude there may be some gain of controller sensitivity, but the most important consideration is probably a change of receptor organs, from the well-damped CO_2 sensors of the fourth ventricle to the oxygen sensors of the carotid bodies, intimately bathed by arterial blood.

CHAPTER 7

THE SENSORY RECEPTORS

IN THIS CHAPTER, WE SHALL LOOK briefly at the sensory systems of the skin, mouth and nose, and will consider the receptor organs of the ears and eyes in greater detail. Particular attention will be directed to normal perception in industry and problems arising from an excessive or an inadequate input of sensory information.

GENERAL CONSIDERATIONS

Specialization of Receptor Organs

Specific receptors deal with individual modalities of sensation. The skin contains nerve endings for the detection of touch, pain, heat and cold. The taste buds of the mouth report the flavor of ingested foods, and the olfactory receptors of the nose the smell of respired air. But specialization is incomplete, and excessive stimuli can be misunderstood. Thus heavy pressure on the eyes can be interpreted as a flash of light, and an infection of the middle ear may give rise to a buzzing or a ringing sensation.

The characteristics of the impulse transmitted along any given nerve pathway are largely independent of the information carried. Interpretation as a specific conscious sensation is based upon the central connections of the nerve fiber in question, and patterns of recognition established within the association areas of the cerebral cortex (page 82).

The human brain has a rather limited capacity for processing information (page 85). Only a small proportion of the total sensory input reaches consciousness at any given instant. Familiar data require no more than an *automatic* response; it is thus filtered out, and such selectivity on the part of the cortical computer is important to its effective functioning.

The Phenomenon of Adaptation

Let us suppose that a lumberjack is operating a chain saw. A constant intensity of vibration is applied to receptor organs in the hand and arm. Initially, trains of impulses pass up the corresponding sensory nerves, reporting an irritating tingling in the affected part. But as the

116

work continues, the discharge of the receptors diminishes and may even cease. Adaptation has occurred.

Some receptors, such as those concerned with touch, adapt rapidly. Others, such as those modiating painful stimuli, adapt little if at all. The relevance of adaptation to the design of warning signals has been discussed elsewhere (page 365). What of adaptation to a restricted sensory flow, as in a man who has lost his sight? Braille is used very successfully by some blind people, but many of those who are less intelligent or less well motivated fail to attain a useful reading skill. Somewhat surprisingly, a blind person does not have any greater basic tactile sensitivity than a man with normal sight. However, the more intelligent blind patients make improved use of the available sensory flow; the potential of the association areas in the cerebral cortex is realized, and there are substantial gains in tactile pattern recognition.

No systematic studies have yet reported on the possible use of taste and smell as additional channels of sensory input. However, the smell of burning commonly gives the aircraft pilot an early warning of fire, and it is usual practice to ensure that gas has sufficient odor for detection of leakage even if the eyes and ears are fully occupied elsewhere.

Techniques to Improve Sensory Capacity

Despite a filtering out of familiar signals, the operation of complex machinery such as high performance aircraft tends to be limited by the finite rate at which the cerebral cortex can process vital information.

One approach to this problem is to improve the instrument panel (page 25). Several related items of information are grouped in an analogue display, leaving individual signals accessible to inspection on demand (page 28). But on occasion, such an approach is either impossible or impractical. Interest thus attaches to the possibility of speeding information flow through the use of various combinations of visual, auditory, tactile and even olfactory signals.

The introduction of tactile signals does not impair visual or auditory acuity as measured by simple laboratory tests. Nevertheless, the interpretation of information received by the eye and ear becomes sometimes less accurate. Thinking in terms of a cerebral computer (page 84), we may visualize a rapid switching of the input selector to accommodate the three modes of sensory flow. As in some physiological recorders, additional information is handled, but detail tends to be blurred by the switching process.

Prolonged isolation leads to increasing anxiety, with hallucinations and finally an elaborate visual imagery (Bexton *et al.*, 1954; Zuckerman and Cohen, 1964). Perceptual constancy is gradually lost (Doane *et al.*, 1959); the solitary observer begins to interpret sensory information in a bizarre manner, particularly with respect to the parts of an object that are not immediately visible. Thus, the pilot returning from a solo mission may see that part of the hangar he is entering in a normal fashion, and yet misinterpret the arrangement of the unseen portion.

Prevention of the sensory deprivation syndrome involves the use of the techniques previously suggested for maintenance of arousal (page 65). In addition, the bush pilot, oil explorer or deep-sea diver should never work alone. The astronaut also should have a co-pilot, or if the pay load does not permit such an arrangement, contact with ground control should be maintained at all times. If an industrial process monitor must work alone, he should be required to telephone a central control building at regular intervals.

TASTE AND SMELL

Taste and smell are theoretically distinct modalities of sensation. The taste buds of the tongue can distinguish four discrete sensations (sweet, sour, bitter and salt). Nevertheless, in practice, the interpretation of flavors depends upon the combination of taste information with aromas detected by the olfactory nerve endings in the upper and anterior part of the nose.

Application of tasting skills is found in crafts like wine, tea and coffee blending; in a more general sense, the skilled chef also has undoubtedly developed the associative capacity of his cerebral cortex for sensations of taste and smell. *Taste blindness* is a well-recognized inherited defect that leads to difficulty in detecting a bitter flavor. It is unlikely that an affected person would enter the practical side of the food industry. However, formal recognition of the condition and a classification of its extent is quite simply achieved; the patient merely tastes various dilutions of a standard bitter product (usually phenylthiourea), and reports those that seem bitter to him.

Four modalities of smell have been described: acid, fragrant, acrid and capric (Crooker and Henderson, 1927). Each of these modalities has up to six levels of intensity: absent, threshold of detection, threshold of identification, definite, strong and very strong odor (Kaiser, 1962). The initial sensitivity of the olfactory receptors is very great. Pungent substances can be detected at a concentration of no more than one part per thousand million (Dravnieks and Krotoszynsky, 1966). However, adaptation is rapid. An odor may be readily detected on entering a room, but if the individual remains for a minute or so, it is no longer apparent. The probability of detecting an odor also varies with its psychological connotations. The familiar smell of household pets or of a tobacco-laden room may pass unnoticed by the regular inhabitants of a dwelling, but the acrid smell of burning rubber causes alarm and is reported at very low concentrations. Memory serves to sharpen some olfactory sensations, while it dulls others.

Smell is important not only in the detection of fire and gas hazards, but also as an index of ventilation. The accumulation of chemical or body odors within a building may present no measurable hazard to health, and yet an atmosphere can become sufficiently stale to affect working efficiency. The standard remedy is an adequate and well-distributed supply of fresh air. The recommended minimum ventilation is 3000 cubic feet per person per hour; this should be provided at a comfortable temperature (65° F for an active worker, 70° to 75° F for an office building) without undue noise or draughts. No problem arises in meeting such requirements in an office or factory, but difficulty is encountered in spacecraft, submarines and the well-sealed homes of the Canadian arctic. The author's one voyage in a submarine was well remembered by his family for several months—two suits had become firmly impregnated with the smell of engine oil. The usual treatment of a closed environment is to rely upon scrubbing and filtering systems. Occasionally, recourse is also made to the masking effect of more pleasant odors. Research is concentrated on several possible methods of destroying malodorous compounds, including catalytic heating (usually a rather expensive proposition), oxidation and the combination of several odors to yield a more pleasant overall smell (Del Vecchio and Mammarella, 1970).

HEARING, SPEECH AND COMMUNICATION

The ergonomist must consider the biological implications of a variety of low frequency oscillations, including not only the wave bands concerned with speech (20 to 10,000 c/sec), but also suprasonic and infrasonic vibrations. However, this section considers simply the problems of hearing, speech and communication.

The Mechanism of Hearing

Sound is transmitted through the air in the form of pressure waves. These waves are collected by the external ear, and impinge upon the drum (tympanic membrane); a chain of three small bones (auditory ossicles) transmits energy from the drum to the specialized receptor organ of the inner ear, the cochlea. This contains in essence a long and tapering vibrating reed, in which are embedded the detector *hair cells* of the organ of Corti. The reed acts as a selective vibrator; the frequency of sound is determined from the region that vibrates, and the intensity from the amplitude of vibration. The detectable frequency range depends upon age, but normally extends from the low-pitched rumble of an organ or distant traffic (16 to 20 c/sec) to a high-pitched squeak (10 to 15,000 c/sec).

The ear is most sensitive to variations in sound energy and pitch over the range 1000 to 3000 c/sec. The average person can distinguish some two thousand tones, and in the more sensitive range a difference in pitch of 0.3 percent is readily detected. Intensity is measured on a logarithmic scale (decibels). If E_o is the reference level, and E_1 is the intensity under investigation, then the decibel level is given by

$$N_{db} = 10 \log \frac{E_1}{E_o}$$

The energy of sound is proportional to the square of the pressure. Hence we may also write

$$N_{db} = 20 \log \frac{P_1}{P_o}$$

The reference pressure P_o, corresponding to 1 decibel, is 2×10^{-4} dyne/cm^2. If presented in the frequency range 1000 to 3000 c/sec, a one-decibel sound is just audible to a young person. A whisper amounts to some 20 decibels, normal office conversation to 50 decibels, and heavy street traffic to 80 decibels.

Normal auditory acuity may be disturbed if gas pressures within the middle ear are reduced relative to ambient air; this is a situation that can arise during diving or descent of a partially pressurized aircraft. Middle-ear pressures can be equalized by swallowing, but a differential pressure of 20 cm H_2O is fairly normal before the need to swallow is appreciated; such pressure differentials can lead to an 8 to 10 dB impairment of hearing (Melville Jones, 1949). Larger pressure differences may arise if the internal ear passage is congested by upper respiratory infections; these can produce structural damage in the tympanic membrane and thus long-lasting hearing problems.

Loss of auditory acuity with aging is a significant consideration in the employment of the older worker (page 281). It is sometimes difficult to distinguish a physiological from a pathological hearing loss. Normally, deterioration of acuity is first seen at very high frequencies, and it is not until the age of 50 or 60 that losses in the frequency range 4000 to 5000 c/sec begin to affect the intelligibility of speech.

The Mechanism of Speech

Speech is produced by raising the gas pressure within the lungs and allowing a vibrating column of gas to escape slowly through the approximated vocal cords. Appropriate movements of the lips, tongue and palate supplement the normal resonance of the thoracic cavity and paranasal sinuses, giving the characteristic modulations of the spoken word.

Speech is an expiratory phenomenon, and thus leads to a shortening of the inspiratory phase of respiration, more obviously at rest than during physical work (Ernsting and Gabb, 1962). This is of particular significance when oxygen is to be inhaled from a mask, as in military flying. During talking, the rate of inspiration may become so fast as to exceed oxygen supply, causing an inward leakage of ambient air. Furthermore, demand valves may fail to follow the rapid decceleration of gas flow at the end of inspiration, and if a compensated expiratory valve is used, the resultant phase lag may impose a heavy expiratory load, hindering both respiration and normal speech.

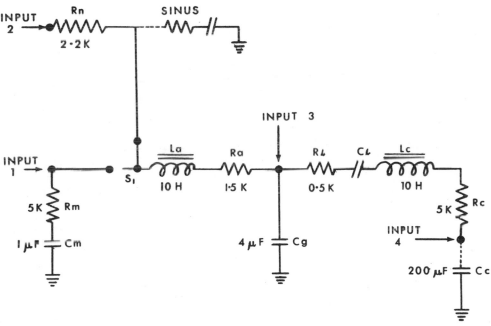

Figure 39. Electrical analogue of respiratory system. (Redrawn with permission from R.J. Shephard, Airway dynamics in unstable breathing systems, *J Aerosp Med*, 37(10): 1014–1021, 1966.)

Detailed analysis of mechanical, electrical and mathematical models of the respiratory tract (Fig. 39, Shephard, 1966) suggests that we are dealing with a complex under-damped system. In a simple linear mechanical analogue, the damping ratio h would be given by

$$h = \frac{R}{2} \sqrt{\frac{C}{L}}$$

where R is the resistance, C is the compliance and L is the inertia of the system. The overall damping ratio for the respiratory tract is less than unity, and airflow is thus inherently unstable. Attempts to reduce the work of breathing by lowering the resistance of demand valves exaggerate the tendency to instability, and the inspiratory valve of a facemask may readily develop an irritating chatter as it vibrates upon its seating. The simplest engineering solution to both the instability and also the phase lag of the demand valve is to increase compliance. The pilot is required to inhale from an oxygen reservoir, commonly his *pressure-breathing waistcoat.* By this simple modification of the oxygen system, the overall damping ratio is increased to unity or above, and at the same time it is no longer possible for inspiration to exceed the rate of oxygen delivery during speech.

Expiratory gas flows must be modulated rather precisely while talking. This may be quite difficult when piloting a low-level attack aircraft or driving a tracked vehicle over rough terrain. The respiratory tract is a complex oscillator (Shephard, 1966), but its principal resonant frequencies lie within the range (1 to 10 c/sec) encountered by passengers in such vehicles. If the imposed accelerations exceed $\pm \frac{1}{4}$ g, much of the respiratory air flow becomes attributable to vibration rather than deliberate expiratory effort, and the intelligibility of speech is markedly reduced.

Vocal resonance depends upon the density of respired gas. Clear speech is thus very difficult when gas pressures within the airway are modified. At altitude, the diminution in ambient pressure reduces the overall energy of the transmitted sound. Fortunately, oral consonants are affected less than vowels, and since intelligibility depends mainly upon the consonants, mistakes are less frequent than might be anticipated from the reduction in sound energy (Kryter, 1940). In diving operations, low levels of illumination often preclude the use of hand signals, and poor communication can seriously impede working efficiency. The problem of increased ambient pressure is commonly compounded by an alteration in the chemical composition of respired gas. Low density gases such as helium are added to the breathing mixture; these pass freely through the vocal cords, while the poorly ventilated paranasal sinuses contain appreciable quantities of air that resonates at a different frequency. Intelligibility of speech may be further reduced by the pressure of water or gas upon the cheeks and of a helmet upon the jaw. It may be necessary to grip a mouthpiece, or to wear a resonant facemask (Davies, 1962). There are also many interfering sounds such as the hiss of compressed gas. Finally, the psychological tensions of complex maneuvers, whether at altitude or in diving, militate against careful diction and careful listening. Communication with the diver can be improved by the use of high fidelity microphones, fitted with selective filters, and other more complicated techniques have been suggested to reduce the overall frequency of the transmitted sound (Bennett and Elliott, 1969).

Effective Communication

The efficiency of most team tasks drops unless speech can be both heard and understood. The effectiveness of communication depends upon

1. intrinsic anatomical limitations of the transmitter and receptor organs
2. defects in any intervening equipment
3. the signal to noise ratio and
4. the extent of pattern recognition in the cerebral cortex (particularly the anticipation of specific word frequencies, and the association of ideas).

While the frequency range of normal speech is quite broad (90 to 5000 c/sec), the spectrum can be restricted to a much narrower band (1000 to 3000 c/sec) without serious loss of intelligibility (Fletcher, 1922). This question has received extensive study by those concerned with the development of telecommunications equipment; it is of some practical importance, since the average telephone receiver is hardly an item with high-fidelity characteristics. When testing equipment, the problem of individual variations in pattern recognition is overcome by the use of *articulation scores*. These may be based on a formal study of the speech spectrum, band by band, but are normally calculated more simply from the percentage of unrelated words understood when read quietly at a fixed distance from the listener.

Transmission of speech is seriously impaired when wearing a device such as a respirator (gas mask). The basic respirator acts as a low-pass filter (Fig. 40), cutting off most transmitted vibrations with a frequency greater than 1000 c/sec (Shephard, 1962). The loudness of speech is largely unimpaired, but intelligibility is poor. The remedy adopted by the designers of gas masks has been to introduce a voice plate. This increases the transmission of higher sonic frequencies, and if suitable lightweight materials are chosen the plate is virtually aperiodic, so that it imposes little of its own personality upon the transmitted sound.

An alternative solution found in typical Air Force oxygen masks is to install a microphone. The need for cable connections is very restricting unless a sedentary task is to be performed, and because the microphone is mounted within a resonant cavity, there is a danger

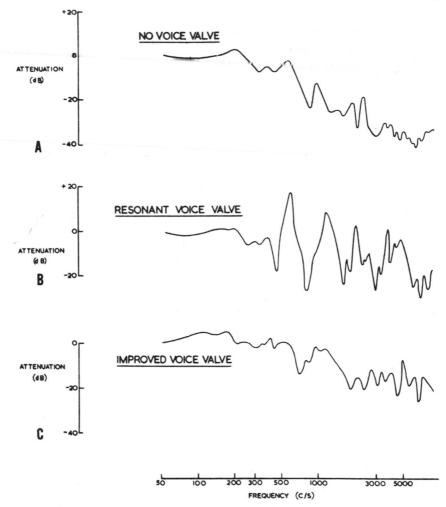

Figure 40. Transmission of sound through a respirator (gas mask). (A) basic pattern, (B) resonant voice plate, (C) improved voice plate. (Reprinted with permission from R.J. Shephard, Ergonomics of the respirator. In C.N. Davies (Ed.), *Design and Use of Respirators*. New York, Pergamon, 1962.)

that low frequency sounds and valve hiss may be exaggerated at the expense of normal speech frequencies. The usual aircraft microphone is designed to respond to sounds within the range 500 to 3000 c/sec. Instabilities of electronic equipment and valve hiss preclude the acceptance of higher frequencies, but fortunately these contribute no more than 15 percent to overall intelligibility (Ernsting and Gabb, 1962).

Laryngeal speech transducers have been used experimentally. A light vibration sensor is strapped over the larynx. This has the advantage that the face is left free, but the detected speech has a peculiar quality, since vowel sounds originate mainly from the larynx, while consonants are formed by the tongue, palate and lips. Vibration sensors could theoretically be applied to such bony areas as the forehead and the mastoid process of the jaw. Unfortunately, the energy of vibrations emitted from these regions is low, and extensive amplification would be needed to produce usuable sound.

If recourse is made to radio-telephonic communication, the ergonomist must advise on an appropriate earphone design. Other factors being equal, the weight of the headset should be as light as possible, with its center of gravity low yet above the axis of the vertebral column. Lateral bulk should not be excessive. In some instances, special earphones may provide protection against high intensities of noise, or may form part of a general system to protect the head from injury.

NOISE

Noise and Communication

Articulation scores are commonly used to define intensities of noise that interfere with communication (*speech interference levels*). The intensity of the interfering sound is conveniently calculated as the arithmetic mean of intensities recorded over three frequency bands (600 to 1200, 1200 to 2400 and 2400 to 4800 c/sec). Rosenblith and Stevens (1953) have shown that if a speaker is separated from his audience by as little as 12 feet, a normal voice is barely audible with a background noise of 43 dB. A raised voice can be understood at 49 dB, and by shouting a message can be conveyed at 61 dB. These figures refer to ideal conditions of communication. Much of the spoken word is normally interpreted from secondary cues—lip and facial movements and gestures play a significant role in the transmission of ideas. Thus, much lower tolerances are found if the speaker cannot be seen, as when messages are sent over a noisy telephone line. Furthermore, although some measure of communication is technically possible at the specified sound levels, annoyance may lead to a rapid impairment of articulation. In many circumstances, partial articulation can be dangerous; it is more obvious to a speaker if his message is unheard than if it is misunderstood.

Interior designers should seek to attain the following standards of background noise (Hickish, 1970):

    20 to 30 dB—large conference room or executive suite

    30 to 35 dB—small conference room (15-foot table)

    35 to 40 dB—office (6- to 8-foot table)

    40 to 50 dB—telephone area

If these standards cannot be met, it may be necessary to modify such items as telephones (see page 129).

The Problem of Noise

Noise is an intensely personal experience, and is regarded by some people as one of the more serious environmental problems of our present day. Personal reactions to noise can be judged by the techniques of the social scientist, although the most useful studies to date have been based on objective criteria. The industrial classification of noise is based upon its periodicity and its spectral characteristics. Steady, broad-band noise is typical of machine shops, furnaces and fans. Steady, narrow-band noise arises from such sources as circular saws, machine planes and gas valves. Intermittent noise arises from traffic, aircraft and engine testing bays. Impulsive noise is produced by hammers and power presses, and repeated impacts are encountered in rivetting and engineering shops. For laboratory purposes, it is convenient to synthesize *white* and *pink* noise. *White* noise is distributed with equal energy over the entire audible range, while *pink* noise is so arranged that equal intensities of energy are presented in successive octave bands (40 to 80 c/sec, 80 to 160 c/sec, 160 to 320 c/sec).

Noise has four principal biological effects. We have already discussed interference with communication. We shall now look specifically at annoyance, deterioration of skilled performance, and physical damage to hearing mechanisms.

Annoyance

The characteristics that make a sound annoying are elusive. Often, the motives of the most strident complainants are devious; political office or some other form of notoriety may be sought rather than industrial efficiency or community health. Observers have used such techniques of evaluation as opinion polls and the presentation of recorded noise patterns to a representative sample of the affected community. Among possible adverse characteristics, Guignard (1965) notes the following:

1. Quality: loud, high pitched, harsh or with distinct *tones*.
2. Frequency: irregular or intermittent, with sharp onset or cessation, and prolonged or frequent exposure.
3. Predictability: unexpected.
4. Localization: poorly directed or diffuse.*

---

* This emphasis can be understood in simple mechanical terms; if the source is localized, difficulty in conducting a conversation can be minimized by varying the direction in which the speaker is facing.

5. Individual susceptibility: for example, ill health or disturbance of current activity; particularly attempts at mental concentration.
6. Information content: for example, a partly heard public address message.
7. Emotional content: for example, fear of an aircraft accident or dislike of neighbors.
8. Deliberate and apparently avoidable nature.

Effects upon Performance

The immediate effect of noise is an increase of arousal (page 215). Depending upon the complexity of the working task and the personality of the individual, performance may either improve or worsen. Broadbent (1957) found little deterioration in performance of a laboratory task (five-choice serial reaction rate) with a noise level of 90 dB. Higher intensities (up to 100 dB) gave an increase of error scores, more marked for high (> 2000 c/sec) than for low pitched (<2000 c/sec) sounds. Other investigators have shown a loss of vigilance, a slowing of responses, and an impairment of time judgment. The effects of noise supplement other adverse features of the working environment (page 65).

If the threshold for loss of performance is as high as Broadbent's experiments suggest (90 dB), then any effects should be of academic rather than practical concern. Continuous industrial exposures to such sound levels are inadmissible on grounds of hearing conservation. Nevertheless, Broadbent was able to find factory situations with noise readings of 90 to 100 dB, and to demonstrate that in such areas errors of production occurred as predicted.

Gross abuses of this type are likely to be eliminated in the foreseeable future. But the problem of intermittent bursts of noise will continue. Sounds such as the sudden roar of an aircraft take-off not only increase arousal, but also distract attention; the latter can cause additional errors in work requiring sustained vigilance. If a complex task such as aircraft control is in process, a sound lasting one second may disturb normal performance for 15 seconds or more (Woodhead, 1959).

Pathological Changes

Physicians interested in hearing conservation are becoming increasingly cautious when specifying permissible noise exposures. Much depends upon the duration, the band width and any unusual peaks of noise. Low frequencies are tolerated better than high frequencies, and over the critical range for speech (1000 to 3000 c/sec) harm may result from prolonged exposure to no more than 80 dB. Unfortunately, it is difficult to disentangle the effects of aging from those of prolonged

noise, and many workers accept slight deafness as an inevitable concomitant of their later years.

Commonly accepted safety standards are exposure to 85 dB for long
portions of a working day, 110 dB for one hour, and 120 dB for a
few seconds. If such limits are exceeded, the hair cells of the organ
of Corti become dislocated from their basilar membrane and
degenerate. Hearing loss occurs primarily over the frequency range
where over-stimulation has occurred. Industrial deafness is often seen
first at frequencies of approximately 5000 c/sec (Fig. 41).

Very high intensities of sound can be felt, and may even cause pain,
hemorrhage and rupture of the eardrum.

Counteracting Noise

A wide range of approaches to noise abatement have been suggested.
The most obvious remedy is to control the noise at its source.
Sometimes, control takes the form of legislation—limitations upon the
power or rate of climb of aircraft, restriction of airport landing hours
and the like. A better method of control is to improve design. Both
noise and vibration reflect energy wasted in a mechanical system. Efficiency is thus gained as vibration of the moving parts is reduced. The
designer may alter the oscillation mechanics of a machine so that it
is critically damped (page 121). Lubricants may be sealed into bearings

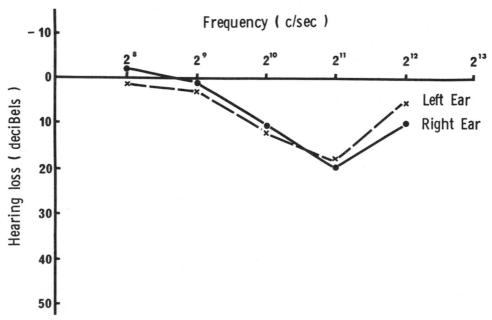

Figure 41. Typical audiogram with moderate industrial hearing loss (a normal person
would show 0 dB loss).

so that vibration is minimized irrespective of the frequency of maintenance. Attention may be directed to basic principles of hydraulics, thereby quietening the flow of air and liquids through conduits; in particular, ducts and nozzles can be shaped to reduce the proportion of turbulent flow.

Transmission of noise away from machinery can be greatly reduced by rubber mountings, flexible couplings and silencers. Resonant casings should be avoided. Noise barriers may be devised, both for individual items of industrial machinery (small cubicles) and for larger noisy areas such as airfields and expressways. Current expressway barriers (trees, earth and concrete walls) are rather ineffective, reducing noise levels by no more than 10 percent; the most useful devices seem slatted walls that deflect the offending sound waves.

Appropriate design of buildings and of cities can minimize the impact of inescapable noise. Where possible, airports should be sited in rural areas. Parking bays for individual aircraft should be set well back from the control tower and other *sensitive* positions, and the planes should be so angled that engine noise is directed away from the terminal facilities. Similarly, in planning a factory, noisy machinery should be kept as far as possible from office areas.

Interior design can contribute much to noise suppression. Soft carpets, heavy drapes (curtains), acoustic tiles and padded cloth furnishings all supplement the effects of double glazing and even underground construction. The additive nature of noise deserves emphasis; unnecessary sounds such as protracted public address messages should be eliminated. Unfortunately, many workers do not appreciate the importance of avoiding unnecessary noise, and programs of both education and hearing conservation are needed in industry.

If noise levels cannot be reduced to an acceptable level, personal equipment may need to be redesigned. Thus, telephones can be fitted with both high and low frequency filters to remove unwanted sounds. Automatic *gain* controls may be required, since the first reaction of many people when telephoning from a noisy room is to shout louder. Filtration of high frequencies may improve intelligibility, since vowel sounds are selectively eliminated. Under very adverse conditions, miniature stereophonic earphones may give additional help, particularly if the message can be oriented at right angles to the interfering noise.

Personal protective measures should be a last resort. Well-fitting ear plugs or muffs with a liquid seal can reduce sound energies by 20 to 30 dB, and if greater protection is required, a full helmet can be provided.

VIBRATION

Audible sound spans but one part of a continuous spectrum of pressure waves. Frequencies greater than 10,000 c/sec are classified as *ultrasound*. Vibration can be felt over much of the sonic range, but also extends to infrasonic frequencies (< 20 c/sec). Problems arise from the resonance of body parts, particularly over the range 1 to 10 c/sec, and exposure to frequencies around 1 c/sec tends to produce motion sickness. This last topic will be discussed on page 138.

Ultrasound

Ultrasound is finding increasing application to both industry and medicine. Echo-sounding devices are used in oceanography and in mapping the contours of the heart. The energy content of ultrasonic vibration can also be used to generate fine aerosols and to warm body parts (as in rehabilitation medicine). Accidental exposure leads to adsorption of the vibrant energy, with heating of the exposed part; providing heating is not excessive, there is no evidence of any adverse effect upon health or performance.

Nature of Vibration

Vibration may be defined as a periodic pressure fluctuation that produces a sustained oscillation of body parts; normally, the vibration is transmitted through some rigid structure, but it can also pass through air or water. It is usual to distinguish free vibration (where the body part moves at its natural frequency) from forced oscillation (where the natural frequency is exceeded). It is also customary to classify vibration in terms of the principal axes of movement; these include three linear translations (a vertical heave, a forward surge and a lateral sway) and three types of torsional movement (forward pitching, lateral rolling and lateral yawing). Whole body vibration is encountered in many vehicles, while vibration of individual body parts is a characteristic of machine tool operations.

Assessment of vibration has traditionally been subjective in type. As with noise, much depends upon the expectations of the assessor. A driver of a tank will tolerate more vibration than a passenger in a first-class railway compartment. Objective readings showing the frequency and amplitude of vibration can be obtained by mounting accelerometers at normal crew positions.

In land vehicles such as cars and railway carriages, the nature of the vibration depends upon the stiffness of the springing and the damping or amplification introduced by seating. The stiff springing of older cars transmitted little energy in the motion sickness range (~1 c/sec) but gave an excessive and tiring vibration at higher frequencies. Modern car designs represent a compromise between comfort and the likelihood of motion sickness. The present-day railway passenger is exposed to substantial vertical vibration at 3 to 4 c/sec, with lateral vibration at 0.5 to 1.5 c/sec (Begbie et al., 1963). Vibration is particularly severe in tracked vehicles (tractors, earth-moving equipment, snow transport and tanks).

The problem of vibration in civil aircraft is usually subordinate to that of noise. However, some older multi-engined piston-driven planes produced enough vibration to cause fatigue, with difficulty in both writing and the reading of instruments. The problem is now largely overcome by improved engine design and rear mounting. Low-flying military aircraft that attempt to beat radar screening devices encounter a serious problem from ground turbulence; pitching and yawing movements are superimposed upon vertical accelerations of 0.5 g or more. The effects depend upon the speed of the plane and the location of the crew stations; rotational accelerations are greatest at the front and rear of the aircraft, while at the center, movement is mainly in a vertical direction. Stratospheric and supersonic flight does not minimize vibration as much as might be thought; although there is little turbulence at 65,000 to 70,000 feet, the absence of air gives no basis for damping once an oscillation has been initiated.

Some of the newer forms of transport are plagued by serious vibrational problems. Helicopters commonly show a marked vibration at the speed of their main rotor. Hovercraft pick up vibration in rough water, much as does a fast motorboat; however the frequency spectrum is likely to cause motion sickness rather than a decrement of psychomotor performance. Space vehicles may encounter marked vibration coincident with maximum aerodynamic drag (about a minute after take-off) and also during re-entry of the earth's atmosphere.

Body Mechanics

Vibrations in the frequency range 1 to 100 c/sec can cause annoyance, impaired performance and even structural damage to the body. Much of the reported work is based upon exposure in industry and during normal travel, although devices are now available for experimental vibration of the human body; these include very powerful loudspeakers (*electrodynamic vibrators*) and shaking platforms (mechanical vibrators). The body forms a rather complex resonator (Fig. 42) and experimental study is made difficult by variations in damping with posture and muscle tone. Bent knees and slouching shoulders damp out vibrations received via the feet—an item of bioengineering well appreciated by ancient charioteers. In more modern vehicles there is quite commonly a dynamic interaction between the body and the

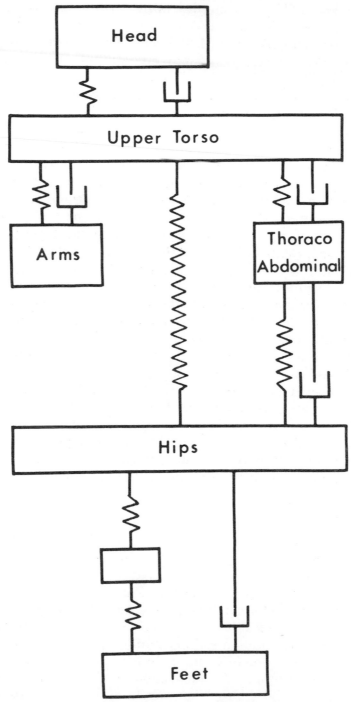

Figure 42. Simple mechanical analogue of the human body, considered as a vibrating system (based on a design by Coermann, 1970). Note that the mechanical properties of the system are greatly modified by a bending of the knees.

seat, and this makes it very difficult to devise an accurate mannikin for the assessment of vibrational stress.

The response of the body depends upon the direction and site of application of the vibrating force, its intensity and its distribution. If a driver is seated, a front to back vibration causes resonance at approximately 2 c/sec; however, little vibration occurs in this plane when standing, because the legs provide an excellent damping system. Vertical vibration at frequencies of less than 3 c/sec produces displacement of the whole body without local resonance. Local movement of the shoulder girdle and the thoracoabdominal viscera first appears at 4 to 5 c/sec, and a second major resonance of the torso is seen at 11 to 14 c/sec (Guignard, 1965). There is local resonance of the head at 17 to 25 c/sec, of the hand at 40 c/sec (Békésy, 1939) and of the skull and jaw at frequencies above 100 c/sec; fortunately at 100 c/sec, 95 percent of vibrant energy is damped out between the feet and the head.

Arm mechanics are of particular interest because many workers must support vibrating machinery (pneumatic drills, chain saws and various machine tools). Both the resonant frequency and the amplitude of oscillation vary with muscle tone. Ballistic and tapping movements coincide with the natural frequency of the relaxed part (for example, finger tapping at 10 c/sec). Under normal working conditions, two peaks of resonance are seen, at approximately 5 c/sec and 30 to 40 c/sec; most of these vibrations are damped out in the arm and shoulder girdle (Coermann, 1970). Further control of vibration is possible by tensing the muscles involved, but this is very fatiguing to the worker.

Perception of Vibration

The slow rolling motion of a large passenger liner is readily perceived down to frequencies as slow as 0.1 c/sec. However, at such frequencies the postural receptors of the inner ear and vision contribute more information than stimulation of cutaneous receptors; indeed, the rolling sensation is usually reported as occurring *in the head*.

Over the range 1 to 20 c/sec, the threshold of stimulation varies from 0.002 to 0.01 g, the most sensitive range being 3 to 6 c/sec (Goldman and Von Gierke, 1960). Guignard (1965) gives a good description of the effect of increasing frequencies. Moderate vibration at 1 to 2 c/sec can have a relaxing, soporific effect. At 3 to 4 c/sec, differential movement of the abdomen and shoulder girdle becomes noticeable. Problems of breathing and speech are observed at 5 to 8 c/sec (page 122). Fluttering of the lower abdomen is seen at 8 c/sec, and a fluttering of the cheek at 9 to 10 c/sec. At frequencies above 15 c/sec, stimulation of the local mechano-receptors in skin, muscle and tendon is dominant, and at frequencies of 20 to 30 c/sec no more may be felt than a slight tingling of the buttocks. A tickling sensation persists to some 900 c/sec, the most sensitive range being 200 to 300 c/sec. Frequencies of 1000 to 10,000 c/sec can cause stinging or burning, and frequencies around 20,000 c/sec produce a velvety sensation, progressing to intense heat.

Annoying Vibrations

Representative panels of subjects are used to rate the intensity of vibration, particularly when judging the qualities of a given vehicle. Some authors (for instance, Janeway, 1948) have prescribed safe working limits in terms of jolting (change of acceleration, span 1 to 6 c/sec, limit 40 ft/sec²), acceleration (span 6 to 20 c/sec, limit 11 ft/sec²) and velocity (span 20 to 60 c/sec, limit 0.1 in/sec). However, a continuous scale is more convenient. Dieckmann (1957 and 1958) has proposed calculation of an amplitude/frequency product:

$$K = 5 \; af$$

where a is the amplitude, and f is the frequency of vibration. A K reading of 0.1 gives just perceptible vibration, 10 is unpleasant, and the tolerance limit is 1000.

Others have plotted the limits of acceleration against frequency. Much depends upon the circumstances (Fig. 43). On a six-hour railway journey, vertical vibration greater than 2 ft/sec² and lateral vibration greater than 1.5 ft/sec² are unacceptable (Loach, 1958). However, brief exposures to 64 ft/sec² (vertical) and 10 ft/sec² (front to back) cause no serious harm. Annoyance arises from difficulty in reading and eating, and (in low-flying aircraft) problems are encountered when speaking and operating delicate controls.

Vibration and Performance

Performance of skilled tasks may deteriorate due to direct interference with vision, speech and postural control; annoyance, arousal and alarm also affect performance.

If the dial to be read or the target to be followed is moving with a sine wave vibration of 1 c/sec, it is a simple matter for the body to compensate by moving the eyes with a similar rhythm. Visual tracking breaks down if the movement is erratic or has a higher frequency of oscillation. Paradoxically, instruments can often be read better at a frequency of 6 to 8 c/sec than at 3 to 4 c/sec; with the higher frequencies of oscillation, the eyes make no attempt to follow the gauge, but merely read it at its extremes of movement (Edholm, 1967). Small but widely spaced letters or digits are sometimes easier to read than those that are larger; small symbols form discrete images at the extremes of their movement, whereas larger or more closely set symbols

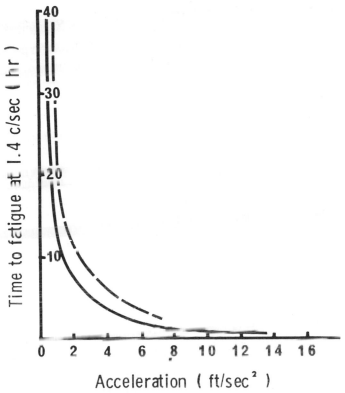

Figure 43. The relationship between duration of vertical vibration and acceptable accelerations. Continuous line, lateral vibration; interrupted line, vertical vibration. Data for railway passengers, after Loach, 1958.

still overlap one another. Because it is difficult to use the most sensitive parts of the visual receptor surface consistently, and because of blurring caused by the movement, visual acuity is decreased by vibration. The illumination and contrast of instrument dials must thus be proportionately increased in order to maintain performance. Visual acuity deteriorates markedly if the frequency of vibration is greater than 15 c/sec, and impairment of vision is particularly serious over the frequency band 60 to 90 c/sec, due to local resonance of the eyeball within its socket (Coermann, 1970).

Problems of speech have been discussed elsewhere (page 122); intelligibility is poorest with vibrations of 3 to 15 c/sec; in this range it is very difficult to understand messages at vibrant accelerations exceeding 0.5 g. (Guignard, 1965).

Stimulation of proprioceptive nerve fibers by vibration leads to a loss of control when performing delicate movements. Sustained forces, such as the grip of the hand upon a control lever, are well maintained, but difficulty arises with tasks requiring proprioceptive supervision. Many tasks that are normally performed *automatically* (page 87) revert to conscious control if there is much vibration; examples include the movement of the foot upon an accelerator pedal and the fine hand movements required in writing. Difficulty is also encountered when a steady hand is required (as when using certain navigational aids). The tracking ability of the hands has been studied by pursuit meters, the subject being required to move a light or heavy lever with an oscillatory motion. The hands, like the eyes, can follow a regular sine wave to a frequency of about 3 c/sec; irregular vibrations, on the other hand, cannot be followed beyond 1 c/sec. The tracking error on a simulated driving task is greatest with lateral vibration at 1.5 c/sec (Hornick *et al.*, 1961). When standing at a console, additional problems arise from oscillation of the shoulder girdle at 4 c/sec (Forbes, 1959). Writing is most difficult with a vibrant frequency of 5 c/sec; unfortunately, amplification within the body is such that the head, fingers and paper often vibrate with independent rhythms. The effects of sustained vibration upon postural control persist on leaving a vehicle, and it is thus possible to assess its riding qualities from subsequent tests of body sway or local hand steadiness. For a more dramatic demonstration of the same phenomenon, it is possible to take a pneumatic drill operator to a tavern at the end of a hot day's work and to find his proprioceptors so disorganized that he cannot raise a tankard to his lips (Poulton, 1970).

Low frequency oscillations have a soporific effect; rocking chairs have been used by many generations of nursing mothers, and a ride in a modern car often has a similar advantageous effect upon a screaming infant. The danger that the driver will fall asleep adds one more piece to the jigsaw of impaired vigilance upon long journeys (page 217). Higher frequencies of oscillation increase arousal (page 215); whether this helps or hinders performance depends upon the complexity of the task and the personality of the worker. Arousal increases muscle tone, and performance may thus show a secondary deterioration after several hours, due to fatiguing muscular tension. If oscillation is sustained (for instance, a long sea voyage), habituation occurs. Habituation is less likely if vibrations are irregular or intermittent. Cerebral effects of arousal may be supplemented or opposed by vibration-induced hyperventilation. The latter produces an excessive washout of carbon dioxide from the body, with a consequence diminution of blood flow to the brain.

Pathology of Vibration

Very severe generalized vibration leads to internal hemorrhage, loss of weight and even death of laboratory animals. Claims of industrial injury are somewhat difficult to separate from unrelated accidents and the normal effects of aging. As a general rule, accelerations of more than 30 ft/sec² are painful, and injury can result from accelerations of more than 60 ft/sec² One Air Force officer was exposed experimentally to 320 ft/sec²; he experienced anginal pain at 10 c/sec, and gastrointestinal bleeding at 25 c/sec (Guignard, 1965). Spinal radiographs have shown a high proportion of abnormalities in young tractor drivers. Older workers with a history of vibration tend to be afflicted with vertebral problems (such as lower lumbar pain and prolapsed intervertebral disc), anorectal conditions (particularly pilonidal sinus) and duodenal ulcers.

The fingers of pneumatic drill, chain saw and machine tool operators may experience both local vibration and severe chilling. A proportion develop the Raynaud phenomenon, characterized by an unusual sensitivity to cold, with intense and painful vascular spasm at each exposure. Other possible effects of vibration include a cystic decalcification and osteoarthritis of the wrist, a thickening and contracture of the palmar fascia, and ulnar nerve injury. The maximum permissible amplitudes of vibration (Agate and Druett, 1947) are 0.2 mm at 20 to 30 c/sec, and about a third of this figure over the most sensitive range (40 to 125 c/sec). Frequencies in the band 125 to 200 c/sec do not give rise to a clear-cut Raynaud phenomenon (Dart, 1946), but there may be a painful blue swelling of the affected hands (erythrocyanosis).

In some types of drill, additional mechanical support is provided by inserting a rest into the armpit (Huzl *et al.*, 1970). This arrangement is likely to give rise to edema and injuries of the brachial plexus.

Minimizing Vibration Effects

As with noise, vibration is wasted energy. It implies inefficient machinery, and can damage not only the worker but expensive items of equipment (for instance, a whole fleet of Comet aircraft). The most effective basis of prevention is at source, through improvements of design, adequate maintenance, use of shock mountings and couplings, and development of the minimum power consistent with objectives.

Unfortunately, flexible linkages are not always a success. In the case of the chain saw, vibration is reduced, but accurate control becomes impossible. The original springing of vehicles was provided by the knees, either of the driver (charioteers) or the porter (sedan chairs). The use of formal springs began in the eighteenth century. The choice has rested between heavy suspensions with stiff damping, and soft suspensions with lighter damping. Modern cars, buses and trains tend to the latter pattern, having a natural frequency of 1 to 1½ c/sec, and a damping ratio of 0.5 to 0.7 (Guignard, 1965). High frequency vibrations (~5 c/sec) are now filtered out quite efficiently, so that fatigue is lessened upon a long journey. However, low frequency vibrations (~1 c/sec) are less well counteracted, and susceptible individuals thus tend to suffer from motion sickness.

Soft cushions are sometimes helpful, but they interact with the body, amplifying certain frequencies. It is also difficult to operate fine controls from a cushioned base. A suspended seat (such as a hammock) minimizes body movement, but gives a large displacement of the operator relative to his controls; devices to *float* instrument panels with the operator are currently under study. Body resonance is generally reduced by rigid bindings such as safety harnesses. On the other hand, arm and head rests, long control columns and working desks may by-pass some of the normal damping mechanisms of the body. The length of the back-rest in a car seat materially influences the transmission of vibrations to a driver; a good seat amplifies 3 c/sec vibrations by no more than 30 percent while a poor seat may give 100 percent amplification (Coermann, 1970). If vibration is severe, it can be helpful to provide some support for the arm or leg that is operating the controls.

The operator of any vehicle should work at that point in his craft where vibrations are minimal. Fingertip, servo-operated controls are more effective than those needing powerful arm movements under vibrant conditions. Good illumination and certain arrangements of lettering are helpful (page 25); it is also an advantage to mount instruments so that pointer and control movements are at right angles to the main plane of vibration, with the head of the operator tilted forward while reading the gauges.

MOTION SICKNESS

Receptor Organs

The vestibular apparatus of the inner ear detects linear and rotational forces. There may be some overlap of functional responsibility, but the otolith organs (saccule and utricle) provide the main indication of linear accelerations, while the semicircular canals sense angular movements.

Overstimulation of the Receptors

Individuals vary in the sensitivity of their vestibular receptors. Postural sensitivity is an advantage to certain occupational groups, such as the *spider-men* who erect modern skyscrapers, and professional figure skaters. But the main problem of the average citizen is overstimulation during rough air and sea voyages. In this respect, the deaf-mute or the person who has recovered from a surgical extirpation of the vestibular apparatus has a substantial advantage over the average traveler.

Motion sickness is essentially similar, whether caused by land, sea or air travel. Negative acceleration can cause a direct mechanical expulsion of the stomach contents, but the usual disturbance results from over-stimulation of the vestibulum by movements in the frequency band 1 to 10 c/sec. Angular acceleration is more unpleasant than linear, and because of the effect of radius upon acceleration, head movement is more disturbing than a movement of the entire body (Johnson *et al.*, 1951). Vague uneasiness is followed by nausea and actual vomiting. In addition to subjective complaints, the affected person is often pale, with an increased heart rate and a cold, sweaty skin (Whiteside, 1965); the entire mind of a severe case becomes preoccupied with the sensation of sickness, to the exclusion of more important tasks.

Although primarily a vestibular response, other systems may become involved. In some instances, sickness becomes worse as travel is repeated; a bad initial experience has conditioned the traveler to anticipate trouble, and sickness may even be induced by an associated stimulus such as a barely detectable smell of gasoline or the sight of an airsickness bag. Liability to motion sickness varies with the individual's overall level of anxiety. Thus, undue susceptibility to rough weather may be one of the first signs that an anxiety state is developing in an experienced pilot. The usual response to repeated exposure is a loss of susceptibility. This is partly a true habituation to intense stimulation, and partly a learning of techniques to avoid excessive head movement. The latter, at least, is vehicle specific, and the captain of a small fishing trawler may be acutely sick if he takes a cruise on a large ocean liner.

Effects on Performance

Performance of any skilled task inevitably deteriorates if sickness progresses to the point where the mind is unduly exercised by the sensations of sickness and the mechanics of vomiting. Formal tests of volunteers on swinging platforms and rafts have shown a 30 percent decrement in the performance of mental arithmetic relative to people protected by hyoscine (Brand *et al.*, 1967).

Problems are naturally most common with inexperienced travelers. Prevention of sickness can effect substantial economies in the time and thus the cost of training aircrew. Some 90 percent of passengers on oceanliners, and 60 percent of soldiers in landing craft are sick in rough weather (Hill, 1936; Tyler, 1946). Again, 0.6 percent of all passengers on commercial airliners are sufficiently sick for this fact to be reported (Whittingham, 1950). Among crew members, navigators are affected more than pilots, presumably because it is more difficult for them to keep their heads still. Passengers are also more susceptible than drivers because they are unable to anticipate and counteract movements of their vehicle.

Prevention of Sickness

Improvements in vehicle design can reduce the likelihood of motion sickness. Ships can be fitted with stabilizers. Car springing can be arranged to minimize movement at 1 c/sec. Crew positions on ships and aircraft can be placed at *nodal* points. Modification of vehicle movement patterns may also be helpful. Car sickness often arises from sudden accelerations and decelerations; a constant speed of driving with some *softness* in the braking system is thus to be recommended. The speed of a ship should be reduced in rough weather, and aircraft can alter either altitude or course to avoid local patches of turbulent air.

Individual protection is best achieved through head immobilization; this is realized at the expense of some increase in the transmission of vibrations. Habituation and learning also contribute to a lessening of symptoms, and sickness is less likely if a heavy meal has been avoided for several hours prior to travel. Drugs are best administered at least an hour before exposure. Once sickness has begun, they are difficult to retain and therefore much less effective. The antispasmodic hyoscine (0.7 mg) is recommended for short journeys, and the antihistamine dimenhydrinate (Dramamine,® 100 mg) for longer voyages. Unfortunately, the side effects of these drugs, such as drowziness, blurred vision, dizziness and tinnitus severely limit their usefulness to vehicle operators.

APERIODIC ACCELERATION AND DECELERATION

Positive accelerations are encountered mainly in flying. Small, high performance military aircraft readily develop stresses of 3 to 5 times the normal gravitational force (3 to 5 g) during tight turns, and forces of up to 5 g are encountered when naval aircraft are launched by catapult. Larger forces of up to 15 g are sustained by both the astronaut who is leaving the earth's atmosphere and the pilot who finds it prudent to use his ejector seat (Glaister, 1965).

The aviator and astronaut may also be exposed to severe decelerations. On entering the earth's atmosphere, a space capsule sustains a deceleration of about 15 g. The pilot who parts company with his aircraft encounters an immediate combination of tumbling and rapid deceleration (up to 20 g) with a further severe jolt (up to 30 g) as his parachute is deployed. The main civilian exposure to sudden deceleration is in vehicular accidents (page 355); forces of at least 30 g are encountered in that popular blood sport of the twentieth century, a head-on collision. Construction workers may also be exposed to sudden decelerations if they fall from tall buildings or bridges (page 145).

The topics of acceleration and deceleration are considered here because the performance of skilled tasks is affected by the unusual stimulation of the vestibular apparatus. However, acceleration has many other effects upon the body, physiological and mechanical (Howard, 1965). Even if the affected individual is securely strapped to his seat, the arms, legs, jaw and eyeballs are forcibly displaced, while a substantial portion of the circulating blood volume is shifted in the direction of the applied acceleration.

A sudden upward movement (*positive* acceleration) robs the eyes and brain of blood, through both a fall of arterial pressure and also

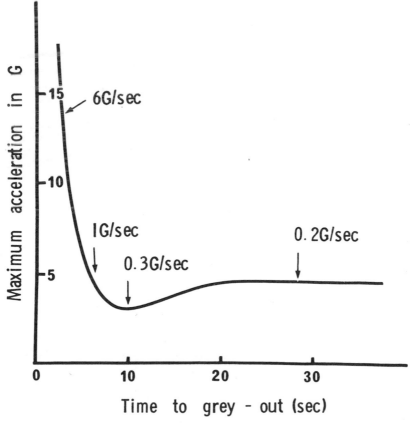

Figure 44. Probability of *grey-out* in relation to intensity and duration of applied gravitational force (based on experiments of Beckman *et al.*, 1960).

a forcible emptying of the venous side of the circulation. The eyes are commonly affected more than the brain, so that a *grey-out* or *black-out* of vision precedes loss of consciousness (Fig. 44 & Fig. 45). Blood in the pulmonary circulation is directed toward the base of the lungs, and the air spaces in this region tend to collapse owing to the in creased pressure of pleural fluid. A defect of blood oxygenation thus compounds the problems of poor venous return, reduced cardiac output, falling arterial pressure, and restricted cerebral blood flow. If the acceleration is maintained for more than 5 to 10 seconds, there is a compensatory increase in the tone of both arterioles and veins, so that displacement of blood into the legs is reduced. However, circulatory pressures within the blood vessels of the lower limbs remain high, and a progressive outflow of fluid from the capillaries into the tissue spaces further compromises an already inadequate blood volume.

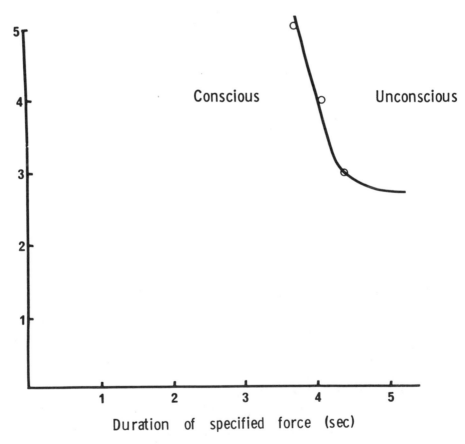

Duration of specified force (sec)

Figure 45. The influence of maximum acceleration (ordinate) upon time to *grey-out* (based on data of Stoll, 1956). Note that the rate of acceleration becomes progressively faster on moving along the curve from right to left

Downward thrusts are less well tolerated than those occurring in an upward direction, and vehicles are best designed to avoid *negative* accelerations. Such stresses drive blood towards the head, giving an intense congestion of the eyes (*red out*), followed by loss of consciousness. There is danger of brain damage if accelerations exceed 3 g.

Positive accelerations lead to decrements of visual acuity and poor performance of various psychomotor tasks such as tracking and choice reaction times, even if consciousness is not lost (Frankenhauser, 1958). The deterioration of performance has a complex basis. It reflects not only the reduced blood flow to the brain but also visual problems (distortion of the eyeball and restriction of ocular blood flow), mechanical displacement of the limbs and attempts to compensate for the apparent increase in mass of the moving parts by the use of visual rather than *automatic* proprioceptive control loops (page 87). The minimum acceleration necessary to disturb cerebral function remains uncertain. Frankenhauser (1958) demonstrated a 12 percent slowing of a predominantly mental task at 3 g; however, his test was quite simple and a lower threshold might be found for more complex activities. Space research workers have currently lost interest in specifying the threshold, since in any event the astronaut must be exposed to accelerations at least five times as great as threshold accelerations (Chambers, 1963).

The visual illusions that accompany sudden accelerations can cause a dangerous disorientation of aircraft pilots. Aircrew must therefore be trained to appreciate the nature of such illusions and to rely upon their instruments when these conflict with the visually perceived orientation (Clark and Nicholson, 1954; Nuttall and Sanford, 1956). In order to understand the *oculogyric illusion*, let us consider a beautiful ice skater pivoting on one foot. As she begins to turn, her eyes stop fluttering and fix upon some stationary external object; as this moves out of range, the eyes are flicked quickly to a second and then a third fixation point. Such adjustments of eye position are entirely automatic; this facility makes an important contribution to clarity of vision when the head is moving. However, when a man is enclosed in the cab of a rotating vehicle, the normal feedback loops are no longer appropriate, since both the head and the instrument panel are displaced by similar amounts. The eyes still make compensatory movements at the behest of the vestibule, and the instrument panel appears to be displaced in the direction of acceleration. Unfortunately, the illusion persists even if the driver has learned to hold his eyes still; he cannot avoid the unusual feedback of impulses from the oculomotor muscles (Byford, 1963). The apparent movement causes difficulty in reading instruments

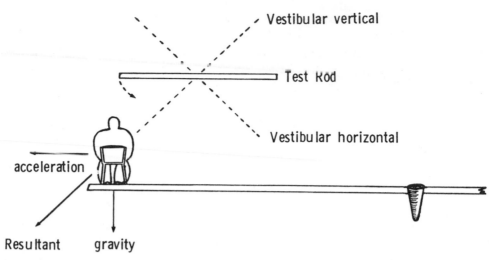

Figure 46. To illustrate the oculogravic illusion. A man is firmly strapped to a seat fixed vertically upon a turntable. The force acting upon his vestibular apparatus is the resultant of radial acceleration and gravity. Hence, the vestibular horizontal is displaced, and if the man is shown a horizontal luminous rod it appears to rotate in a counterclockwise direction.

(Meiry, 1965), and when calculating the possible work load of a pilot, due allowance must be made for the stress of correctly interpreting the orientation of a rotating or accelerating aircraft. The oculogravic illusion (Graybiel and Clarke, 1965) has a somewhat similar basis (Fig. 46); radial acceleration leads to a displacement of what the vestibular apparatus reports as the vertical. A man who is seated upright on a turntable is normally aware of this misinformation, but if the room is darkened, a luminous test object appears to rotate. A dramatic example of this illusion can be encountered by a pilot who makes an unsuccessful attempt to land his aircraft in misty weather. As the speed of the aircraft is increased, the pilot senses that the nose of the aircraft has risen; in fact, it is still directed downwards onto the runway, and the situation is not improved by the open throttle and wheel retraction (Fig. 47).

In vehicular collisions, decelerations are usually sustained in a backward direction. The body is surprisingly tolerant to insults of this type. Death and injury are often due not to the deceleration itself, but rather to a gross displacement of the body relative to the vehicle. A driver may be thrown out of a door, against a steering column, or through a windshield. Experimental studies have been conducted on the sudden stoppage of rocket-powered sleds. If a deceleration of 25 g is sustained for one second, there is a temporary loss of vision (associated with forward displacement of the eyeballs) and mental confusion sug-

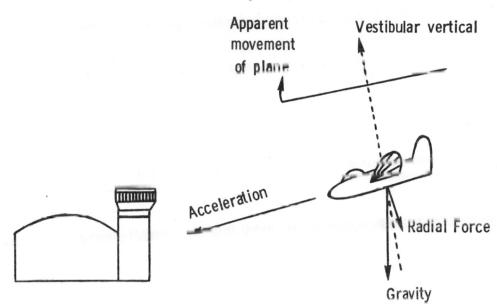

Figure 47. To illustrate the role of the oculogravic illusion in aircraft accidents.

gestive of cerebral concussion (Stapp, 1955). Over briefer periods, men have survived decelerations of 60 g (backwards, for 0.01 sec) and 80 g (forwards for 0.04 sec). Survival following a fall depends upon the firmness of the ground surface. A 50-foot drop onto concrete can produce a deceleration of 5000 g but on the other hand men have withstood falls of 1000 feet. If the initial impact is borne by the feet, much damping occurs in the legs and spine; Swearingen *et al.* (1962) have estimated that a 250 g-force applied to the feet is reduced to 7 g at the shoulders.

The adverse effects of sudden accelerations and decelerations are best minimized at source. Thus, the rate of acceleration of such devices as aircraft catapults and ejector seats can be slowed, and space rockets can be ignited in several stages, giving a moderate and sustained thrust rather than a violent lurch into outer space. The personal equipment of the military aviator includes an inflatable pressure suit that covers the legs and abdomen; the pressure within the suit is controlled by a gravity-operated switch, thereby minimizing the peripheral displacement of blood during positive accelerations. Severe accelerations are best withstood in the prone position, and astronauts are commonly provided with couches molded to their body form. Phases of space travel where severe accelerations are anticipated can be pre-programmed, so that the intended maneuver is performed satisfactorily even if consciousness is lost. Decelerations are least likely to cause injury if passengers sit facing towards the rear of a vehicle; the chair back then provides support for the body. This concept has been adopted in some British aircraft, but is not particularly popular with fare-payers, since they tend to fall from their seats at take-off. If the passenger is facing forward, his harness or safety belt should yield a little, without permitting the body to hit objects such as a meal tray, steering column or dashboard. Motor vehicles can usefully incorporate impact-absorbing features such as water-filled bumpers. Vulnerable roadside furniture such as lamp posts should be either protected by guard-rails or constructed of thin hollow metal rather than reinforced concrete or wood. Barriers separating divided highways should yield somewhat without allowing miscreant vehicles to enter opposing lanes or bounce into the path of following traffic. The only person likely to ponder landing techniques while falling from a great height is the paratrooper; he can reduce impact forces by landing on soft ground with his knees bent.

WEIGHTLESSNESS

The state of weightlessness implies the absence or counteraction of normal gravitational forces. When an astronaut is circling the earth in a satellite, the rotation of his vehicle generates an inertial force that exactly balances the attraction of the earth. Similarly, a diver may encounter no effective gravitational force if the volume of displaced water is equal to the weight of his body plus any equipment he is carrying.

Some of the problems of the weightless state are discussed elsewhere in the context of sensory deprivation (page 117). In this section, we shall consider briefly the disorientation that can result from the absence of normal gravitational stimuli.

The otolith organ of the auditory vestibule is designed to report the position of the head. However, the information flow is much reduced when the head is horizontal or upside down (Brown, 1961); furthermore, gravitational forces are necessary to displace the otolith particles and thus stimulate the receptor fibers. Under weightless conditions, the normal differential stimulation of pressure and muscle tension receptors is also lost, and the only valid information on body position is provided by the eyes. If the transition from normal to zero gravity conditions is rapid, even visual cues may be misleading. The eyeballs either rotate upwards or an unaccustomed balance of tensions develops in the external ocular muscles; in either case, fixed objects appear to fall (Warren *et al.*, 1964). The hands aim too high at any target because tension is lacking in the muscles that normally support the weight of the outstretched arm (Gerathewohl *et al.*, 1957). Compensation is possible if the movement is monitored visually (Whiteside,

1961), but *automatic* movements (page 87) must be relearned. This inevitably places some limitation upon other concurrent cerebral functions.

The diver faces many additional handicaps. For a detailed discussion, other books must be consulted (Miles, 1962; Lambertsen, 1967; Bennet and Elliott, 1969; Shephard, 1972a). The field of vision is usually restricted and light rays are distorted by some form of face mask. Pathophysiological problems such as over-compression of gas in the thorax and middle ear during descent, and oxygen poisoning, inert gas narcosis, and CO_2 accumulation while at depth are compounded by a gross increase in the work of breathing, due to the increased density of respired gas mixtures. Communication is difficult (page 122), and during ascent there are risks of lung rupture and decompression sickness. Physical labor underwater is made more difficult because the worker has no counterforce or support while using tools. He tends to rotate in the opposite direction to any levers that he may turn (page 71), and is also buffeted by the waves. Even simple tasks like inserting and tightening bolts in a screwplate require 50 percent more time for completion (Baddely, 1966)

Astronauts face many of the same problems. Vision is restricted by personal equipment. Locomotion presents serious difficulty, since walking is normally dependent upon friction between the sole of the foot and the ground surface (Margaria, 1971). Adhesive-soled shoes have been used experimentally (Graybiel and Kellogg, 1967), but the most effective pattern of movement for the spaceman seems a gliding motion. This is a new skill, and the best technique of training has yet to be evolved. Because the rarefied atmosphere provides almost no friction, motion continues until some alternative support can be grasped. Unfortunately, the body is likely to develop a rotary motion unless the initial propelling thrust passes through the man's center of gravity. The inflated pressure suit of the astronaut adds further difficulty in maneuvering the body. The eyes provide orientation with respect to the capsule, although Russian astronauts have reported a feeling that they were upside down whether their eyes were open or closed (Graybiel and Kellogg, 1967); this sensation could be corrected by pressing the feet firmly on the floor of the capsule. The effects of weightlessness upon cerebral function have yet to be evaluated. Attempts to simulate weightlessness in aircraft are not particularly successful, since the duration of the weightless state is not long enough to allow the effects of the preceding acceleration to wear off.

The physical problems of space travel (Hess *et al.*, 1967) fall outside the scope of this review. Prolonged immobility and a restricted diet cause many of the physical changes associated with prolonged bed rest (Saltin *et al.*, 1968)—a loss of working capacity and muscular strength, anemia and decalcification of the long bones.

The disorientation of the weightless state can be minimized by staying in well-lighted areas. The instability of an astronaut can be overcome by strapping him firmly in his seat, but it is more difficult to arrange suitable counterpressure for the diver. The loss of physical condition with prolonged space travel can be minimized by improved diet, coupled with physical exercise that includes compression of the long bones (Astrand and Rodahl, 1970; Bassett, 1971).

VISION

The eye is undoubtedly the most important of the sensory receptors, and a long section could be devoted to the physiological problems inherent in the receipt and interpretation of visual signals. However, we shall deliberately restrict our focus to the problems of the working man, including such facets as the intensity of illumination, the acuity of the eyes, glare and contrast, flicker, night vision, perception and the interpretation of colors.

Intensity of Illumination

Illumination is measured in foot-candles. A foot-candle is the light intensity at a distance of one foot from a standard candle. It is the minimum light needed for such tasks as reading and writing. Productivity is increased with the provision of more light, and the standards expected by the average worker have risen markedly in recent years. Thirty foot-candles is now regarded as a comfortable level for most tasks, and as much as 50 foot-candles may be provided for close work, with a minimum of 15 foot-candles in darker areas of a factory. When positioning lights, the aim should be to provide relatively uniform illumination, with a minimum of shadows. In this connection the inverse square relationship between intensity I and distance D must be kept in mind:

$$I = \frac{1}{D^2}$$

Brightly painted reflecting wall surfaces help to eliminate the problem of dark corners. Some shading of lights is desirable to reduce glare, but unduly thick or dirty lamp shades can lead to much wastage of electrical power. Areas for reading and other close work should have additional lighting, but glare is likely if the contrast with the remainder of the room is excessive.

The intensity of daylight much exceeds that of even the most modern artificial illumination. On a clear day, figures of up to 1000 foot-candles are common even at sea level; ascent to altitude is associated with both an increase in overall brightness, and also a selective increase in the shorter wavelengths of light. The eyes are drawn towards any bright source of light, and the very intense unidirectional *natural* lighting provided by the modern glass-sided office is thus undesirable. The worker's eyes are distracted from his desk, and he is left with a visual after-image, usually corresponding in shape with the window frame. Incidentally, he is also likely to complain that he is roasted in summer and frozen in winter, due to an excessive radiant heat exchange through the vast expanse of glass.

Visual Acuity

Overall Acuity

The ergonomist expresses visual acuity in terms of the minimum angular separation that can be perceived. A young worker standing in a large and well-lit warehouse may see an object one inch wide at a distance of 100 yards. This object subtends an angle of one minute at his eye. Clinicians rate visual acuity relative to population standards, using Snellen test type. If the left eye of a person standing 6 meters from a chart can only read the large type intended for 18 meters distance, then the visual acuity of that eye is expressed as 6/18.

The acuity of the individual depends partly upon the efficiency of his lens system, and partly upon the sensitivity of the receptor cells in his retina.

Spherical and Chromatic Aberration

Incident light is refracted at the cornea and at the anterior and posterior surfaces of the lens. Problems thus arise from both spherical and chromatic aberration. The human lens has an added convexity at its center, and for this reason, the central rays of light are refracted more than those at the periphery. Light at the blue end of the spectrum is also refracted more than that at the red end. As in man-made cameras, the optical properties of the system are regulated by varying the aperture of the iris. The pupil is widely regarded as a light regulator, but in fact the possible range of diameters (16-fold) is small relative to the range of light intensities encountered in normal life; even in bright sunlight, the pupil gradually dilates to about 4 mm, and this optimum setting seems determined by the optical properties of the lens. Transient changes of pupil diameter merely give the eyes time to adapt to a brighter light. Chromatic aberration can give a blue object a red fringe, or vice versa; however, the visual areas of the brain have a facility for suppressing unwanted images, and even if a chromatic aberration is deliberately doubled by provision of a poor quality lens it is normally not perceived (Fincham, 1951).

Refractive Errors

The total resolving power of the average eye is about 68 diopters (i.e. it is equivalent to a convex lens with a focal length of 1/68 meters). The majority of refraction is contributed by the cornea, and problems can arise when this membrane is congested by exposure to irritant gases, or is affected by hay fever or the common cold. Refraction at the corneal surface is lost when working under water, and vision grossly deteriorates unless the air interface is restored by the use of goggles or a visor; if goggles are worn, the lens material requires careful selection for optical quality.

Personal errors of refraction arising from an abnormal length of eyeball lead to short- or long-sightedness. Minor refractive errors can be corrected through the activity of the intrinsic eye muscles, altering the curvature of the lens (accommodation). Close work involves both accommodation and convergence of the eyes (an activity of the extrinsic eye muscles). Fatigue of the intrinsic and extrinsic eye muscles is thought responsible for the burning or aching sensation of visual strain (page 75). It can be greatly reduced by provision of suitable reading glasses and a due allowance of rest pauses (page 54).

Problems of Accommodation

If the pupil diameter is 3 to 4 mm, the relaxed lens will focus light from 30 feet to infinity. This is a convenient range for car-driving or flying. If the visual field is empty (as when driving or flying in fog), involuntary accommodation occurs, and the eyes focus upon objects at a distance of 1 to 2 meters (Whiteside, 1957). Distant objects become blurred in outline, and can only be seen if they are larger than the normal threshold size. The long-sighted person is at an advantage under such conditions and the performance of the average pilot can be improved by training techniques that involve the projection of fine patterns at an *infinite* distance upon a blank screen. Instruments likely to be used at night (telescopes and binoculars) are commonly given a slight negative lens setting to compensate for involuntary accommodation. The opposite type of problem arises when a driver or pilot wishes to consult his instruments; the eyes are focused upon the distant horizon, and it is necessary to read a gauge perhaps two feet from the face. The simplest arrangement is to make the gauge easy to read even with incompletely focused eyes. Collimation provides a more expensive solution; the gauge is covered by a lens that in effect sets it at an infinite distance from the eyes. Collimation is essential when instruments are to be read simultaneously with forward vision; it is also easily introduced into rearview mirrors

Regional Acuity

The acuity of the receptor cells varies with the individual, deteriorating with age. It is also greatly influenced by illumination, contrast, glare, flicker and the duration of presentation of a test object. The anatomical arrangement of the eye is such that electromagnetic radiation in the visible band (4000 to 7000 angstroms) must pass through several layers of retinal tissue to reach the transducers (the cone and rod cells). The optic nerve fibers in turn penetrate the retina en route to the brain, giving rise to a *blind spot*, covering an area 3° by 5°. The most sensitive region of the retina is the fovea. Although a few of the cells in this region have a rod shape, all are functionally color transducers (cone cells). Expressing foveal acuity under daylight condition as 100 percent, there is a progressive decline of angular discriminatory power on moving either medially or laterally (Fig. 48). However, the cones are unable to function in dim light; this facility is reserved to the monochromatic rod cell. Night vision is thus most acute when looking 15° to 20° to one side of an object. Special training programs are needed for night workers in order to overcome the natural tendency to look directly towards a target. In the foveal region there is a 1:1 relationship between receptor cells and transmission lines to the brain, but in more peripheral areas, as many as 5000 transducers may converge upon a single ganglion cell. This type of spatial summation is well suited to operation when lighting is dim, and increases the probability of firing in a given ganglion cell (page 350). The observed acuity substantially exceeds what would be anticipated from the number of receptor pathways, and various models have been developed to explain the sharpening of focus within retinal nerve networks (page 349).

Field of Vision

The natural field of vision is adequate for most tasks, but in some working situations it may be artificially restricted, either by cab design, or by the use of protective equipment such as a respirator (Shephard, 1962) or goggles. Considerable compensation is possible if the employee is taught to increase the frequency and range of head movements. At one time, aircraft pilots wore goggles to protect their eyes from both glare and the windblast of an emergency. However, any covering of the face becomes hot and uncomfortable with prolonged wear, and military pilots are now equipped with a color-tinted visor; this gives protection against glare, and snaps shut over the front of a pressure helmet in an emergency.

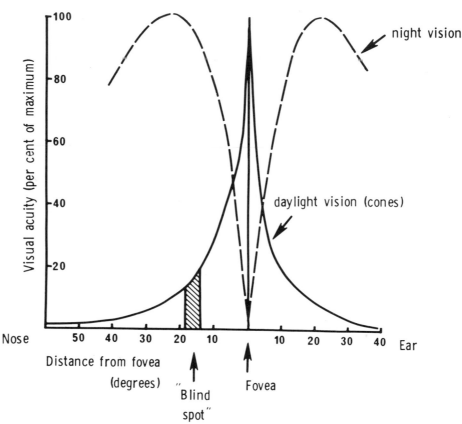

Figure 48. Visual acuity at varying angular distances from the fovea.

Glare and Contrast

Contrast and Tracking

An object is difficult to see unless there is adequate contrast between it and its background (Fig. 49). White paper reflects 80 percent of the incident light (Poulton, 1970). Thus if the intensity of illumination is 30 foot-candles, the brightness of the paper is 24 foot-lamberts. Well-printed black type reflects no more than 4 percent of the incident light, giving a contrast ratio of $\frac{(80 - 4)}{80}$ 100 or 95 percent. However, a publisher may decide a book looks prettier if red type is set on pink paper; the contrast ratio is now much reduced because the pink paper reflects less light than the white, and the red type reflects much more light than the black. Reading becomes difficult if the contrast ratio drops below 50 percent, and the problem is only partially relieved by an increase in the brightness of illumination. Modern hardcover books commonly use 10-point type; this is one seventh of an inch

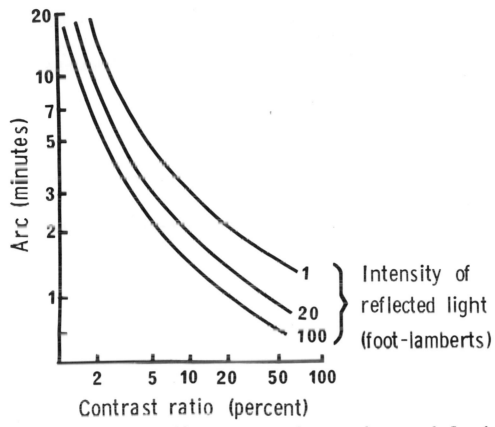

Figure 49. The perception of detail as a function of contrast and intensity of reflected light. The arc refers to the angle subtended at the eye by the gap between the two bars of the Roman numeral II (after an experiment by Cobb and Moss, 1928).

tall, and if the book is held at a normal reading distance of 12 to 15 inches, the gaps between characters subtend an arc of three to five degrees at the eye. It would appear from Figure 49, that such type would be readable in very poor light (<1 foot-lambert). However, in the experiment illustrated Cobb and Moss required their subjects to do no more than distinguish the orientation of a single letter over a period of 0.17 seconds. Speed reading requires more light—at least three foot-candles for good 10-point type, increasing progressively to 20 to 30 foot-candles with smaller lettering, poor paper and badly smudged carbon copies (Tinker, 1963).

The normal technique of reading is a *saccade*; the eyes focus for an instant upon a word or a phrase, and then jerk rapidly to the next fixation point. A moving chart or other target can be followed at speeds of up to 30 degrees per second, measured at the eyeball. This facility is sometimes used in the navigation of high-performance aircraft.

Unless the plane is at high altitude, a conventional map is useless; instead, a strip map is propelled at a speed proportional to that of the aircraft. Discrimination of detail decreases with the angular velocity of the chart (Ludvigh, 1955). Some help can be obtained by rotating the head as well as the eyes, but at speeds above 50 degrees per second there is a marked deterioration of performance; the eye tends to lag behind its target and then catch up with a jerk.

The tracking ability of the human eye has traditionally been valued in the context of English sports such as tennis and cricket, where the ball may move with respect to the eye at speeds of more than 30 degrees per second. Not only must the eye follow the ball, but the information must be processed in the brain and translated into an appropriately coordinated movement of arm and wrist within about 1½ seconds. The problem of visual tracking and reaction times is now encountered in a more lethal form when piloting a military aircraft. An enemy plane may be perceived while flying at three times the speed of sound (Mach 3). The time for perception in the periphery of the visual field is 100 msec, some 60 msec is occupied in rotating the foveal receptors to receive the critical information, and these receptors in turn impose a further 50 msec lag (Strughold, 1951). By this time, the plane has flown over 200 yards. The brain now starts to interpret the information, and before recognition is achieved, 1000 yards may have been covered. The enemy, in the meantime, has also approached by 1000 yards. It is necessary for pilots to spot fast-moving enemy aircraft at a distance of several miles if they are not to fall victim to a surprise attack. Unfortunately, contrast is reduced at altitude, since the brightness of the sky is almost directly related to ambient pressures; the pilot is searching for the dark shadow of a plane against what seems an equally dark sky. Even if the enemy aircraft is detected, there is a danger that its distance will be overestimated, since there is no basis for estimating size in an empty sky. The apparent increase in size of the moon when it first appears over a barn is a similar phenomenon; at this stage, the size of either moon or plane can be compared with that of a familiar object.

Glare

Three categories of glare are distinguished: veiling, dazzling and blinding glare. Veiling glare is a common experience of the English winter driver who faces mist and fog. It may also arise from dirt or scratches on goggles, visors or other equipment. Light is scattered, and fails to form a distinct image. Many English vehicles are equipped with yellow headlights in the belief that such selective filters improve visibility. There is no good evidence that this is the case, although the reduction in intensity of light may lead to less scattering, and thus greater comfort for both the vehicle operator and those traveling in the opposite direction.

Dazzling glare is caused by the scattering of excess light within the eye. It may arise from any smooth and brightly illuminated surface such as industrial machinery which lacks a mat finish. Often, the workers concerned do not complain, because man has a substantial ability to adapt to adverse conditions; nevertheless, performance improves if the glare is corrected. The motorist on older highways is often troubled by glare from the headlights of vehicles traveling in the opposite direction. The problem increases rapidly as vehicles converge, since the intensity of illumination varies with $1/D^2$ (page 148); however, glare diminishes again at about one car length, since the opposing vehicle is now (hopefully) in its correct lane, some 10 or more feet to the side of the driver. The momentary nature of blinding, seen also in the formal *glare-recovery* tests used at some driving instruction centers, emphasizes that difficulty is caused by a scattering of light rather than a loss of night vision. Modern expressways minimize glare by wide separation of opposing lanes, or the erection of glare barriers where this is impractical. Much can also be achieved by regular vehicle inspection (to detect poorly aligned headlamps), and training drivers so that they do not allow their eyes to be drawn towards on-coming lights. Most drivers regard the use of dipped headlights as a glare-reducing courtesy, although at least one formal study has shown that if all road-users adopt a normal beam position, the added illumination more than compensates for any increase of glare (Johannson *et al.*, 1963). Aircrew are troubled by dazzle both day and night. Since the brightness of the sky varies with atmospheric pressure, the greatest intensity of light comes from below an aircraft and the eyebrows are unable to fulfill their normal protective role. The standard remedy is a tinted visor, darker below than above. At night, dazzle may be caused by runway lights; the problem is due not so much to the absolute intensity of illumination as to the rapid change of intensity as the aircraft approaches the landing strip.

Blinding glare implies exposure to light of sufficient intensity to reduce the sensitivity of the retina for a period. Arc-welding, photography and the watching of an eclipse are common sources of difficulty. The problem may also present as *snow blindness* when men work out of doors at high altitudes. The thinner atmosphere provides less attenuation of the sun's rays (particularly of the shorter wavelengths). The liability to glare is compounded by an increased contrast between shadow and sunlight; the sky provides less light to illuminate the shadows, and the sun itself casts a sharper shadow as its surrounding *aureole* is reduced.

Flicker

If a light source flickers at a rapid rate, the eye does not perceive the flickering. However, as the rate is slowed, it becomes apparent; depending upon the level of arousal (page 215), flickering is usually seen first at frequencies lower than 15 c/sec.

The usual reaction to flicker is annoyance. There may also be a modest deterioration in the performance of skilled tasks. In susceptible individuals, an epileptiform attack may be provoked (for instance, by driving a truck down a tree-lined avenue in bright sunlight). Reactions to a flickering tungsten light source are illustrated in Figure 50. The intensity of illumination was deliberately varied three times per second while subjects were reading. Fifty percent of the population were aware of flickering when the fluctuation reached 3.5 percent, and with a 5 percent fluctuation some of the group were prepared to complain to the company supplying the electrical power. Nevertheless, there was no deterioration in reading performance (Poulton *et al.*, 1966). The normal reader moves his eyes some four times per second (page 153). The rate of reading may thus be slowed if the speed of flicker is less

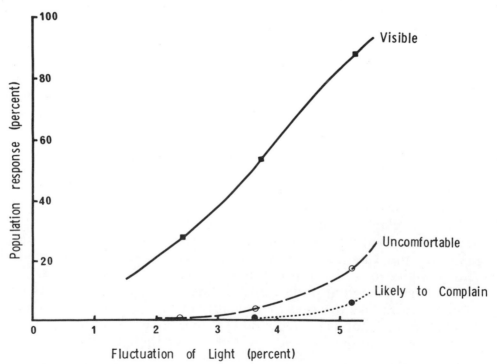

Figure 50 The subjective reactions to 3 cm/sec fluctuations in light intensity (after data of Poulton et al., 1966).

than 4 c/sec, and the intensity of illumination is inadequate for reading during the darker phases of the cycle. Tracking tasks are performed more poorly in a flickering light, because the movement appears jerky (Battig *et al.*, 1955).

Night Vision

Dark adaptation follows a two-component curve (Fig. 51). The more rapid phase of adaptation is complete within five minutes, while the slower phase occupies at least forty-five minutes (Wald, 1954). If adapted eyes are exposed to red light, the subsequent recovery curve shows only the first phase; this seems to correspond with adaptation of the cones, while the slower phase corresponds with adaptation of the rods (which are not materially stimulated by the red light).

The rods are the only effective receptors in dim light, but unfortunately they are absent from the fovea (Fig. 48); the *blind-spot* of daylight vision is thus supplemented by a foveal blind area some 10° in diameter. When searching for a target at night, it is important to keep the eyes moving, and to resist the temptation to fix the vision upon any object that is spotted. During World War II, the British Royal

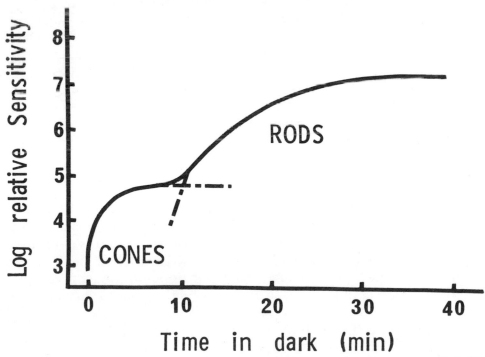

Figure 51. The normal course of dark adaptation (based on experiments of Wald, 1954).

Air Force found it useful to establish special schools to teach techniques of scanning the night sky. It was also found important to ensure that night vision was not impaired by excessively bright illumination of instrument panels and charts within an aircraft or control room. Luminous paint did not provide very satisfactory dial markings. It became rather thin at the edges of figures, blurring their outlines, and a large part of its brightness was lost within a few years. A better approach was to use a fluorescent paint, and to illuminate this with ultraviolet light derived by selective screening of a tungsten filament. Navigators attempted to work at desks illuminated by red lights; an extensive recoding of maps was then necessary, using appropriate shades of grey, blue and green (Whiteside, 1949). Fortunately, most of these measures have become less necessary with the development of radar navigational aids.

Electromagnetic energy is converted to nerve impulses through the breakdown of a pigment (visual purple) within the rod cells. Resynthesis of visual purple is very susceptible to oxygen lack, and pilots of unpressurized aircraft are thus advised to wear oxygen masks at altitudes in excess of 4000 feet.

Because the rod cells respond to the green portion of the spectrum, green objects appear brighter and red objects darker as twilight approaches. Red objects may be mistaken for shadows, and the normal stereoscopic cues derived from shadows are reduced or lost. Detail also becomes blurred due to the lower intensities of illumination, and driving becomes much more hazardous. Artificial street lighting eases the situation, but problems of altered colors and misinterpreted shadows may still arise, particularly if the light source covers a narrow spectral band, as in the usual sodium, potassium and mercury vapor discharge lamps.

Perception

Appropriate performance requires not only stimulation of the visual receptor organs, but also interpretation of signals by the visual association areas of the cerebral cortex. The duration of a stimulus determines whether it is perceived vaguely, clearly or in detail. Brighter illumination shortens the time to recognition. Much depends upon the complexity of an object and its familiarity; the situation is also influenced by the vigilance of the observer (page 217). A familiar object seen briefly from a high-speed vehicle is readily misinterpreted if it changes but slightly; this may explain in part how a driver of an express train, thoroughly familiar with a route, can pass a signal set at danger (page 364). It also explains the occasional errors of proof-reading in this and other books. The average reader looks at a word or even a phrase at a time, and a deliberate effort must be made to check each letter individually. Misspelled words are usually interpreted without thought in the general context of the paragraph. Let us suppose the words *soal* and *whail* appear in the text. If the subject under discussion is marine mammals, a high percentage of readers will perceive *seal* and *whale*, while if the topic is boating, *sail* and *wharf* become more likely interpretations (Edholm, 1967). Much depends also upon the emotional needs of the moment; if the reader is hungry or thirsty, a misspelled word may be interpreted as a form of food or drink, while a dimly printed or blurred picture of food assumes a new brilliance and clarity. Colors interact with emotional state. Red conjures up feelings of warmth, while blue is cold, clean and impersonal. Reactions to color are studied extensively by advertisers and packagers anxious to create an emotional environment where the shopper will feel a strange compulsion to buy their particular product (irrespective of its merits).

The correct perception of *depth* is important to many industrial skills. At one time it was believed to depend mainly upon sensations arising from the extrinsic muscles, as the eyes converged on a target. Certainly, the external muscles of the eyes have a greater nerve supply than other *voluntary* muscles. However, the body uses many cues to assess distance and relative movement of objects, including not only the sensations of convergence, but also the apparent size, clarity, color, perspective and position of shadows. Some highway authorities insist upon testing the stereoscopic vision of prospective motorists; evaluation is based mainly upon sensitivity to convergence. Nevertheless, it is now well recognized that a person who is blind in one eye can drive a vehicle and even land an aircraft with reasonable safety.

For some purposes, it is necessary to go beyond perception. A visual display must be not only seen and recognized, but understood within a broad framework of knowledge. This is particularly true of visual aids used in teaching (page 240). Illustrative material should be simple, free of irrelevant information, and adequately explained. Slow-motion films can be useful in teaching specific skills, but at normal cinematograph speeds actions occur too rapidly for important details to be seen and learned.

Color Vision

Color may be expressed in such terms as the emission of electromagnetic radiation, reflectance and the sensitivity of the retinal cells. However, it is ultimately a centrally perceived phenomenon, and there can be no simple one-to-one relationship between the physical stimulus and the reported color.

Natural sunlight provides rather uniform radiation over the entire visible spectrum (Fig. 52). However, many factories are illuminated by north lights to avoid excessive heating, and a preponderance of blue light is then admitted. Standard tungsten filament bulbs emit mainly yellow and red light, while fluorescent tubes show sharp peaks of emission in the blue, green and pink parts of the spectrum. Discharge lamps glow at the wavelength characteristic of the vapor they contain; thus, a sodium lamp produces a very narrow band of orange/yellow light. The three primary colors of emitted light are red, green and deep blue. Yellow is formed from red and green, blue from green and deep blue, and white from a suitable mixture of all three primary colors. Observers can distinguish not only hue, but also the depth and brightness of a color. The *depth* or *saturation* depends upon purity; one of the mysteries of the Renaissance was the quest for pure colors both in stained glass and in the painter's pigments. The *brightness* of a color depends upon the *intensity* of light. *Subtractive* colors are found in paints and pigments. They are formed by materials that absorb or filter out a portion of the incident light. The primary subtractive colors are red, yellow and blue, and a suitable mixture of these colors produces black. An object that appears blue is absorbing green light, and reflecting mainly blue radiation (Fig. 53).

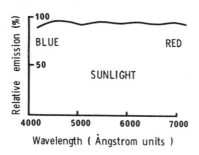

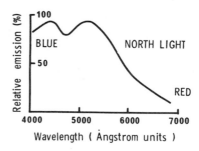

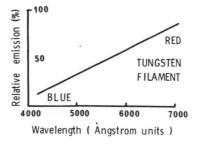

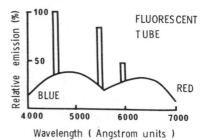

Figure 52. Schematic diagram showing the wavelengths of light provided by different types of illumination.

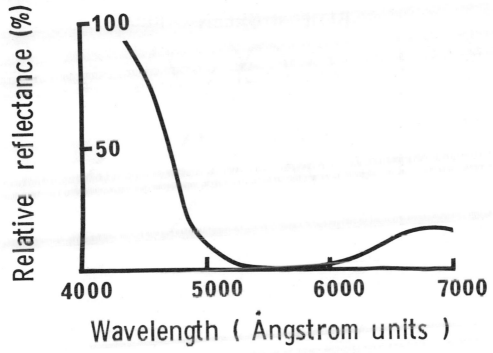

Figure 53. Reflectance curve for a blue book cover.

The relative sensitivity of the receptor cells is illustrated in Figure 54. The rod cells show a maximum at 5100A, with a sharp decline on either side. Many texts show a similar shaped curve for cone receptors, but this refers only to one species, usually the green receptors (peak 5400 to 5500A); if account is taken of all three types of receptor, the sensitivity to colored light is relatively uniform throughout the visible spectrum.

Appearances are readily modified by simultaneous or immediately subsequent contrast. Thus a neutral grey pigment appears bluish grey if viewed on a yellow background, and yellowish grey on a blue ground. Again, blue on a yellow background appears much brighter than blue on black.

The accurate standardization of color is important in many industrial operations ranging from the printing of banknotes and postage stamps to the preparation of foods; many customers will refuse to buy a product unless it is of the *correct* color. International agreement has now been reached upon the average sensitivity curve for the retina, and standard light sources have also been specified. Let us suppose that we wish to match a given unknown color U and to specify its characteristics; the color U is projected upon half of the screen, and varying amounts of red (R), green (G), and blue (B) light are mixed upon the opposite

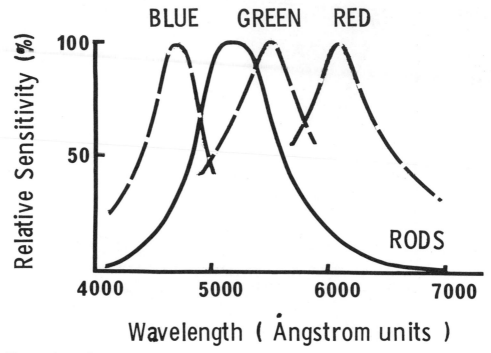

Figure 54. Relative sensitivity of visual receptors for blue, green and red light and for night vision (rods).

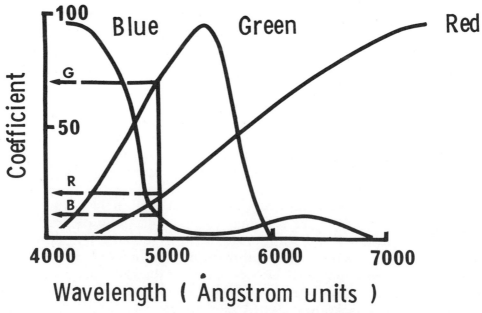

Figure 55. To illustrate the mixture of primary colors needed to produce light of a specific wavelength. Light of 5000 angstrom units is formed by mixing red, green and blue light in the proportions indicated by the coefficients R, G and B.

half of the screen until the color is matched (Fig. 55). This is essentially a judgment, and must be made by a suitable panel of well-trained observers:

$$U\bar{U} = R\bar{R} + G\bar{G} + B\bar{B}$$

In some parts of the spectrum, it is not possible to define a color by simple addition; subtraction is necessary, for instance by feeding blue light to the *unknown* half of the screen.

$$U\bar{U} = R\bar{R} + G\bar{G} - B\bar{B}$$

In order to avoid the inconvenience of specifying colors by negative coefficients, it is usual to transform data to three unreal primary colors X, Y and Z having slightly different spectral properties. Allowance is made for the nature of the light source and/or reflectance (β_λ) and the physiological function of the eyes (P_λ) at any given wavelength. Thus for a complex color or pigment

$$U\bar{U} = X\bar{X} + Y\bar{Y} + Z\bar{Z}$$

$$X\bar{X} = \sum_{4000\,\text{Å}}^{7000\,\text{Å}} P_\lambda\,\beta_\lambda\,Y_\lambda\,\Delta_\lambda$$

It is inconvenient to represent colors in a three-dimensional space, and for many purposes it is preferable to derive two-dimensional coordinates x and y such that

$$x = \frac{X}{X + Y + Z} \quad \text{and} \quad y = \frac{Y}{X + Y + Z}$$

All possible combinations of color then fall within the shaded area of an x/y diagram (Fig. 56). Pure spectral colors lie along the upper boundary of the diagram, while colors formed by admixture (such as purple or white) lie within the diagram. If a color is required for a signal light—or any other industrial purpose—it is possible to define it accurately in terms of its x and y coordinates. In some operations such as the blending of paints and inks, it is also necessary to specify the error of matching, in terms of the dispersal along x and y coordinates. The average person is less certain of matching in the green than in the red and blue parts of the spectrum (Fig. 57). Some individuals are *color blind*. The large majority of these (anomalous trichromats) can match all spectral colors, but use unusual proportions of the primary colors in so doing. Protanopes (Fig. 58) are insensitive to red, and match the spectrum as they see it with blue and green. Deuteranopes are insensitive to green. Tritanopes with an insensitivity to blue, and monochromats with no color sensitivity are rare. The protanope has a series of isochromatic lines radiating from point P in the red corner of his x/y diagram (Fig. 58); in the same way, the deuteranope has

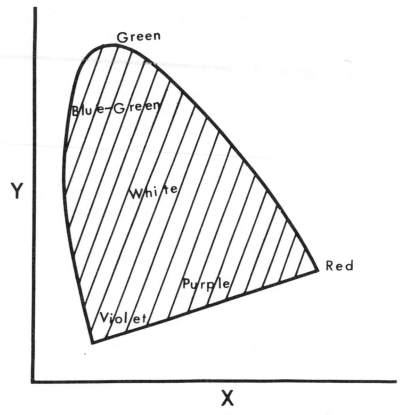

Figure 56. Two-dimensional representation of colors (chromaticity diagram). For significance of x and y coordinates, see text.

a series of isochromatic lines radiating from point D, and the tritanope a series radiating from point T.

Color-blind individuals are most accurately diagnosed by plotting areas of color confusion on an x/y diagram. However, this is time consuming, and requires specialized equipment. Thus for practical assessment, a worker may be required to sort skeins of wool (Holmgren's test) or read numbers from a multicolored chart (Ishihara test). In setting up such screening procedures, a standard light source must be used, and the colors presented must be so adjusted that an appropriate proportion of the total population is diagnosed as abnormal.

Some 10 percent of males have an appreciable defect of color vision. A job analysis is thus desirable to specify necessary standards of visual acuity and color vision in any particular occupation. Unfortunately, decisions are commonly made on an emotional rather than a rational basis. For example, *normal* vision is still mandatory for navigators within the Royal Canadian Air Force. This may have had relevance to flying when extensive use was made of color charts, but is plainly

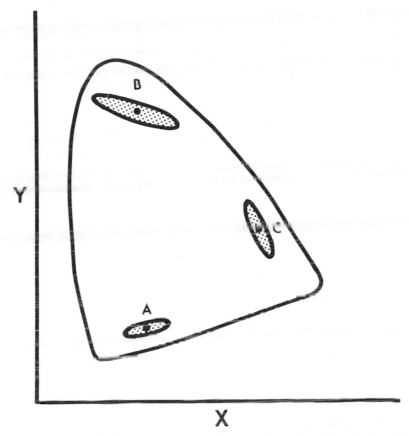

Figure 57. Average error in matching colors at three selected points A, B and C within the x/y chromaticity diagram. Note that errors fall within an elliptical area, and are larger in the green region of the spectrum (see also Fig. 58).

much less important now that the main basis of navigation is the use of a monochromatic radar screen.

Most color signals are based on red, yellow and green. Considerable research effort has been devoted to specifying the x and y coordinates of colors that can be recognized by a large proportion of the population. In consequence, the blue content of many green traffic signals has been increased. Although adequate recognition may be possible with good visibility, the proportion of the population making mistakes increases with adverse weather conditions such as fog or drizzle. There is thus much to commend supplementing the color code by size (larger and brighter signals for red) and shape (for example, green = circle, yellow = diamond, and red = square).

The matching of colors in terms of x and y coordinates is time consuming, and when choosing paints, dyes and the like it is convenient

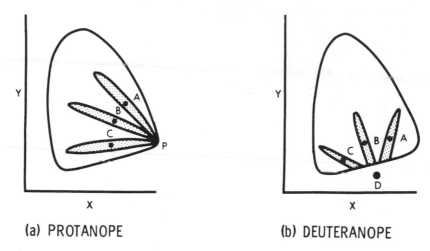

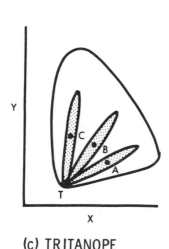

(c) TRITANOPE

Figure 58. To illustrate areas of confusion for *color-blind* individuals. (a) protanope, (b) deuteranope, (c) tritanope. All data plotted on a chromaticity diagram (see Fig. 57).

to compare colors to a standard book of tints such as the Munsell or the Ostwald systems. In any area where colors are matched, be it a paint shop or a tailoring warehouse, suitable illumination is most important. Where possible, incident light should have a flat spectral distribution, as in natural sunlight; alternatively, workers should have specific training to overcome problems related to the nonuniform spectrum of illumination.

CHAPTER 8

THE BIOMECHANICS OF HUMAN MOVEMENT

T HE ERGONOMIST IS CONSTANTLY striving to minimize the physical work load encountered in industry and the home. Ultimately, he may well suggest modifications to tools or techniques and even replace human functions by an automaton. However, a necessary first step is to study the biomechanics of a given operation, to note the limitations of the human operator, and to obtain ideas for the design of purposeful *humanoids*.

This chapter thus makes a brief survey of the biomechanical principles governing gross body movements, with specific reference to the descriptive techniques of the kinesiologist and the time and motion engineer (Dyson, 1970; Rasch and Burke, 1972; Brunnstrom, 1972; Mundel, 1970). Applications that are considered include the lifting of heavy loads, walking, and the design of prostheses and walking vehicles.

BIOMECHANICAL FUNDAMENTALS

Mass, Space and Time

The physical properties of a biomechanical system can be described in units of mass, space and time. *Velocity* (distance/time) is characterized by direction, and may be linear or angular; most work involves repetitive and nonuniform angular movements. Motion may be resolved into components parallel with standard *body planes*. A convenient starting point is the *anatomical position*, when a man is standing erect, with his palms facing forward. The *sagittal plane* then passes vertically through the body from left to right, the *coronal or frontal plane* passes vertically from front to rear, and the *transverse plane* passes horizontally from front to rear. Notice that when a movement is carried out in one of these three planes, the remaining two planes describe the axis of rotation. A point within the body (such as the center of gravity) can be described by reference to two axes or three planes. The planes passing through the center of gravity are known as the *cardinal planes*.

Acceleration (distance/time^2) can occur in any plane, but is typically developed with or in opposition to gravitational acceleration. In any given muscle system, maximum speed is attained at zero external loading, and maximum force at zero speed.

167

The Laws of Motion

The fundamental laws of motion were propounded by Sir Isaac Newton in the seventeenth century. They may be related to biomechanics when stated in the following form:

Law I. Inertia and conservation of momentum. A body segment remains in a state of rest or uniform motion unless acted upon by some force external to the segment. If at rest, there is inertial resistance proportional to the mass of the segment and any attached load. If in motion, there is a momentum proportional to the product of mass and the linear or angular velocity of the part.

Law II. Acceleration and applied force. Any change in the motion of a body segment is proportional to the effective external force (that is, force = mass × acceleration). The disturbance may arise from the normal gravitational acceleration (page 194), an agency external to the body such as a gust of wind or a mechanical impact, or a force generated by muscular activity in an adjacent body segment.

Law III. Reaction. Every action induces an equal and opposing reaction, thereby conserving the energy of the system.

The work performed upon a system is given by the product of force and displacement. Energy (the capability for performing work) can be neither created nor destroyed. A man may boast a certain reservoir of *potential energy* (such as that due to climbing a flight of stairs) or *kinetic energy* (due to body motion), but once these resources have been exhausted the person who is producing the displacing force needs energy replenishment (usually given in the form of food, page 43). Movement is characteristic of man, and indeed the human organism can be regarded as a complex and rather inefficient system for the transfer of energy.

The Human Machine

As a simple example of the human machine, let us examine a rudimentary elbow joint (Fig. 59). The upper arm is held at the angle illustrated because certain of the shoulder muscles are exerting a fixation force to counteract the influence of gravity. The forearm segment has a finite mass, effectively concentrated at its center of gravity, and a finite inertia, proportional to mass.

Unless there is intervention from some external agency, movement of the forearm is attributable to the resultant of gravity and forces developed by the flexor and extensor muscles about the elbow joint acting against the inertia of the segment. Part of the potential muscle force is expended in overcoming the viscous resistance of the fibers; such internal work develops exponentially as the speed of movement is increased.

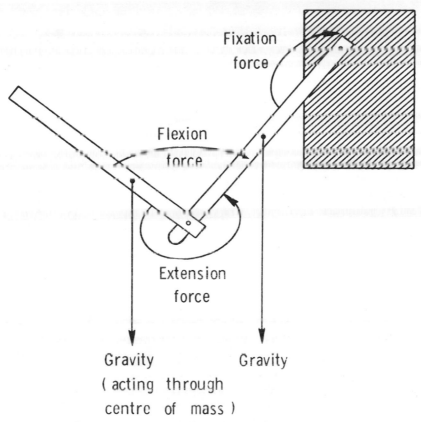

Figure 59. A diagram illustrating the balance of forces about the elbow joint.

Center of Gravity

The concept of the center of gravity is fundamental to biomechanics; the mass of the body may be considered as concentrated at this point without alteration of forces. The center of gravity of the average man is set in the horizontal plane at 56 to 57 percent of standing height. In women, a combination of heavier hips and thighs with a lighter upper half of the body give an appreciably lower center of gravity. The center of gravity moves with displacement of individual limb segments, and in certain athletic maneuvers such as the jackknife dive or the *Fosbury flop* of the high jumper it may be displaced outside of the body.

Twin scales and a beam may be used to determine the center of gravity for the body as a whole (Fig. 60). The center of gravity of individual limb segments is generally calculated by the use of equivalent coniforms, cylinders and spheres, making *reasonable* assumptions regarding the densities of bone, muscle and fat.

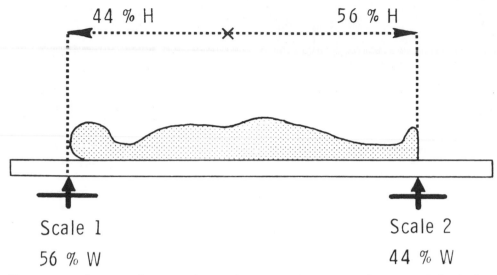

Figure 60. The use of twin scales to determine the center of gravity of the body.

Equilibrium

Equilibrium may be static or dynamic. If the body is in static equilibrium, then a perpendicular line drawn from the center of gravity (the *gravitational line*) passes within the standing base; the likelihood of achieving equilibrium thus varies directly with the size of the standing base, and inversely with the height of the center of gravity from the ground. The degree of stability depends upon the distance separating the gravitational line from the edge of the standing base, the combined inertia of the person and any load that he may be carrying, and the magnitude of any external forces (such as gusts of wind). During rapid marching (page 196), the body is tilted forwards, carrying the gravitational line close to the toes, and loads are best carried as a backpack that restores the center of gravity to a more central position.

If the body is in dynamic equilibrium, then the sum of forces acting about the center of gravity produces zero torque; normally, gravitational forces are being counterbalanced by muscular activity. The equilibrium position varies from moment to moment, due not only to varying external forces (such as wind gusts), but also to anatomical displacements induced by cardiac and respiratory forces. Since muscular effort is needed to maintain a dynamic equilibrium, it is economical to devise methods of achieving static equilibrium. A worker may be provided an external support such as a chair with an adequate back, or he may carry out his work from the prone or the supine position. In building construction, both thermal comfort (page 69) and maintenance of equilibrium can be helped by temporary windshields.

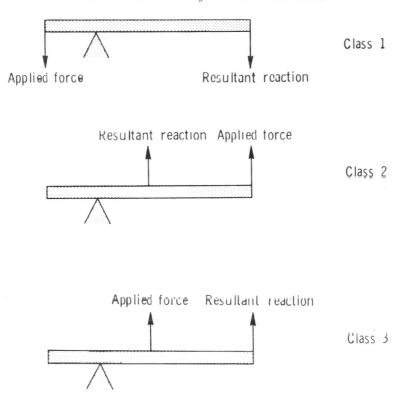

Class 1

Applied force   Resultant reaction

Resultant reaction Applied force

Class 2

Applied force   Resultant reaction

Class 3

Figure 61. Three classes of lever.

Levers

Three classes of lever may be recognized (Fig. 61). The *first class* has a central fulcrum, so that the resultant force operates in the opposite direction to the applied force. This is exemplified by the action of the triceps at the elbow joint. In the *second class*, the distance separating the applied force from the fulcrum is greater than that for the resultant. The movement is thus powerful but slow. Examples within the body are not too plentiful; plantar flexion of the ankle and closure of the jaws may be instanced. Applied and resultant forces are interchanged in *third class* levers, so that the movement is fast but lacking in power; elbow flexion provides a good example.

Calculations of leverage in the body are complicated, since forces are summated about a large number of joints. Further, the effective angle of insertion of most muscles changes throughout movement, and this inevitably alters the resultant force (Fig. 62). In some instances, a joint surface may be used as a pulley (Fig. 63). In other situations, the applied force may induce angular rather than linear motion about a joint. Notice that inertia is then determined by the distribution of body mass relative to the axis of rotation. This phenomenon is exploited

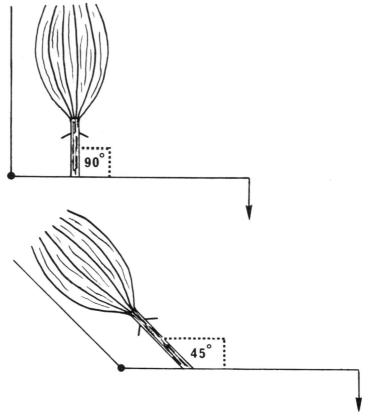

Figure 62. To illustrate the change of leverage as a movement develops.

by the skater (who speeds a spin-turn by bringing his arms to his side) and by the laborer (who is careful to lift a heavy load by holding it as close as possible to the trunk). In the first case, there is a transfer of momentum from the arms to the trunk, and in the second there may be a transfer of momentum from the body to the object to be lifted (page 168).

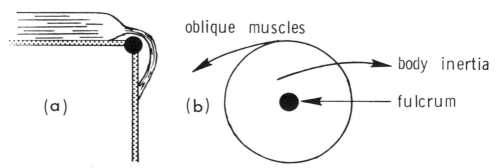

Figure 63. To illustrate (a) pulley arrangement of tendon, and (b) wheel-type arrangement of oblique muscles of trunk.

Ballistic and Controlled Movements

Once movement has been induced in a limb by contraction of an appropriate muscle group (the agonist), it follows from Newton's first law that the movement will continue unless arrested by gravity, an external resistance, the internal viscosity of the moving parts, or the activity of an opposing muscle group (the antagonist). Unresisted (ballistic) movement proceeds at a rate determined by the natural frequency of the part; it is appropriate to fast and relatively unskilled activity, such as the displacement of a light control through the middle range of travel. When greater accuracy is required, alternate bursts of activity appear in agonists and antagonists (Fig. 61); movement becomes slow and energetically costly but precise. If a very high speed of movement is needed, a costly forced oscillation can be induced by vigorous and sustained contraction of the agonists throughout the intended range of movement.

Movement Description

Description of the quality of movement includes a note of its rhythm (smooth or jerky), tempo (fast or slow), range, precision and many other characteristics (Roebuck, 1966). Considerations of psychology (mood, personality, body image and aspirations), sociology (cultural background, economic stratum) and environment (for instance, extremes of heat and cold) alter the manner in which a given task is performed, affecting not only the metabolic cost to the worker but the type of machinery he may be capable of operating.

Coordination

The coordination of the worker is progressively improved through training, until he acquires a smooth and well-integrated pattern of muscular contraction. To paraphrase some words of Bard, the central nervous system learns to "distribute messages to the muscles in such quantities and with such a dispersion in time and space as to bring about an appropriate sequence of integrated motor events."

Most movements include elements of volitional, reflex and automatic activity. However, in a well-learned industrial task automatic movements predominate. The decision to carry out a given operation no longer requires the provision of detailed instructions to individual muscles. Through a process of motor learning (page 244), an appropriate sequence of γ loop settings has been mastered and stored. As Adrian has put it, "the mind orders a movement, but leaves execution to lower orders of the central nervous system." While a large part of the acquisition of automatic movements is undoubtedly a function of careful training, some authors have claimed that there is also a significant genetic component (Cooper, 1971); this view is based upon similarities

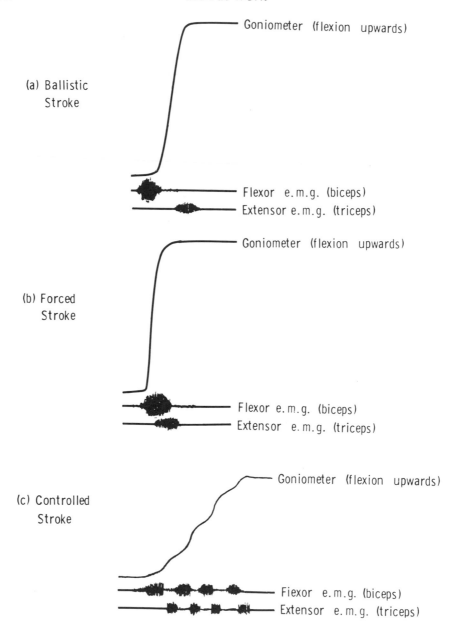

Figure 64. A comparison of muscular activity in ballistic movements, controlled movements and forced oscillations, showing goniometer records of limb displacement and electromyograms obtained from agonists and antagonists.

in the movement patterns of identical twins even when learning has proceeded independently.

Many older training programs first taught individual components of a movement, and later attempted to synthesize the overall activity.

However, this is not the normal pattern of learning, and for this reason is probably a faulty approach. A worker should rather be allowed to attempt the task as a whole, with the instructor concentrating upon the correction of serious errors in technique. Major faults should be set right at an early stage, since it is quite possible to learn bad technique through frequent repetition of a wrong movement. The learning process can sometimes be speeded by mental rehearsal of the task (page 86).

TIME AND MOTION STUDY

Criteria of Success

The ergonomist who carries out a time and motion study of human movement on the factory or office floor has as his prime objective assistance of the managerial process. Depending upon the *purpose* of the system, his criteria of success may range from a reduction of human loading to the achievement of economies in men and materials. In some work situations, existing activity patterns may be creating intense physical or psychological stress, with intolerable fatigue. In other instances, the safety of the worker may be endangered, or there may be a limited reservoir of manpower with the needed skills (Singleton *et al.*, 1967; Mundel, 1970).

More often, the prime goal of an operation is profitability. Account must then be taken of labor usage (direct and indirect), and a proper balance established between the two types of labor requirement. If the skills of individual workers can be matched carefully to the demands of the tasks they must perform, then the total need for skilled labor may diminish. A second possible strategy is to seek a large gain of productivity through the introduction of new and more complicated machinery, matching this where necessary by the hiring of more skillful staff. However, expensive machinery is all too frequently idle or underemployed. The reasons for this can be documented by application of queuing theory (page 350); the remedy may be to hire more or better staff, or alternatively to reduce the sophistication and extent of the equipment inventory.

The average work space (page 317) is both inconvenient and wasteful of floor area. Careful observation of body movement patterns may thus suggest measures that will combine greater convenience with the saving of space. Changes in working procedures may also be justified in terms of functional improvements in the end product, particularly if these increase acceptability to the customer. If the raw materials are expensive, either wastage or usage may be reduced, and in some cases a less costly alternative material may be substituted.

Anticipated gains of productivity from any revised working method must normally be set against the costs of work analysis, training procedures, and physical changes to equipment and work space. However, if a problem of health or safety is involved, a relatively high cost of implementation should be acceptable to management.

Worker Acceptance

Casual inspection may suggest that current methods of production are inefficient, that machines or staff are underemployed, and that workers are becoming physically or psychologically tired over the course of the working day. Nevertheless, one must be cautious in attributing increases of productivity to any changes of working techniques that are introduced. The classical study of Pennock (1930) at the Hawthorne works demonstrated dramatic increases of assembly-line production with each of many changes in schedule (introduction of rest pauses, abolition of Saturday work, commencement of payment by group results and the like). Pennock was forced to the conclusion that the main determinant of productivity was neither the time schedule nor the method of payment, but rather the attitude of the girls towards their work. The personal interest of the supervisors conducting the experiments helped the girls to find a sense of personal freedom and self-actualization (page 237), and it was this that provided motivation to greater productivity.

The worker often regards a time and motion observer in a much more negative sense, suspecting that he will generate unemployment or impossible work loads. It is not sufficient for management to explain that unemployment is a temporary and inevitable accompaniment of technological change. Work study must help the employee, and he must realize this to be the case. There must be an atmosphere of mutual understanding, and an acceptance that change is compatible with the interests of both workers and management. If the unit number of employees is to be reduced, then it is advisable to wait until this can be accomplished by natural turnover and/or expansion of operations. Worker education programs should cover not only the advantages of the new technology, but also consequences (such as loss of competitiveness) that would result from a maintenance of the *status quo*.

Some Typical Examples

An uncorrected and very inefficient shop-floor operation is illustrated in the upper panel of Figure 65. A man is normally working at a bench, but periodically he must take items from a conveyor belt and put them in a sandblasting room. His accepted routine involves a sequence of nine operations:
1. he walks to the unloading area,
2. he drags a skid to the conveyor belt and loads it,

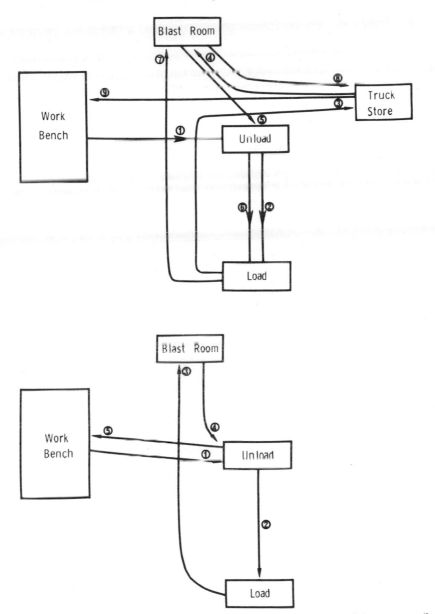

Figure 65. An example of improved time and motion in a sandblast room (based on observations of Mundel, 1970).

3. he walks to the truck storage area,
4. he wheels a hand-lift truck to the sandblast room,
5. he lifts the skid of blasted parts onto the truck and wheels it to an unloading area,
6. he lowers the skid and then wheels the truck to the loading area,

7. he lifts the newly filled skid onto the truck and wheels it to the sandblasting room,
8. he deposits the skid in the blasting room and returns the truck to a storage area,
9. he finally returns to his work bench.

By re-routing the operation and providing a wheeled buggy, as shown in the lower panel, the walking distance can be reduced from 244 to 128 feet, with a 47 percent saving in the time of the worker (Mundel, 1970).

Many industries have used time and motion study with great success. Fish (1953) has described a simple stopwatch analysis of citrus fruit picking in Israel. Time was divided between picking, rest periods and delays caused by shortages of boxes and ladders; after training foremen in improved methods, employment was reduced from 9000 to 7000 and spoilage was also lessened, yet there was a 40 percent increase in the quantity of exportable fruit. A similar study of a company producing diesel engines raised output from 20,000 to 60,000 units per month, without hiring additional staff or making significant expenditures on automation (Mundel, 1970).

Scope

A well-designed time and motion study should examine all phases of an operation, ranging from the input of materials, through the orderly usage of physical and human resources to the final stage of output design.

INPUT. The input of material (whether supplies or data) may be modified with respect to specification, format and speed of delivery.

RESOURCE USAGE. The design of the work station and associated equipment may merit improvement. Advantage must be taken of the potential for body movement, perspective and cognitive activity without imposing excessive physiological or psychological stress upon the worker. It may be advantageous to change the order of operations, combining some steps and separating or eliminating others in order to avoid unnecessary delays and permit a better interaction with other machines or other workers.

OUTPUT. The output, whether of products or of completed services, may be arranged in a more conveniently processed form. Thus, if the end product is a magazine, it may reserve cover space for mechanical address labeling rather than rely upon traditional envelope mailing.

Techniques

Many very effective work studies have been based simply upon a stopwatch and notebook, using the techniques of work activity analysis, work sampling and memomotion analysis. Such approaches can be supplemented by film and videotape records and by measurements of physiological and psychological stress. If possible, a typical work cycle should be identified, and representative elements of the cycle should be studied. Unfortunately, in some forms of work —particularly clerical and supervisory activity—there is no clearly defined cycle.

WORK ACTIVITY ANALYSIS. The worker is required to compile a chronological summary of the various tasks undertaken over one or more days, with a note of the time allocated to each. Thus, a desk clerk might indicate the periods allocated to typing, answering the telephone, finding other employees, photocopy and stencil work, filing, delivery of materials, personal affairs and any idle time.

There is some danger that the completion of work analysis sheets may serve as a *conscience* (Shephard, 1967), increasing the proportion of the time devoted to gainful employment. The close cooperation of worker and supervisor is thus necessary to avoid bias. Typical problems revealed by work activity analysis include a failure to combine errands, an unwillingness to *farm out* simple tasks to less-skilled personnel, inadequate stockpiling of material for any given operation (such as stencil reproduction), poor organization of desk space, and poor arrangements for filing and the provision of necessary supplies.

WORK SAMPLING. With this technique, a large number of observations are made at random intervals (Tippett, 1935; Morrow, 1946). Whereas work activity analysis is intensive, preserving the natural sequence of events over a short period, work sampling is extensive and well-suited to an operation that changes from day to day. Detailed study is subsequently directed to those parts of the total operation that either occupy a large segment of time or are in obvious need of change.

As with work analysis, the sampled time is distributed between identifiable operations. Thus, a factory floor study might note percentages of time allocated to production, packing, clerical work, cartage and idling. If idling is excessive, it is useful to classify work loss arising from shortages of material, lack of instructions, breakdown of machinery, other avoidable causes and personal factors.

Detailed study (Mundel, 1970) may be helped by drawing up operation and man/machine charts. The *operation chart* lists the sequence of activities required of the worker, including *suboperations* such as taking hold of a part, lining it up or assembling it, *movements* of the body and any products that must be carried, *holding* operations and *delays*. Holding is usually fatiguing (page 72), and where possible the hand should be replaced by a vice, a bench stop or a clamp operated by the more powerful leg muscles. A task can sometimes be found for the left hand while the right hand is active. Superficially, this is more efficient, and minimizes delays. However, it also requires more coordination, thus increasing entry requirements and lengthening the training period for new employees.

Individual suboperations are often facilitated by reducing visual demands (for example, by raising bench height), by provision of better tools and containers, and by changing leverage to make full use of inertia (for instance, providing a long and heavy balanced lever arm to spin when tightening wheel nuts). The technique may be improved by a change in the direction of movement (page 62), the use of fewer muscles, and the avoidance of a jerking motion.

Man/machine charts provide information on the interaction between the worker and one or more machines. It is noted when the man is waiting for the machine, and vice versa, and this information is related to the costs of man and machine, measured in terms of the objectives of the system. Delays incurred in the loading and unloading of machines may be minimized by the introduction of automatic supply systems, and the ratio of workers to machines can also be varied so that both men and equipment reach an acceptably short percentage of idle time.

MEMOMOTION ANALYSIS. Memomotion analysis describes families of gross movements (Mundel, 1958). It is particularly helpful when arranging employment for the physically handicapped (page 289). Movements involving missing members or injured joints are located, and if possible are eliminated from the task structure. An operation is filmed at slow speed (perhaps one frame per second) and is projected at a much higher speed (for example, 16 frames per second). A one-hour operation can then be viewed in four minutes or less, and the distribution of activity within this time span can be broken down into either job or movement categories.

Micromotion analysis is a more detailed variant. Hand movements are classified into seventeen *Therbligs*, including grasp, pre-position, position, use, assemble, disassemble, release load, transport empty, transport loaded, search, select, hold, unavoidable delay, avoidable delay, rest for fatigue, plan and inspect (Gilbreth and Gilbreth, 1920).

OTHER FILM TECHNIQUES. The human eyes, ears and hands all impose serious limitations upon a time and motion study because they can only process a finite quantity of information over a given time span. Capacity can be extended somewhat by storing a verbal commentary on a cassette tape recorder, but for detailed examination of a repetitive task a cine-camera or a videotape recorder is invaluable. Frame-by-frame analysis permits careful study of the velocity and acceleration of individual body segments, together with displacements of the center of gravity of the body and any load that must be lifted.

The *cyclegraph* and the *chronocyclegraph* (Clark, 1954) allow complex movements to be recorded on a single photograph. The general intensity of illumination is kept low, and a small lamp (steady or flickering in a rhythmic manner) is attached to the hand. By slow or repeated exposure of the film, the path of limb movement can then be traced over a complete work cycle. The *stroboscope* provides a comparable picture of whole body movement by rapid cyclic illumination of the entire work space. Each of these three techniques is invaluable when redesigning a restricted working area.

PHYSIOLOGICAL AND PSYCHOLOGICAL STRESS. The intensity of physiological stress engendered by a given task is conveniently assessed by the displacement of such variables as oxygen consumption, heart rate, respiratory minute volume, integrated muscle action potentials, body temperature and blood lactates from their resting values (page 53). If the work is repetitive, then the chosen index of stress is best expressed as the average value for the final hour of the working day, but if short bursts of more intense activity are anticipated, then it is also necessary to record peak readings.

If oxygen cost is used as a measure of stress, it should be remembered that in tasks requiring gross body movement, the energy expenditures required vary almost directly with body weight (Godin and Shephard, 1973); for this reason, stresses calculated for an average employee cannot be applied directly to a more obese worker. If heart rate is used as the index of stress, account must be taken of the decline in maximum heart rate with age; a pulse rate of 155/min is a relatively mild load for a young man, but is close to maximum effort for the worker who is nearing retirement.

Psychological stress is difficult to evaluate objectively, although some data, both qualitative and quantitative, can be obtained from galvanic skin resistance, catecholamine excretion and the like (page 215).

Synthesis of Information

The first step in the synthesis of time and motion data may be to prepare a simple *possibility card,* specifying possible alterations to the job, equipment and product in the form of a checklist. If the task is simple, it may become obvious that certain steps in the operation should be combined, separated or eliminated. Likewise, suggestions may be made to redesign, combine, separate or eliminate certain tools, and possible changes in the shape, size and packaging of the input and output may be listed.

With more complex operations, *process charts* and equivalent mathematical modeling procedures (page 327) may be necessary to identify critical paths limiting productivity.

It is often useful to specify *standard times* for the various operations that are studied (Swan, 1956). This may indeed be the first practical step towards clarifying requirements of labor and equipment, devising appropriate sequences of activity, and comparing the effectiveness of different work techniques. Standard times also provide an objective basis for schedules, piece-rates and wage incentives. Data conform to a bell-shaped curve (Fig. 66). Rudimentary worker selection narrows the distribution, and the ratio of performance between the substandard and the superior worker is rarely greater than 2:1. The performance scale is usually established so that the productivity of a typical worker is 125 to 130 percent of *standard*. There is thus a rather small group of employees in the 80 to 100 percent productivity range who receive only their base pay, and a much larger group receiving premiums proportional to their additional productivity.

Relaxation allowances can be incorporated into the standard time concept to adjust for personal needs and the additional stresses imposed by the lifting of heavy weights (page 185), adverse posture, and adverse environmental conditions (page 65). Workers who fail to achieve standard times despite adequate training have probably been poorly selected (page 224). Workers who exceed production by too wide a margin are equally in the wrong job; they should be offered more

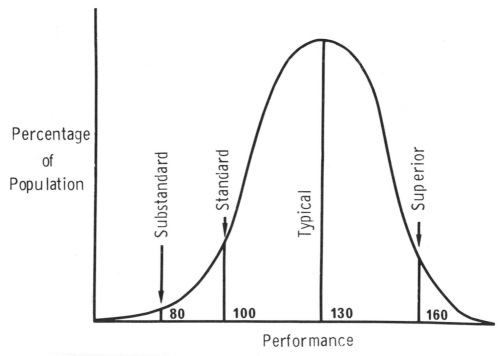

Figure 66. Distribution of performance in an industrial operation.

demanding employment, not least because they are perceived as a threat by their less productive colleagues.

Although the concept of standard time seems simple, application can be quite difficult. In order to determine the bell-shaped curve accurately, it is necessary to time the performance of a substantial number of workers under a representative range of operating conditions. Commonly, recourse is made to the simpler substitute of judging a worker's apparent effectiveness on a numerical scale such as 70 to 120. Such a procedure can lead to very inconsistent results, but accuracy is improved if individual observers are shown filmed sequences of acceptable work rates and are then required to draw their own distribution curves, so that they can judge whether they are compressing or expanding the generally accepted range of speed for the task. Moores (1970) suggested that time-study engineers who use this approach frequently underestimate the potential output of the worker. He related the standard rate accepted by 105 trained observers to energy expenditures as follows:

Task	*Standard Rate*	*Caloric Cost*
Emptying dust (garbage) bins	64/hr	6 Kilocalories / min
Barrowing cement	2.6 mph	8.4 Kilocalories / min
Walking upstairs	3.8 flights / min	12.2 Kilocalories / min
Stacking concrete	18.3 piles / hr	7.5 Kilocalories / min

Such tasks would be heavy if sustained all day. However, Moores pointed out they were normally interspersed with much lighter work, and the theoretical maximum loading (40% to 50% of aerobic power for an 8-hour day, page 55) was thus rarely attained by men who worked at the standard rate.

PROBLEMS OF LIFTING AND CARRYING

The prevention of accidents when man is interacting with complex machines will be explored in the context of transportation (page 355). In this section, we shall discuss the much simpler interactions between a man and a heavy load.

Epidemiology and Economics of Back Injuries

As in any discussion of industrial injuries, account must be taken of (1) the total claims, (2) their number relative to the employed population, and (3) their severity (days of partial or complete disability).

Injuries from lifting and carrying heavy loads are common. In the United Kingdom, the incidence of industrially reported *handling accidents* remained relatively constant at six to eight cases per 1000 employees per year over the period 1930 to 1956 (Brown, 1972), although during the war years (perhaps as a consequence of poor lighting and makeshift working conditions), the figure increased temporarily to 10 per 1000 employees. About 25 percent of all reported industrial accidents in Britain are attributable to *handling;* carrying, pushing and pulling account for three quarters of this total, and lifting for the remaining quarter.

Statistics for Ontario are almost equally disturbing. During 1967, 21 percent of all claims and 19 percent of all payments made by the Workmen's Compensation Board were related to *over-exertion.* It was indeed the third most common category of industrial accident in the Province, with 52 percent of *over-exertion* claims relating to handling and throwing, and 32 percent to lifting. Total provincial disability payments for this accident category are now in excess of $15 million per year, with $6 million being directly attributable to lifting injuries.

Injuries arising during pushing and pulling may be dealt with quite briefly. Frederik (1960) has shown that pushing is mechanically more efficient than pulling, and has recommended that pulling be reserved for conditions where visibility is impaired (for example, a truck laden with bulky packages). He has set the optimum force of horizontal traction at 15 kg, with the preferred handle height 30 to 40 inches above the floor. Accidents involving trucks can be minimized by fitting automatic brakes, placing bumpers at ankle and hip heights, and eliminating sharp corners from the vehicle.

Back injuries are less readily prevented. The most common age of injury, both in Canada and the United Kingdom is 30 to 40 years (Brown, 1972). We may speculate that at this age seniority ensures lifting is no longer a regular occurrence, but the worker is not yet old enough to admit he is unable to lift heavy loads. Other factors may be a decline of personal fitness and (for the new entrant to a given industry) difficulty in learning proper lifting methods. Guthrie (1963) drew an interesting contrast between the high incidence of back problems in white supervisors at a South African explosives factory and the low incidence in African laborers. He suggested that the back problems of the white man might be related to poor posture, obesity, lack of physical fitness, and the use of unduly soft chairs and mattresses.

One difficulty in analyzing compensation statistics is how to eliminate from our consideration those patients who complain of low back pain by reason of an inadequate personality (hysterical symptoms) or a greed for unmerited financial support. Objective data can be based on the demonstration of intervertebral disc displacement (injection of radio-opaque dye into the vertebral canal) or at a later stage the observation of pathological changes in the vertebrae proper. Schroter (1958) has used the latter approach to show the adverse experience of manual workers relative to office employees (Table II).

TABLE II. THE FREQUENCY OF PATHOLOGICAL CHANGES AT SELECTED JOINTS, IN RELATION TO TYPE OF OCCUPATION (MANUAL HANDLING VERSUS BANK EMPLOYEES). BASED ON DATA OF SCHROTER (1958).

Joint	Frequency of Pathological Change	
	Manual workers	Bank staff
Vertebral column	98%	37%
Elbow	35%	3%
Knee	32%	13%
Hip	28%	0%
Shoulder	12%	5%

The stereotype of long and intractable disability is fortunately by no means true of all compensable back problems—indeed, the back accounts for a rather similar percentage of both claims and dollar disbursements, so that the average period of disability is comparable with that for other industrial injuries. In the Province of Ontario, some 82 percent of low back injuries are classed as sacroiliac strains; these have a relatively short average period of disability (28 days). A further 12 percent present as visceral hernias, and these merit an average of 43 days disability payments. Only 6 percent have prolapse of an intervertebral disc, but in this group (1) disability is extended (average 135 days) and (2) arrangement of alternative employment is often necessary, with attendant problems of retraining (page 248).

Mechanics of Lifting

The conventional wisdom of the past forty years has taught that heavy loads should be lifted by a rigid back, bent-knee technique (C.I.S., 1962). Much effort has been devoted to the production of training manuals and the teaching of approved methods of lifting, but the incidence of injury has shown remarkably little reduction (Brown, 1972). Several possible explanations could be advanced for this paradox:

1. injuries arise mainly in the untaught,
2. the muscular strength of the population has diminished and this has offset any gains from improved lifting techniques,
3. the straight-back, bent-knee method is impractical, and
4. the straight-back, bent-knee method is incorrect.

With regard to the first two possibilities, Canadian industries where heavy lifting is frequent (such as meat handling) have a relatively good claims experience; injuries are more frequently encountered among sheet metal, construction and transportation workers, where lifting tasks are intermittent and poorly formalized. The latter three groups often contend with uneven standing surfaces, inadequate loading bays, minimal tuition, and awkward loads that obscure visibility. The back problems of the Canadian construction industry are in marked contrast with British experience, where building operations have shown a consistently low rate (1 to 2 back injuries per 1000 employees per year). Until quite recently, construction in the United Kingdom has been characterized by an absence of both prefabrication and mechanical tools; traditional brick-laying methods have been used, with emphasis upon the repetitive carrying of moderate to heavy loads. One may speculate that frequent repetition has formalized the task, and at the same time has maintained the physical condition of the bricklayers.

Although inexperience, poor working conditions and lack of physical fitness undoubtedly contribute to back injury, observation of workers experienced in the handling of heavy loads shows that the straight-back, bent-knee method is rarely used. Davis *et al.* (1965) have suggested this is because the proposed method is impractical. The leverage exerted by the quadriceps muscle is most unfavorable if a heavy weight is lifted from the floor with the knees bent, and the average worker cannot develop sufficient force to raise legally permitted loads by the official technique.

How useful is it to persist with teaching a method that may prove impossible to adopt in practice? The concept of lifting with a straight and rigid spine apparently originated with measures of intervertebral disc strength. The discs were found to be more resistant to compression than to tension, shear or torsion; maximum compression forces were set at 30 kg/cm^2 over the age range 20 to 35, 25 kg/cm^2 from 35 to 50 years, and 20 kg/cm^2 in those over 50. According to accepted principles of leverage, it was reasoned that the force imposed upon the lumbar vertebrae increased with the angulation of the trunk (Table III), and the back was therefore most stable when converted to a rigid vertical pillar by simultaneous contraction of the flexor and extensor muscles. Floyd and Silver (1955) demonstrated electromyographically that when lifting was carried out from a position of extreme flexion, the back muscles were relaxed, thus apparently placing the intervertebral discs in danger. In the unskilled worker, the problem was compounded by the increased forces associated with rapid and jerky movements.

Recently, the straight-back, bent-knee technique has met increasing criticism—from the United States National Safety Council in Chicago, from Himbury in Australia, from Davis in the United Kingdom, and

TABLE III. STRAIN ON FIFTH LUMBAR INTERVERTEBRAL DISC IMPOSED BY BENTBACK LIFTING (BASED ON DATA OF INTERNATIONAL OCCUPATIONAL SAFETY AND HEALTH INFORMATION CENTER, 1962).

Inclination of trunk (degrees)	Load			
	0 kg	*50 kg*	*100 kg*	*150 kg*
0	50	100	150	200
30	150	350	600	850
60	250	650	1000	1350
90	300	700	1100	1500

from Joseph in Canada. The method of the safety manuals is in marked contrast with dynamic, natural techniques of lifting, learned from one's parents in childhood and from skilled workers on entering industry. An analysis of forces due to Jones (Figs 67 and 68) suggests that stresses on the vertebral column are much smaller in dynamic than in straight-back lifting. With the straight-back method, counterforces are provided almost exclusively by the erector spinae muscles. Rasch (1971) has suggested that the risk of forward displacement of the lumbar vertebrae and injury of the erector spinae muscles is enhanced by a combination of pelvic tilting, hyperextension of the spine, and weak counterpressure from the anterior abdominal muscles. An imbalance of strength between the flexor and extensor muscles of the trunk is certainly associated with low back disorders (Flint, 1958), although Chaffin and Moulis (1968) have found no relationship between the inclination of the sacrum and back pain. On the other hand, they note a significantly smaller sagittal diameter of the fifth lumbar vertebra in those men having back problems. Chaffin thus supports the view that back injury arises from application of excessive pressure to the intervertebral discs, and he comments that women, with smaller vertebrae, are necessarily at greater risk for a given load.

Klausen (1965) has suggested that while the short muscles of the back control individual intervertebral joints, the stability of the spine as a whole depends upon the long muscles of the back and/or the abdominal muscles. If a load is carried high upon the back, the tendency is to lean forwards, increasing the load on the back muscles. However, when lifting from a stooping position, an increased proportion of the needed counterforce is developed by the anterior abdominal muscles.

In many instances, back injury is associated with torsion of the spine while lifting or carrying a heavy load. Due to bad industrial house-keeping, the worker stumbles over an uneven floor or slips upon an icy or greasy surface. The straight-back, bent-knee method can thus be criticized from a pedagogical viewpoint. The worker has failed to learn dynamic movement patterns, and in consequence is unable to correct a momentary loss of equilibrium without injury. Certainly, it

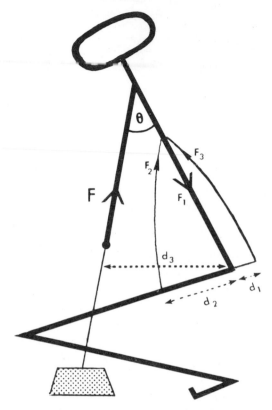

Figure 67. Mechanical forces for the straight-back, bent-knee method of lifting (based on analysis by Jones, 1969). Compare with Figure 68.

$$F = W + ma$$

where W = weight of load, m is its mass = W/g, and a is acceleration.

$$F_1 = F_3 = F \; Sec \; \Theta + F \; \frac{d_3}{d_1}$$

where F_1 is the force in the spine, F_3 is the force exerted by the spinal muscles, Θ is the angle between the spine and the applied load, d_3 is the distance between spine and load, and d_1 is the distance between the spine and the resultant of forces developed by the spinal muscles.

One method of reducing the loading on the spine is to reduce the ratio (d_3/d_1) through an increase in the effective length of d_1. Only small changes can be brought about by altering the resultant of forces in the erector spinae muscle group, but if the task can be transferred to the abdominal muscles (F_2), the much longer lever arm (d_2) can be used. A third possibility is to reduce d_3 by alterations of body position and changing the direction of F_1, as in the dynamic method of lifting (Fig. 68).

seems more important to maintain the *relative* positions of adjacent vertebrae than to avoid all curvature of the spine while carrying a heavy load.

Figure 68. Mechanical forces involved in dynamic method of lifting. Based on an analysis by Jones (1969). Compare with Figure 67. Bending the back forwards moves the resultant force F further up the spine, where the force can be transferred to the abdominal muscles. The effective load is also halved by tilting it until it rests upon the knee. Thereafter, the effective force F is moved even further forwards as the resultant of forces developed by the arms (F_a), the forward thigh (F_4) and the rearward leg (F_5). The momentum thus developed relieves much of the compression force on the vertebral column while the back is bent.

Anatomical and Physiological Studies

Davis and his associates (1965) used chronocyclophotography (page 181) to evaluate spinal movements during lifting. They concluded that there was little displacement of the thoracic vertebrae with either straight-back or dynamic lifting. The lumbar spine showed a slight initial flexion, followed by continuous extension, movements being greater with dynamic than with straight-back lifting. However, when the attempt was made to lift a heavy load from the flexed-knee position, the hips rose faster than the shoulders, automatically converting the flexed-knee to a more dynamic pattern of lift. It was suggested that conversion reflected partly weakness of the quadriceps muscle, and partly the greater inertia of the loaded trunk.

Substantial pressures are developed in both the abdominal and the thoracic cavities during the early stages of lifting; these pressures help the spine to resist the large flexor forces that develop as acceleration of the load begins. Asmussen and Poulsen (1968) found that a forward stoop increased the intra-abdominal pressure, confirming the view that effort was transferred from the back to the abdominal muscles by this maneuver. However, the increase of intra-abdominal pressure did not account for the entire decrement in activity of the erector spinae muscles, and Asmussen postulated that additional counterforces were generated through a stretching of the annulus fibrosus of the intervertebral discs.

The intensity of stress resulting from different lifting techniques can be assessed physiologically from increments of oxygen consumption, pulse rate and electromyographic activity. Bedale (1924) studied eight methods of carrying a load, and concluded that oxygen consumptions were smallest when wearing a yolk; postural strain was minimized by the natural body position, and there was a minimum displacement of the center of gravity of the body. Many demonstrations of the physiological costs of poor posture can be found in the literature. Bedford and Warner (1955) showed that walking with a stoop increased the energy cost of progression by 30 to 50 percent; this is a frequent necessity when mining shallow seams of coal. Droese and his colleagues (1949) carried out an ergonomic study of housewives; they found that whereas the energy cost of standing was only 2.25 kilocalories/min, walking and carrying household items had a cost of 3.09 kilocalories/min, reaching upwards cost 4.07 kilocalories/min, reaching down 4.44 kilocalories/min, and bending down 5.57 kilocalories/min. Brown (1971) examined the influence of bench height upon the oxygen consumption of an industrial worker sorting screws and lifting weights of 1 to 10 kg through a height of 10 cm. Results followed a similar pattern for each of these tasks, and may be illustrated by the effort required to raise 1 kg nineteen times per minute. The energy cost was 3.02 kilocalories/min at the optimum working height of 93 cm. When the bench was raised so that the hands had to be raised above the shoulders (162 cm), the cost increased slightly (to 3.14 kilocalories/min). If it was necessary for the body to slump until the hands were at knee level (bench height 44 cm) the cost increased further (to 3.61 kilocalories/min). The least economical approach to this mode of activity was the use of an intermediate bench height (69 cm) with a 45° forward angulation of the trunk (4.19 kilocalories/min). On the other hand, there are advantages to a forward tilt of the body when additional pressure must be brought to bear upon a task (as in hand planing).

When lifting or carrying, the body weight of the worker often accounts for a substantial fraction of the total energy expenditure. Godin and Shephard (1973) recently demonstrated that whereas in sedentary activity the net oxygen cost of activity was independent of body weight, in tasks that involved a lifting and lowering of the body mass, the net cost of movement was almost directly proportional to body weight. The cost of lifting is further augmented if a worker must balance on a moving vehicle (Vos, 1966).

Static work is both fatiguing and inefficient. Dr. Harry Davies, of Rochester, New York, compared the cost of 20-inch lifts, starting from the floor and from a height of 20 inches. Three loads were tested, each at a rate of six lifts per minute and the straight-back, bent-knee technique was used throughout (Table IV). There was a large (2.4

TABLE IV. ENERGY COST OF LIFTING BOXES WEIGHING 20, 40 AND 60 POUNDS (BASED ON EXPERIMENTS OF DR. H. DAVIES, ROCHESTER).

Load (pounds)	Energy cost (kilocalories/min)	
	0–20 inch	20–40 inch
20	5.6	2.8
40	6.4	3.8
60	8.2	5.8

to 2.8 kilocalories/min) increase in the energy cost of lifting when the load was initially at floor level. Under such conditions, whole body motion rather than arm movement was required, and mechanical efficiency was low (<2%). There were large inter-individual differences in both pulse rates and oxygen consumption readings; high values were recorded from those who were obese, and low values from those aged 30 to 40 years. Presumably, the latter group were skillful, and developed a relatively small static effort when lifting.

Brown (1972) compared physiological responses to straight-back, dynamic and free-style lifting. His subject was a student without particular industrial experience. The pulse rate was similar with all three techniques. During straight-back lifting, muscular activity was apparently distributed rather evenly between the erector spinae and the rectus femoris, but with free-style lifting the greatest stress was carried by the erector spinae muscles. At all loads from 5 to 40 kg, the oxygen cost was greatest for the straight-back method, and lowest for the free-style method.

Acceptable Loads

Commonly reported standards of acceptable loading have been based upon opinion or the supposed behavior of the intervertebral discs. Unfortunately, it is difficult to prescribe definitive legal limits because many variables, both human and operational, are involved:

Human variables	*Operational (task) variables*
Age	Size of load
Sex	Range of heights for lifting
Size	Frequency of lifting
Posture	Speed and steadiness of lifting
Training and experience	Steadiness of base
Physical fitness	Need for twisting motion
Health	

The Swiss Accident Insurance specified safe limits for occasional lifting in 1961. A man lifting with a straight back was permitted to raise
a load diminishing from 400 kg to 50 kg as the back was inclined from
the vertical to the 90° position. Further downward adjustments were
made for bulky loads, jerky or infrequently repeated movements,
aging, and spinal deformity. Quoted figures for bent-back lifting (based
on the theory of disc compression) were only half as great. Women
were assumed to have 60 percent of the physical capacity of men;
even lighter loads were prescribed for adolescents, partly because of
the danger of damaging developing bone structures, and partly because
they were thought to tire more quickly.

The Swiss study suggested that an inexperienced worker might well
use the worst (90°) technique, and the limiting load for a young man
was thus set at 25 kg. More recently, the I.L.O. (1964) has proposed
a rather higher limit (40 kg) for men working in jobs that normally
involve some carriage of heavy weights. Asmussen and Poulsen (1968)
suggested that the effort required of the back muscles should not
exceed 40 to 55 percent of maximum isometric force. On this basis,
a maximum load of 32 kg was recommended for 35-year-old men lifting
with a 45° forward stoop. A third alternative is the subjective approach
(Snook and Irvine, 1967), allowing workers to vary loads until a maximum acceptable level is reached; limits are similar to those proposed
on theoretical grounds:

Range of lift	Acceptable to 50% of population	Acceptable to 90% of population
Floor to knuckle	30 kg	24 kg
Knuckle to shoulder	28 kg	23 kg
Shoulder to arm reach	24 kg	22 kg

Heavy people are willing to lift heavier loads, perhaps partly an expression of muscularity and partly attributable to the relationship
between personality and body build. On the other hand, lightly built
people apparently have more dynamic strength, and are prepared to
accept a faster rate of lifting than those who are heavier. As with other
forms of time and motion standard, the chosen loadings must be acceptable to the majority of the population (80% to 90% rather than 50%).

Another possible criterion of acceptable loading is elevation of heart
rate; the ceiling for repetitive lifting should be perhaps 40 to 50 percent
of maximum. Snook has suggested a figure in this range (115/min for
the young worker). Brown's data (1971) for repeated and fairly rapid
lifting of a 30-kg load showed average pulse rates ranging from 125/min
for free-style lifting to 145/min for straight-back lifting.

Prevention and Treatment

Back injuries frequently recur. It is thus important to screen out workers with a history of disability at the preemployment medical examination. Subsequent prevention is directed partly to the employee (through the encouragement of personal fitness; good posture and dynamic lifting patterns; avoidance of sudden twisting movements) and partly to the task (design of more compact loads with adequate handles; improvement of loading docks and standing surfaces; and provision of adequate rest pauses).

Prime treatment often involves prolonged immobilization of the back, and thus a carefully planned program of rehabilitation is needed before the worker returns to carrying heavy loads. Surgery is not a panacea of treatment, and before recourse is made to drastic intervention It is important to be certain that an organic lesion is responsible for disability. If the injury recurs, there may be a need to arrange alternative employment, with the usual problems of retraining (page 248).

WALKING AND MARCHING

Many workers devote a substantial percentage of their time to walking, and until recently the success of armies has depended upon their ability to march long distances while carrying relatively heavy loads.

Mechanics of Walking

Walking has been aptly described as a controlled catastrophe. In essence, a heavy mass (the trunk and head) is supported by a series of levers, linked through four joints (the lumbar spine, hips, knees and ankles). Balance is preserved by muscular activity, but equilibrium is precarious, being disturbed by unexpected movements such as slipping on an icy surface. If posture is adjusted to bring the center of gravity in front of the knee, this joint can then be locked; however, in many activities such as bent-knee lifting or skiing, the center of gravity is behind the knee, so that a controlled jackknife position is adopted.

For at least 30 percent of walking time, the body weight is supported on a single limb. In order to maintain equilibrium, the summed turning moments about each of the four joints must be brought to zero by an appropriate combination of muscle tone and leaning. Normally, the activity of the quadriceps muscle prevents buckling of the knee joint, and soleus activity prevents buckling at the ankle. The several stages involved in walking a single step may be classified as follows:

Stance
1. Heel stage—weight acceptance.
2. Midstance—forward glide of trunk.
Swing
1. Push-off and balance assist.
2. Limb pickup.
3. Reach.

The mechanics of walking has been studied by many techniques ranging from the simple (walking on a wet floor, observation of a weighted string attached to a body segment) to the esoteric (chronocycle and cine-photography, force platforms, accelerometers, recording goniometers, floor-level photo relays, microswitches built into shoes, and electromyography).

Individuals vary widely in their walking patterns. The toe-out pattern requires more energy than the toe-in, but there is no real evidence that either type of motion is *wrong* relative to parallel-toed walking; indeed, the Eskimo finds a pigeon-toed position a significant advantage when crossing an icy surface. Natural stride length and frequency also vary widely. Older people take shorter strides than those who are younger, partly because they are shorter, and partly because they have poorer joint mobility; mechanisms of postural control are also poorer, and an elderly walker thus oscillates further about his equilibrium position. Women generally take shorter strides than men. Many wear high-heeled shoes; in consequence, there is a very guarded initial contact between the heel and the ground, followed by a rapid plantar flexion of the ankle in a desperate search for a firmer basis of support. However, even when wearing low-heeled shoes, the fast stride length of a woman is no more than 93 percent of stature, compared with 106 percent in a man (Murray *et al.*, 1967).

The proportion of the time devoted to the swing of the leg increases from 30 to 50 percent of the total cycle as walking speed is increased. Grieve and Gear (1966) have shown that the maximum step frequency is proportional to $1/\sqrt{H}$, where H is the standing height. However, the time allocated to a swing rarely reaches the theoretical maximum of a quarter of the natural period for the part; calculations based upon the behavior of a compound pendulum indicate that the lower leg has a theoretical periodicity of 1.1 seconds. When the swing occupies almost 50 percent of the walking cycle, corresponding to a stride length of about 1.2 meters, most individuals find it more convenient to break into a run.

Ralston and his associates (1969) have calculated the total energy (potential and kinetic) for the various body segments. During much of the walking cycle, there is an interchange between kinetic and potential energy, so that the total energy level of the head, arms and trunk remains essentially constant. However, there is a significant input of work associated with contraction of the hamstring muscles just before the heel makes contact with the ground. Other muscles at the knee and ankle joints play an important controlling role. Thus, the tibialis anterior restrains the foot so that it does not slap the ground as a reaction to heel impact; the muscle then undergoes a graded relaxation, allowing the sole to make firm but gentle contact with the walking surface. In diseases where the normal proprioceptive feedback is lacking (as in tabes dorsalis), walking is transformed to a clumsy, stamping gait. Among other adverse consequences, additional impact stresses are thus imposed upon the talofibular joint, the meniscal cartilages of the knee, and the lumbrosacral joints. During *midstance*, there is relatively little muscular activity, and the incorporation of a natural *rest pause* is one reason why a long distance can be walked without undue fatigue. The long muscles of the foot normally provide some stabilization of the anteroposterior and lateral arches during stance, but in those who have developed flat feet there is an attempt by these same muscles to provide active support. This can be very tiring.

During the *pickup phase*, there is activity on the part of the hip and knee flexors, with the quadriceps playing a dominant role (Becket and Chang, 1969). However, the action is more than a simple heel kick, and eccentric activity of the knee extensor muscles ensures that the heel does not make inadvertent contact with the buttocks, particularly during running.

The *balance assist phase* largely conditions the length of pace, as the walker tips forward upon his toes. An amputee, or a patient with tabes dorsalis lacks the proprioceptive feedback needed at this stage, and in consequence moves with small and inefficient steps.

Much of the *reach phase* is probably a ballistic stroke (page 173), but an eccentric contraction of the hamstring muscles gives a controlled deceleration of the leg swing so that the movement is brought almost to a halt some two inches above the floor. Further lowering of the foot is a closely controlled movement; the sole is rotated outward and makes contact with the ground along its lateral border, from the heel forwards.

Energy Cost of Progression

Passmore and Durnin (1967) have made extensive measurements of energy expenditures during normal walking. Over the range 2 to 4 mph, the oxygen consumption is apparently a linear function of speed, although as Margaria (1968) has pointed out, the energy cost of moving from point A to point B is almost independent of the speed of walking over this same range. The cost of walking varies with the weight of the body and any added load, the grade, the extent of training and the nature of the terrain. Givoni and Goldman (1971) give the following equations, applicable to young soldiers:

$$M(\text{kilocalories/hr}) = \eta(W + L)(2.3 + 0.32 [V - 2.5]^{1.65} + G(0.2 + 0.07 [V - 2.5]))$$

where M is the energy cost of movement, W is the body weight, L is any added load in kg, V is the speed in km/hr, G is the grade in percent, and η is a terrain coefficient—1.2 for a hard road, 1.5 for a ploughed field, 1.6 for hard snow, and 1.8 for sand dunes. Additional costs are incurred if loads are carried on the hands or the feet rather than the back, and a 40-mph wind almost trebles the energy cost of progression.

The speed of troops is normally controlled by setting a marching pace of 90 to 120 steps/min; with an average pace length of 30 inches, this corresponds to a speed of 3 to 4 mph. A substantial distance can thus be covered in one day without exhaustion. Under good conditions, oxygen consumption (0.75 to 1.00 liters/min STPD) is about a third of the maximum value for a sedentary young man. Adverse conditions of terrain or climate, including uphill marching and strong head winds may necessitate a much lower speed. The back pack should not exceed 40 percent of body weight for level marching, and 30 percent for uphill marching. It will necessarily lead to at least proportionate increases of oxygen consumption.

During uphill walking, mechanical efficiency increases progressively to a maximum of some 25 percent during ascent of a 22 percent grade. With downhill walking, it decreases to -120 percent on a 9 percent grade. Under normal gravitational conditions, the vertical force exerted in lifting the center of gravity of the body far exceeds the horizontal force required for propulsion, and the angle of the resultant is such that slipping is unlikely except when running at high speed on a dusty surface. Under low gravity conditions (as in moon walking), the vertical component necessarily falls with the diminution of body weight, and the risk of slipping is proportionately greater. Margaria (1968) has suggested that the most appropriate method of locomotion for the lunar explorer is a jumping type of movement.

ARTIFICIAL LIMBS

Design of a Lower-limb Prosthesis

A patient who has undergone an amputation of a lower limb can be trained to move quite rapidly on crutches. However, these are difficult to use on an icy surface. There are also psychological advantages to a limb that looks real and functions in a fairly normal manner. Few people are prepared to accept a non-cosmetic prosthesis, regardless of its efficiency. Indeed, no more than 60 percent of amputees use even the most life-like artificial limbs.

A normal lower limb may weigh 8 to 10 kg, depending upon the level of amputation and the extent of muscular development. A typical artificial limb is much lighter (about 4 kg) and is sometimes difficult to maneuver because of its limited inertia. The proprioceptive feedback from muscle and joint tendon receptors is distorted by the reduced mass, and the patient may lift his prosthesis too high when walking.

A young man can use an artificial leg after as little as three weeks of training, although initially he will find the cost of progression is three or four times that of normal walking. With an older worker, the learning process is much slower. Residual proprioceptors are less sensitive than in a young man, new *automatic* movements are less readily learned, and residual muscles in the stump are weaker. Furthermore, the amputation may have been undertaken to treat gangrene, secondary to diabetes and peripheral vascular disease; in such a case, cerebral ischemia or blindness may hamper learning, while coronary atherosclerosis prevents development of the energy expenditures required for walking with the artificial limb (Kavanagh and Shephard, 1973). Occasionally, the problem can be overcome by a preliminary period of cardiac training in a swimming pool.

At all ages, the main mechanical problem of the artificial knee joint arises from uniplanar movement (biaxial joints are still in the developmental stage). The prosthesis is usually arranged to *jackknife* if weight is applied to the front of the *foot*, and to lock if weight is applied to the *heel*. The patient must thus learn to place his foot on the ground with unusual care. Hydraulic devices that limit the rate of collapse of both knee and ankle joints are currently being investigated; their main drawback is the additional weight that must be carried by the limited musculature of the stump. At the present time, the best arrangement seems a solid ankle with a rubber foot and thickly cushioned heel. This works quite well on a level surface, but the absence of movement in the lateral plane causes difficulty when walking on rough ground.

Some proprioceptive feedback is derived from the skin of the weight-bearing stump, but this is inadequate to allow precise control of limb movements. Experiments are thus in progress to supplement information flow by mounting pressure sensors in the sole of the prosthesis and transmitting signals to discrete areas of thigh skin.

The use of a prosthesis calls for the redesign of many pieces of domestic and industrial furniture. To quote but one example, an increase of chair height (page 291) greatly helps the climb from a sitting to a standing position.

Upper-limb Prostheses

A simple hook is the most effective form of upper-limb prosthesis for tasks such as eating, although many patients find it cosmetically unacceptable. More sophisticated devices can be powered by small electric motors, gas cylinders, springs or hydraulic feeds. However, it is difficult to simulate the versatility of the human hand; even the most elaborate prostheses do not attempt combinations of more than two or three forms of linear and angular motion. Control switches may be placed in the armpit or the opposite hand; sensors are also under development that will detect action potentials in the muscles of the stump and apply this information to control of the prosthesis. The ultimate in design would seem a sensor that is activated directly by the behavior of the motor cortex. The main limitation of current upper-limb prostheses is an inadequate sensory feedback. If sufficient force is developed to apply an effective grasp, then fragile objects such as a glass are liable to be crushed. Ideally, the user should retain control over terminal position, velocity and force. Salisbury and Colman (1969) are thus developing a device that reallocates function between man and prosthesis. *Position* and *grasp* decisions are taken by the human operator, but the force of grasp is regulated by a servo-mechanism that detects slippage by means of a piezoelectric crystal.

THE DESIGN OF HUMANOIDS

The main roles of the humanoid (mechanical man) are to carry out tasks that are dangerous or involve too much drudgery for the human operator. Obvious applications include the *hot room* of a radioisotope laboratory and the assault area of the battlefield. In the latter situation, the infantryman still has two important advantages over an unmanned vehicle—a knowledge of when to take cover and an ability to avoid obstacles. However, there is no fundamental reason why vehicles fitted with image recognition devices (page 349) should not eventually be capable of making the same type of decisions (Cohen, 1969). Requirements would include (1) a *visual* system to detect obstacles and explore their magnitude, and (2) a *brain* that would relate obstacle size to an avoidance path and calculate the subsequent course needed to reach the original target.

WALKING VEHICLES

Current military vehicles are extremely uncomfortable if forced from the road by mines, obstacles or a search for cover. The maximum practical speed drops to around 10 mph, and the average attained speed often degenerates almost to zero (Liston, 1969). This may be unimportant in conventional warfare, where logistic considerations restrict the speed of advance to 10 or 12 miles per day, but it is a serious limitation when dealing with the ever-increasing problem of guerilla warfare.

In considering alternative forms of military transport, it is instructive to note that nature does not use the wheel when crossing rough country; many animals can still move up to an order faster than human machines in this domain. Ground-effect vehicles may ultimately provide the best solution, but at present they give a very bumpy ride. An alternative possibility is to devise a jointed walking vehicle. This can step over obstacles and select the smoothest parts of the available terrain. As in human walking, the main energy losses occur in lifting the center of gravity and accelerating and decelerating the *limbs* at each step; however, frictional losses at the *joints* are larger than in the human body.

Walking machines are particularly adept at climbing steep slopes, negotiating river banks, and crossing fast-flowing streams. The main problem is to design an effective control system within the competence of a reasonably skilled operator. One suggestion has been to develop servo-mechanisms that mimic normal walking. Thus, if the operator feels the machine tipping, he leans in a direction that will restore balance, and sensors attached to his body induce the walking machine to make the postural adjustments needed to restore equilibrium. As in the operation of very large ocean-going oil tankers, human problems arise from the need to adjust the scale of movement. Thus, if the operator is sitting sixteen feet above the ground, the angular acceleration applied to his vestibular organs is three times that experienced in normal standing, and an over-response may occur to a given tilt. However, as in the 500,000-ton tanker, a mechanical or electronic scaling device can be interposed between the pilot and the vehicle.

Current walking machines are still very much in the developmental stage, and are intended to move quite slowly. One plan is for a quadruped that will carry a payload of 500 pounds at an average speed of 5 mph.

PART III

PSYCHOLOGY OF WORK

Chapter 9

ERGONOMICS AND THE MIND

THIS SECTION WILL EXAMINE some of the basic tenets of human psychology as these relate to behavior and performance (Ferster and Perrott, 1968; Ulrich *et al.*, 1966) McGaugh, Weinberger and Whalen, 1967). Although it is possible for the psychologist to argue that a man can be *conditioned* to accept any task, as in other branches of ergonomics the preferred approach is to start with human characteristics and frame the industrial task accordingly. For this reason, we shall review briefly concepts of the human mind, including the manner in which a task is perceived and learned. Consideration will be given to respondent and operant behavior, and the impact of the latter upon motivation will be examined, together with problems of vigilance, boredom and the extinction of unwanted behavior patterns. Lastly, we shall look briefly at the influences of personality and of arousal upon psychomotor performance.

CONCEPTS OF THE HUMAN MIND

It is difficult to give any brief statement of current concepts of the human mind. There seem almost as many concepts as there are psychologists. Perhaps this is inevitable when dealing with a subject that is so complex and resistant to experimental study. Many psychologists, indeed, believe that *mind* is but the description of behavior; in their view, emphasis should thus fall upon patterns of behavior and associated or *causal* environmental changes. In the present chapter, it will only be possible to indicate some of the more significant hypotheses.

Conventional Behaviorism

Contemporary North American psychology traces its roots to several sources. The first impetus to behavioral studies came from German sensory physiologists, men such as Helmholtz and Feckner. More than a hundred years ago, this group began to appreciate the nature of sensory stimuli, and the manner in which responses might be induced. The school of conventional behaviorism, as it is currently termed, was strongly influenced by the views of Pavlov. A rather simple stimulus/response pattern was envisaged (Fig. 69). If a dog was pre-

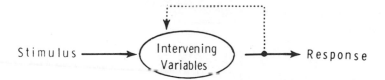

Stimulus ——————→ Intervening Variables ————→ Response

(a) Conventional behaviourism

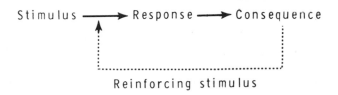

Stimulus ——————→ Response ——————→ Consequence

Reinforcing stimulus

(b) Radical behaviourism

Stimulus ——————→ Traits ——————→ Response

(c) Mathematical behaviourism

Stimulus ——————→ Mind ——————→ Response

(d) Phenomenological behaviourism

Figure 69. To illustrate hypotheses concerning behavior.

sented repeatedly with meat coincident with the display of a light, after a time, salivation could be induced by the light in the absence of meat. A simple *feedback* of information to the *intervening variables* between stimulus and response normally led to a strengthening of response with each repetition. However, the pattern of response could be changed through a process of *association*. Thus, the normal pleasure of the dog on presentation of a particular type of food could be replaced by a cowering fear if the appearance of this foodstuff was linked with an unpleasant stimulus such as an electrical shock.

Radical Behaviorism

To this point, the *intervening variables* were considered as reflex in type. The influence of the mind was largely neglected; indeed, Darwin's essay *On the Origin of Species* apparently left only quantitative differences between man and the higher apes. It seemed that all behavior would ultimately be described in terms of reflex arcs of varying complexity. For a while, some comparative psychologists defended the bastion that the phenomenon of language separated man from lower species. But even this distinction is now challenged. Rudimentary forms of language are present in many animals The fluidity of human speech is lacking. But if a chimpanzee is suitably indoctrinated from an early age, it can acquire up to thirty symbolic gestures, *requesting* items such as water and food, and even distinguishing concepts such as I/you and verb/noun.

Many studies in comparative psychology have used the approach of mazes and puzzle boxes. The end of the maze or the opening of the box provides the animal with a reward, and the satisfying *consequence* leads to repetition of the activity. Such a concept of motivation has developed to the radical behaviorism associated with the name of Skinner (1950). The emphasis of this school of psychology is upon behavior as a product of environment and the consequences of activity. Expressed in Pavlovian terms, the consequence serves as a reinforcing stimulus.

Mathematical Behaviorism

The English empiricists provided a further important drive to experimental psychology. This school of thought believed that the nature of the human mind could be resolved through analysis into its constituent elements. Galton, a relative of Darwin, was possibly influenced by the theory of individual differences, and was stimulated to apply his mathematical expertise to the problems of testing and describing a variable population. He and his followers held the view that characteristics such as intelligence, personality and attitudes were each complexes of normally distributed variables. Thus, by the application of a suitably wide range of tests and the subsequent use of statistical

techniques such as principal component and factor analysis, the underlying structural elements could be brought to light. Intelligence, for
example, could be divided into such domains as *general intelligence,
mathematical ability* and *verbal ability;* in the same way, personality
could be distributed along a number of orthogonal* axes such as
introversion and extroversion, and attitudes equally could be dispersed
along a variety of statistically independent axes. Here is a tradition
that we may call mathematical behaviorism. It holds that the nature
of the response to any stimulus is influenced by personal traits, and
that these traits are susceptible to mathematical analysis.

Phenomenological Behaviorism

A further possible approach is phenomenological—to regard man
as an integrated phenomenon that responds to a given environment in
an observable manner. This school of psychologists developed as a
reaction to behaviorists who either considered single variables in a
way that had little relevance to free-living man, or else subdivided
the mind into a number of apparently meaningless mathematical
abstractions. The phenomenologists were joined by the German Gestalt
psychologists (who wished to study man as an entity) and other workers
who proposed to use a certain measure of introspection to analyze
consciousness and test how far the separate elements of the statisticians
could be integrated into an overall view of behavior.

We may note an important difference of learning theory between
the Gestalt psychologist and the radical behaviorist. The Gestalt man
thinks in terms of a sudden insight, while the radical behaviorist holds
that the strength of response develops gradually from the consequences
of its repetition.

The Role of Freudian Psychology

To this point, no mention has been made of Freud and his school.
In some degree, Freudian psychology embraces elements of all the
above hypotheses. However, the prime emphasis has been medical,
with the objective of treating abnormal patterns of response. Many
physicians of the Freudian school now give substantial time to modification of unacceptable behavior through conditioning therapy.

Psychology for the Ergonomist

Is there a *party line* for the ergonomist? There is obviously an element of truth in each of the hypotheses of behavior that have been put

---

* Orthogonal axes have no correlation with one another.

forward, and at different times ergonomists have endorsed several of the suggested viewpoints. Nevertheless, most ergonomists seek a simple and practical solution to their problems, and for this reason they are attracted to the doctrines of radical behaviorism.

THE MOLDING OF HUMAN BEHAVIOR

Respondent and Operant Behavior

Psychologists rarely distinguish *voluntary* and *involuntary* actions. They prefer to think in terms of respondent and operant behavior. Respondent behavior is elicited directly by the stimulus: for instance, the sudden constriction of the pupil when a glaring headlight crosses the brow of a hill and shines directly into the eye of a driver, or the salivation produced on a dry and dusty route march when an odorless and tasteless object such as a pebble is introduced into the mouth. Operant behavior, in contrast, has its origin within the organism, and in the view of the radical behaviorist is controlled by the consequences of the given pattern of behavior.

Positive Reinforcement of Behavior

Accepting the hypothesis of radical behaviorism, it is obviously possible to strengthen a given pattern of response by appropriate rewards. In a simple case, a pigeon may be taught to peck a certain switch, and this behavior will be reinforced if such activity is rewarded by the provision of needed food.

In man, the reinforcing consequence may itself be conditioned. Desirable patterns of activity are commonly rewarded by metallic or paper tokens (money). The average worker has learned to *pair* these tokens with the satisfaction of his more primitive needs such as food and shelter (page 234). The ability to pair a reinforcing stimulus is seen most frequently in man, but is not a unique human characteristic. Thus, the response of an animal can be quite strongly reinforced by the intermittent use of a pure tone or click that the creature has previously learned to associate with the delivery of food. Most animal training devices are inherently noisy, and the true response cycle may be as illustrated in Figure 70a rather than as in Figure 70b. In a case such as this, a click rather than a supply of food can become interpreted as the reward.

Many real-life situations are inherently noisy, and the correct identification of the reward is important if it is intended to mold behavior. A disobedient child, for instance, may be finding a reward for his

(a) Actual Situation

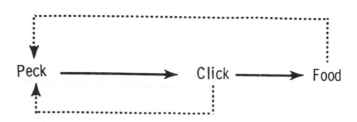

Peck ⟶ Click ⟶ Food

(b) Theoretical Situation for Silent Machine

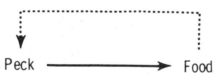

Peck ⟶ Food

Figure 70. The problem of *noise* in a pigeon feeding device. The bird learns to associate the click with the appearance of food.

misdemeanors in either teacher or pupil attention, and classroom discipline will not be improved unless the correct element in the environment is identified and suitably manipulated. Reward identification is sometimes extremely difficult, because a lengthy chain of events intervenes between the primary stimulus and the response. In time, each of the intervening links can itself elicit the same behavioral pattern. Let us suppose that a child is told he will be given a dime to buy some candy. The money is taken from a jar reached by standing upon a particular kitchen chair. Eventually, not only the money itself, but also the jar, and even a particular positioning of the chair may elicit a pleasurable response.

Any environmental change that leads to an increase in the probability of a response can be regarded as a positive reinforcer. In the classroom situation, pupil behavior can be influenced by something as simple as an approving smile on the face of a teacher. Likewise, the teacher responds to very slight cues. If there is greater audience attention when standing to one side of the black (chalk) board, in time, the lectures will be delivered from this vantage point. Traditionally, good grades have been used to reinforce the habit of study, but at the present time the trend is to feed back to the individual student a more personal comment on his own attainments relative to his potential.

Negative Reinforcement

An alternative method of modifying behavior is by negative reinforcement. An animal is taught that *unless* it operates a control in response to a signal such as a light, it will receive an unpleasant electrical shock. In man, frowns and threats are very commonly used as a conditioned form of punishment. A young child is warned that *unless* he behaves himself throughout the visit of Aunt Jane, he will be roundly beaten. An older son is threatened that *unless* he makes an 80 percent average in his grade 13 examinations, he will not be given a motorcycle. An employee is warned that *unless* his productivity increases, he will be fired.

There has been much discussion as to the relative merits of rewards and punishments in producing a desired pattern of behavior. Some psychologists of the Freudian school have hinted darkly that punishment can create deep and long-lasting psychological trauma. This is an important social question, but at the present time the evidence in support of the Freudian hypothesis is equivocal. In practical terms, we can state more certainly that if our objective is to increase an activity (such as industrial production), then either reward or punishment is more effective than indifference.

The virtues of praise and punishment depend in part upon the goals of the hidden persuader, and in part upon the time scale of the operation. On a short-term basis, the threat of punishment is a rather effective procedure, particularly if an acceptable alternative mode of behavior is offered. Differential reinforcement by the praise of good work or behavior is unfortunately a much slower method of achieving the desired results, although in a long-term analysis it is more effective than punishment. Punishment carries several unwanted side effects. Firstly, it is nonspecific. A child confesses to a misdemeanor and is punished; if the sequence is repeated several times, he comes to associate the act of confession rather than the misdemeanor with the resultant punishment. Alternatively, he may associate the consequence too widely, so that the conclusion is drawn that all activity leads to punishment. Some animals react to electrical shock in this way, and many readers will be familiar with the civil servant who finds that the only way to avoid loss of seniority is to make no decisions or other overt signs of activity. Punishment relieves the aggressions of the person carrying out the punitive action. Thus, there is some danger of a positive conditioning cycle; this can reach a serious pitch of sadism not only in prison wardens, but also in teachers and petty officials. Finally, there is the possibility that a positively conditioned reflex may be developed in association with the ending of punishment. Thus, at the end of electrical stimulation, an animal may be cowering in the corner of its cage. In time, it links the act of cowering with the cessation of shock, and much of the day is spent in the corner of the cage. In the same way, an employee may link groveling before an unpleasant foreman with the relief of his misery; an increasing part of the working week thus becomes devoted to abject groveling and other unproductive modes of behavior such as passivity and avoidance reactions.

Extinction of Behavior Patterns

In many practical problems of industry and the home, it is necessary not to develop but to suppress an unwanted form of behavior. This can be achieved either by eliminating the normal reinforcing stimulus (for instance, group approbation of civil disobedience by university students) or by introducing an aversive consequence (expulsion from the university following civil disobedience). It may be possible to weaken the unwanted response by differential reinforcement of more acceptable modes of behavior (for instance, a courteous hearing of legitimate complaints, presented through normal channels). At the animal level, a creature can be rewarded manually every time it turns away from the keyboard. And at the more complex level of human interactions, very subtle rewards can equally place an agitator in a posture where civil disobedience is difficult if not impossible.

Once a conditioned reflex has been established, it is hard to extinguish, particularly if a variable-ratio training pattern has been used (see below). If a pigeon has learned to peck a switch when food is required, its first response when the switch fails to yield food is a more vigorous rate of pecking. If a child is not rewarded for a noisy tantrum, the immediate reaction may be an even greater fit of rage. And if a striking union finds that its demands are not met, the initial consequence is often a more violent and bitter dispute between labor and management. However, in all of these situations the unwanted behavior will abate with patient and consistent removal of the usual reward.

Consistency of treatment is a most important item of management. Inconsistency leads to bizarre and sometimes pathological responses. An example of inconsistent treatment is provided by one of Pavlov's experiments. He trained a dog to salivate whenever it was shown a circle, but not when it was shown an oval. He then progressively modified the shape of the oval until the dog found it impossible to distinguish those signals that were intended to elicit a response. At this stage, the animal became hysterical, struggling fiercely and biting its restraining harness. Children with inconsistent parents and employees with inconsistent management face the same sort of problem. They become uncertain of the cues to which a response is required, and behavior deteriorates to the point where it is unmanageable.

The Shaping of a Response

In normal adults, the appropriate shaping of a particular pattern of response is obtained fairly readily by simple verbal feedback. But there are circumstances such as the training of very young children, the mentally retarded and the deaf and dumb, where more complex shaping procedures must be used.

Let us consider first the problem of teaching a pigeon to peck a keyboard in order to operate a grain feed mechanism. At first, the pigeon is rewarded whenever it faces in the direction of the keyboard. Later, the reward is withheld unless it is also standing near the keyboard. So by a series of successive approximations, the grain-releasing mechanism eventually is operated only when the pigeon pecks the keyboard in the manner prescribed by the operator.

A similar approach is used with very young children. Let us suppose that we wish a boy to wear a pair of spectacles. Initially, he is given a gift of candy whenever the spectacles are picked up. Later, the reward is withheld unless the lenses are near the face. Eventually, the point is reached where the spectacles are worn habitually, and displeasure can be expressed if they are removed. In the human case, it is rather easy to switch the reinforcing consequence as the response develops; for instance, if the boy is suffering from an overdose of candy, the reward can be transformed to a car or train ride.

Dr. C. Webster has described the stages in shaping a more complex response (Fig. 71). Here, a retarded child is required to push a panel near to a black square. Shaping proceeds through a total of some sixty trials, and at each step the task is modified in such a subtle way that a mistake is rarely made. This seems an important virtue of any system of programmed learning. Teaching should be characterized by a positive approach. If a child cannot undertake a task, then a suitable program to teach the necessary skills should be developed. If the student does not succeed, this reflects an error of programming (usually the choice of too large a jump in the sequence of logic) rather than any failure on the part of the pupil.

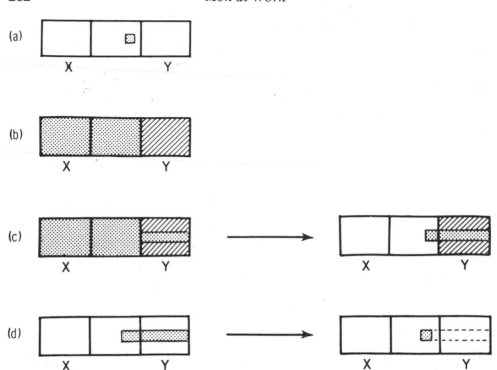

Figure 71. Stages in the shaping of the response of a retarded child. Based on an experiment of Dr. C. Webster.

The Optimum Schedule of Reinforcement

Learning patterns are markedly influenced by the schedule of reinforcement. Reinforcing *consequences* may appear after a fixed or a variable interval, and may bear a fixed or a variable ratio to the number of responses that are made (Holland, 1958).

Let us return to our long-suffering pigeon. Its pecking is rewarded in one experiment by the supply of grain at minute intervals (Fig. 72). The creature quickly learns that there is no virtue in pecking immediately after the grain has appeared, and if a cumulative record of pecking responses is kept, this shows a scalloped stepwise format. In a second experiment, the reward appears at variable intervals. Assuming that the pigeon is still hungry, there are no longer phases of diminished activity. The investigator proceeds to a third variant of feeding pattern; the reward is now provided after, for instance, fifty responses. Again, the appearance of the grain tends to be followed by a pause. However, if the reward/response ratio is varied, the cumulative responses show a continuous increase with time.

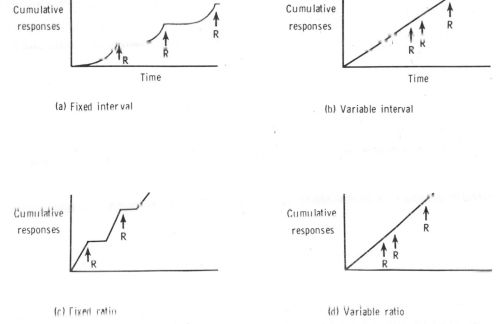

Figure 72. A comparison of several possible schedules of reinforcement. Rewards are supplied at the arrows (R).

Some animals can be conditioned to make an extraordinarily large number of responses before a reinforcing reward is needed. Thus, the chimpanzee can respond to banana pellets if the reinforcement ratio is as small as 1 in 48,000. In shaping such a response, the experimenter starts by alternating the reinforcing consequence, and gradually extends the interval between rewards. Once established, a pattern of this type is very difficult to extinguish. Further, if seen for the first time, it is by no means easy for an observer to spot the infrequent rewards. This has an important bearing on the treatment of apparently irrational or disturbed behavior. We may note also that the high response rate induced by an infrequent and variable reward ratio is exploited in the design of most gambling machines.

In the more usual context of ergonomics, reward schedules have an important bearing upon the productivity of the worker. The constant interval reward may be likened to a weekly wage packet, and the variable interval reward to the occasional bonus payment. The fixed reward ratio is analogous to piecework. In general, it is more effective than the fixed interval reward system, but there tends to be a pause following payment, and the worker is often haunted by the fear that a grasping management may lengthen the reward interval (as in the chimpanzee experiment).

Many production-line tasks call for little more than a key-pressing type of response; accuracy is not involved. However, rather similar principles apply to the teaching of a *correct* response. Thus, if a pigeon is required to choose between a small and a large response key, learning of the task proceeds fastest if a reward is provided after a variable total of correct choices has been made.

The Role of Personality and the Emotions

Some psychologists view personality as an unimportant variable in the context of conditioning. However, there are several possible types of interaction between personality and learning. The extrovert initially responds more readily than the introvert to an external stimulus; on the other hand, he is more easily distracted from learning by *noise* in the environment. Other changes, unrelated to the primary stimulus, soon catch his attention and slow the conditioning process. Personality also has an important bearing upon the choice of reinforcement. An office party may be an excellent Christmas reward for an extroverted employee, but an introvert may be much better satisfied by an equal expenditure upon a book club subscription.

The choice of reward may need to be modified in the light of cultural factors. Concepts of advancement, progress and competence are excellent reinforcers for a solid middle-class citizen, but have less relevance to the gainful employment of a Negro youth born and bred in a Chicago ghetto.

The possible role of the autonomic system in conditioning is receiving increasing attention. Can the state of the emotions influence the learning process? Again, there seem important cultural differences. Some races are encouraged from childhood to note and verbalize autonomic sensations of anger, hostility and fear, while other races keep *a stiff upper lip*, suppressing and ignoring the same sensations. In both types of individual, a strong emotional stimulus may detract from a conditioning program, but whether verbalized anger will enhance or minimize such an effect depends mainly upon whether it relieves or intensifies the emotion. There is currently some interest in conditioning as a means of reducing unwanted autonomic reactions. Athletes such as golfers perform badly when their muscles are excessively tense, and there is evidence that by displaying muscle tension on an electromyograph, a contestant can be taught to relax more completely. In the same way, the heartbeat can be displayed on an electrocardiogram or presented as an auditory signal, and this can be used to condition an executive to avoid an emotional increase of heart rate in a stressful situation (Brener and Kleinman, 1970).

AROUSAL AND PSYCHOMOTOR PERFORMANCE

The Concept of Arousal

Arousal may be considered as an increase in activation of the central nervous system. Arousal is lost as a person becomes sleepy, and reappears as he awakens. The neurophysiological basis of arousal is considered elsewhere (page 99), but we may note here that the level of arousal is fairly readily assessed from the behavior of a number of simple physiological and biochemical variables. Thus arousal leads to an increase of heart rate and blood pressure, an increase of sweating (conveniently measured as the current transmitted through the skin of the palm—the galvanic skin response), an increase of muscle tone (and thus electromyograph signals), a fall of the eosinophil count in the blood (Berkun *et al.*, 1962), and an increase in the urinary excretion of both 17 ketosteroids and catecholamines (Levi, 1965). The galvanic skin response is perhaps the most consistent measure of arousal, but unfortunately it seems relatively independent of the other signs; it is thus useful to note also the mean heart rate and the range of variation in heart rate. Some individuals are able to give a subjective estimate of their arousal, but this does not always coincide very closely with the physiological estimate.

The Performance Curve

The classical curve relating the efficiency of performance to the level of arousal (Corcoran, 1965; Poulton, 1970) has the form of an inverted U (Fig. 73). Efficiency is greatest at point 2 in the figure; in this region a change in the level of arousal has little effect upon performance. At point 1 arousal is insufficient, and an increase of arousal thus improves efficiency. At point 3 arousal is already excessive, and a further increase of arousal has a deleterious effect upon performance.

The location of the curve depends upon the complexity of the task; easy work is performed efficiently at a rather high level of arousal, while a more complex task demands a lower level. With repetition, a task becomes easier, and thus the level of arousal for optimum performance rises. Again, if speed of performance is wanted, arousal should be at a higher level than if accuracy is required.

Anxiety generally increases arousal, and the effect of a stressful situation will thus depend on the relationship between the initial level of arousal and the optimum for the performance of a task by a given individual. A moderate increase in the speed of operation of a conveyor belt may help to increase efficiency, but if the speed appears impossibly

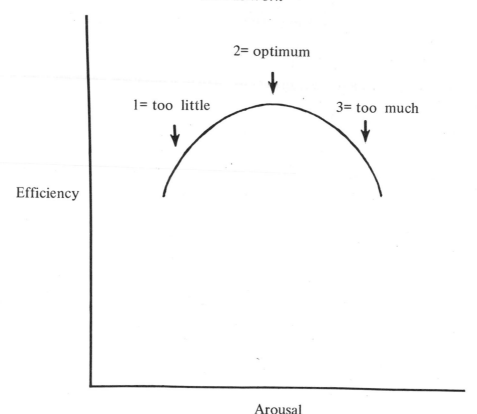

Figure 73. The relationship between arousal and performance (based on experiments of Corcoran, 1965).

fast to an employee, he will become flustered and over-aroused with a drop in useful output.

Introverts are normally more highly aroused than extroverts, and for this reason perform better in a less stimulating environment (Bakam *et al.*, 1960). When considering the ergonomics of his office, an introvert should choose a conservative decor, with thick carpets and acoustic tiles, while the extrovert should seek bright colors and piped-in music.

Maintenance of Arousal

A critical level of arousal is necessary if conditioning is to proceed at its optimum rate. The ideal level varies with the individual and his immediate environmental circumstances, and some people—particularly introverts—can be conditioned more effectively when arousal is decreased by a small dose of a depressant drug.

The maintenance of an adequate level of arousal is particularly important when a person is alone and is required to undertake a rather boring task. Vigilance can often be improved through a deliberate increase of arousal, although in theoretical terms it is difficult to distinguish a purely arousing stimulus from what may be regarded as a conditioned reinforcing signal. Thus, the night driver on an expressway finds benefit from periodic exercise (page 99). This undoubtedly increases the general level of arousal through proprioceptive and neurohormonal mechanisms, but the driver may also have conditioned himself to wake-up in response to this type of stimulus.

Let us now illustrate the problem further by describing a few simple tricks that will help the university student in the difficult and perhaps boring task of mastering his lecture notes. The place of study should be chosen on the basis of personality—more stimulating for the extrovert than the introvert. The rate of working should be paced at the fastest feasible level (for example, five pages to be learned per hour). To avoid over-arousal, the target can be lowered for sections containing difficult concepts or complex mathematical formulae. Reinforcement to continued study is provided by *knowledge of results*. The student should thus choose books that provide quiz sections, or in their absence should devise his own quiz and discuss the material with his colleagues. Boredom will be minimized by varying the task as much as possible (for instance, alternating an hour of ergonomics with an hour of scientific German). Finally, there should be periodic rest pauses, when measures such as physical activity are used to increase arousal.

VIGILANCE

Many tasks call for prolonged watchfulness with only a very occasional response. A simple example is provided by the production-line inspector, watching articles moving along a conveyor belt. Other familiar instances include the engineer monitoring a bank of instruments at a generating station, the airman scanning a radar screen for enemy aircraft or missiles, and a driver on a crowded highway.

The Time Course of Vigilance

With time, all such tasks become boring, and the proportion of errors, missed and incorrect identifications increases. Attention was first drawn to the problem during World War II, when it was realized that submarine patrols were reporting far fewer enemy vessels than were known to be operational. Psychologists were thus asked to make a formal study of human vigilance.

The task of Mackworth (1948 and 1950) is illustrative of their approach. Subjects were required to look at a large and rather old-fashioned clock face. Normally, the hand jumped a distance of 0.3 inches each second, but occasionally (24 times per hour) it would make a double jump. The subjects were supposed to report each such event. Errors developed quite rapidly, averaging 12 percent in the first ten minutes, and 20 percent in the next ten minutes, with further progressive deterioration over two hours of observation.

If a task is self-paced, most people decide on what for them seems an acceptable accuracy, and adjust their output accordingly. However, if prolonged vigilance is required, there is still a decrement of performance (Fig. 74). A *warm-up* occurs over the first few minutes, but this is soon followed by a progressive decline in productivity, interrupted only by a minor recovery after the lunch break and an *end-spurt* as the time of departure draws near (McKenzie and Elliott, 1965). A few investigators have reported a more constant average level of performance under self-paced conditions. Possibly, some workers realize that vigilance has lapsed, and are thus aroused with a temporary gain of output. Certainly, variability of production precedes a frank loss of output (Broadbent, 1953).

If a lapse of vigilance occurs with a machine-paced task, there is inevitably an error, and no compensation can be made by later working at a much higher rate. Thus, productivity can often be improved by self-pacing. Bertelson and his colleagues (1965) illustrated this point by study of a postal sorting depot. By accepting a 2 percent error rate, self-paced mail clerks could sort up to sixty letters per minute. In contrast, the 2 percent error rate could not be sustained with machine pacing unless deliveries were reduced to 40/min. At the higher rate of 60 letters/min, there were 20 percent errors, with two thirds of the incorrectly sorted mail being sent to the address on the immediately preceding envelope.

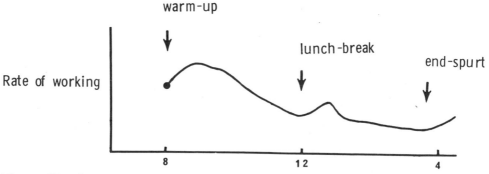

Figure 74. The time course of vigilance—output on a self-paced task (based on experiments of McKenzie and Elliott, 1965).

Improving Human Vigilance

Mackworth (1948) suggested that the loss of performance with pro-longed vigilance was due to a form of reactive inhibition. Vigilance was characterized by an excitatory state, and each response made a finite contribution to the development of an opposing inhibitory state, thereby decreasing the likelihood of a future response; a parallel could be drawn between this reactive inhibition and the external inhibition of classical conditioning. In support of this view, Baker (1959) and Mackworth (1964) emphasized the role of *expectancy*—the longer the interval separating the subject from a previous response, the greater his likelihood of detecting a given signal.

This approach was roundly condemned by Skinner (1950) and his associates on the basis that it was a description rather than an explana-tion of events. The concepts of vigilance, attention, inhibition and the like were essentially behavioral, and because of the difference in the level of observation they could not and should not be used to *explain* the experimental data; the new concepts were in themselves no less mysterious than the results they described.

Deese and Ormond (1953) further confounded the proponents of reactive inhibition by showing that performance improved when the number of required responses was increased. If radar scanners were shown 10 targets per hour, 46 percent were detected; at 20/hr, detection was 64 percent, at 30/hr, 83 percent, and at 40/hr, 88 percent. This seems impossible to explain in terms of accumulation of an inhibitory state! On the basis of such experiments, the current practice is to infil-trate a radar screen with periodic *artificial* signals (Faulkner, 1962). By appraising the viewer of his performance, detection can then be maintained at a high and relatively stable level for a prolonged period. In the same manner, if a factory worker is sorting defective nuts from a conveyor belt, the usual practice is to introduce a few poorly formed nuts into the production line. If the employee is advised of his success in detecting these nuts, the overall accuracy of his output is sustained.

Such observations suggest the hypothesis that both boredom and the resultant loss of vigilance are due to a lack of reinforcing stimuli. For many types of personality, *being right* is a significant reward and indeed is enough to maintain the required response (vigilance). In support of this hypothesis, classical conditioning patterns are seen when fixed interval, variable interval and variable ratio signal presenta-tions are used (Holland, 1960). Furthermore, extinction occurs much as in classical conditioning. If Holland's view is correct, then observing behavior should be controlled rather precisely by a suitable manipula-tion of the external environment—particularly an adjustment of the rate of input of *artificial* signals.

However, the *artificial* signal may be serving an alternative role (Tanner and Swets, 1954). Let us suppose that a manufacturer decides as a reasonable compromise between cost and quality to reject 5 percent of his widgets. Workers are instructed accordingly, but after two hours on the production line their standards change. Joe has become so adept at spotting minor defects that he is rejecting 25 percent (Fig. 75); further, he is spending so much time looking for minor scratches on the base of the widgets that he often misses a gross defect in the super-structure. Bill displaces his criteria in the opposite direction, cheerfully accepting 98 percent of production. His, indeed, is the more common fault on the factory floor. Let us now suppose that defective widgets are deliberately placed on the conveyor belt at selected points (variable interval, variable ratio). The worker spots that these are unacceptable, and his terms of reference are immediately clarified.

Other practical techniques of improving vigilance are to pair operators, to make warning signals as obvious as possible, and to present warnings simultaneously to eye and ear (Loveless, 1957). Thus, when train drivers on the Western Region of British railways confront a red warning signal, a loud bell also rings within the cab of the locomotive.

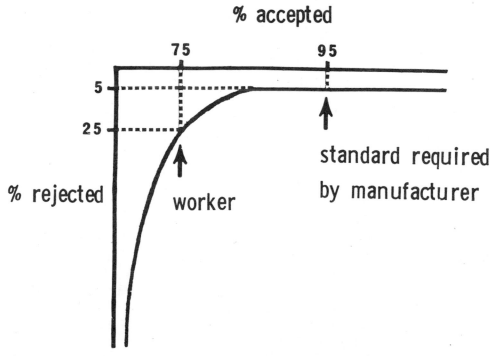

Figure 75. Displacement of standards on production-line task. The worker is supposed to reject 5 percent of production, but by an increase of quality criteria he has raised the rejection rate to 25 percent.

Arousal, Personality and Vigilance

The need for optimum arousal relative to experience, task difficulty and personality has already been stressed. An unfavorable external environment, such as a combination of heat and humidity or an excessive noise level normally leads to a worsening of vigilance (page 217). However, if the individual is initially under-aroused (Fig. 73, position 1), the *unfavorable* environment can lead to an immediate gain of performance. If arousal is insufficient, performance can also be improved by the use of stimulant drugs such as Benzedrine® or caffeine (Mackworth, 1950), nevertheless, this is most undesirable as a working routine. Lack of sleep (page 271) leads to a decrease of arousal, and there is usually an associated loss of vigilance (Corcoran, 1962); adequate sleep is thus important to any task where watchfulness is required. Stress leads to an increase of arousal, and thus (usually) to poor performance. Stress may sometimes arise from a combination of two tasks that seem very simple when carried out individually. A person should undertake no more than one vigilance-type task at a time, unless the second occurrence is most infrequent.

There has been considerable argument as to the type of personality that performs best on a vigilance task. Much of the conflicting evidence can be reconciled in terms of the inverted-U relationship, and associated differences of task difficulty and experience (Fig. 73). Introverts may initially respond more poorly than extroverts, since they have been conditioned to react to internal rather than external stimuli; however, extroverts are also more readily distracted. Furthermore, the usual vigilance task has a progressive depressant effect upon the general level of arousal. Introverts who initially are highly aroused are brought to an optimum mental state, whereas extroverts who initially are somewhat under-aroused suffer a further depression of their reticular activating system (page 99). Obviously, the optimum personality depends on the nature and duration of the task, the experience of the worker, and the degree of arousal contributed by other features of the working environment.

Vigilance and Automation

It may be argued that man is not well-suited to the performance of vigilance work. Whatever maneuvers are adopted, it is difficult for the average worker to maintain a uniform standard of watchfulness over a long period. Attempts to sustain vigilance are stressful, and human failure becomes a serious possibility. Thus, the intermittent vigilance demanded of the normal car driver is not unpleasant, but the constant attention required when operating a vehicle at high speed on an overcrowded multi-lane expressway is very tiring. A similar problem is found in busy airport control rooms.

Some operations can be carried out very successfully by animals. Verhave described an interesting experiment where pigeons were trained to the point that they rejected 99.8 percent of badly formed capsules passing along a conveyor belt. By arranging two pigeons in series with each other, he achieved a level of vigilance far exceeding that of the average worker. Unfortunately, the pressure of unions and the fear of public ridicule has so far prevented exploitation of this approach. Even registration of an appropriate patent has been turned down on the basis that a mental rather than a physical process is involved.

Currently, machines seem a more acceptable substitute for the human operator; the trend is thus to replace man by a suitable machine during routine phases of work, saving human skills and judgments for the most difficult stage of an operation. For example, aircraft are flown across the Atlantic Ocean by an *automatic pilot* and the experience of the aircraft's captain is reserved for instrument failure and problems of take-off and landing.

CHAPTER 10

PERSONALITY AND APTITUDES

A T SEVERAL POINTS IN THE TEXT, emphasis has been given to the importance of personality and aptitude as determinants of human performance. Personal characteristics strongly influence such fundamentals as the pattern of task learning, the ability to develop and maintain an appropriate level of arousal in the working environment, techniques of motivation, and the nature of interactions with other employees.

Tests of both personality and aptitude are widely used by industry, particularly in North America. The increasing social distance between employer and employee has led to a delegation of selection procedures to applied psychologists, and the use of such specialists has become an integral part of industrial management. However, the effectiveness of the procedures currently used is questionable; often, the motivation for introduction of a testing service is not its proven efficacy but a fear of losing ground to a rival company that has a division of applied psychology.

The number of available procedures for personnel testing is very large (>2500); the purpose of the present chapter is not to review details of individual methods, but rather to discuss some general principles governing such measurements.

ETHICS OF TESTING

Many psychological tests include questions of a very personal nature. It is thus necessary to respect professional codes of ethics relative to such matters as informed consent, personal benefit and secrecy (W.M.A., 1964; Shephard, 1967; Baumrind, 1964; Malik, 1969).

The psychologist administering a test battery must ensure that an employee or prospective employee has given his free and informed consent to application of the proposed tests. If testing is a condition of consideration for employment, this should be made clear to an applicant before he has placed himself in a position where the meeting of such an entrance requirement cannot be avoided by selection of alternative work. Information given to the candidate should include briefing upon the general nature of the procedures to be used, the reasons behind the various tests, and the individuals to whom the results will be disclosed. In some instances, as when testing intelligence, it may be impossible to make detailed prior disclosure of the test methods. This point was reviewed by the Canadian Supreme Court, when an irate father demanded to see certain intelligence tests administered to his son by a local school board; a ruling was given against disclosure of the detailed protocol, on the grounds of (1) the accepted efficacy of the test and (2) the likelihood that its usefulness would be destroyed if the questions became widely available. The psychologist must ensure that when the tests have been completed, the results are used for the welfare of the employee—perhaps to determine his suitability for a particular position—rather than for his exploitation. This point is well illustrated by the status of U.S. college athletes. Such students are often subjected to extensive personality testing. The reporting of adverse results to the coach may be helpful in improving an athlete's skill, but if the information becomes widely available there is a danger that the individual concerned may be dropped from the college team and thus lose his athletic scholarship or other emoluments necessary to his education.

Personal privacy is currently a matter of public concern. Computers have the capacity to store vast quantities of information on large segments of the population, and psychologists have rightly resisted pressure to transmit their data to memory banks. Not only is privacy an important foundation stone of our society, but it is unlikely that an employee will answer a questionnaire frankly if he suspects that his responses will be widely disseminated. Questionnaires that contain personal information should be held in strict confidence, and where possible should be summarized in numerical form with destruction of the original answer sheet. In this way, the possibility of embarassing disclosures in response to a court subpoena are avoided.

TEST DESIGN

Personnel tests may be grouped according to the characteristics under evaluation, or alternatively according to the method of evaluation that is used. In the former case, one may distinguish three main categories:
1. ability: intelligence, mechanical aptitude, motor skills and learning ability (for example, Raven, 1956, and Fig. 76).
2. motivation: attitudes and interests (for example, Allport *et al.*, 1951).
3. personality (for example, Eysenck, 1959).

In terms of test methodology, one may distinguish:
1. subjective self-reports and questionnaires,

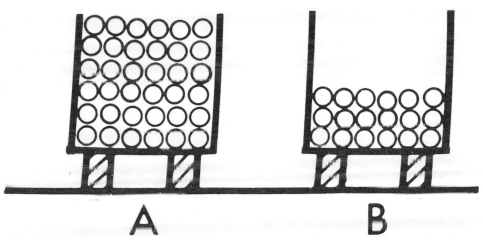

Figure 76. Example of a simple paper-and-pencil test of mechanical aptitude. Which truck is more likely to tip over, A or B?

2. objective self-reports, including biographical data and measures of capacity such as intelligence tests and work samples,
3. evaluation by peers or supervisors,
4. observation by others, and
5. use of archival data.

The great bulk of personnel tests fall into the category of subjective self-reports. It is generally conceded that such tests are not particularly effective, but the hope persists that by use of a new and improved questionnaire the desired information will be obtained. *Paper-and-pencil* tests remain popular partly because they are easy to administer, and partly because there is a latent belief that the verbal output of the individual must bear a significant relationship to his personality. Critics emphasize that many of the questions asked are offensive to some subjects, and that elimination of the offending questions leads to a further deterioration in the effectiveness of the questionnaires. Alternatives to the questionnaire (Lovell, 1967; Radloff, 1968) are examined briefly on page 230.

Certain assumptions are inherent in most designs of test. The investigator must normally accept that the variable to be measured—whether vocabulary, mechanical knowledge or mathematical ability—is normally distributed in the population. Statistically predictable percentages of people should then attain specified scores above or below the mean value. It might be decided, for instance, that a Wechsler intelligence score of 130 was essential for a university professor. This is two standard deviations above the population mean for Wechsler's test, and assuming a normal distribution of data, only 2.27 percent of the population would be expected to reach this standard.

It is also assumed that behavior is causally determined. The purpose of testing is thus to narrow the possible range of explanations for a given behavior pattern. Let us suppose a girl shows a poor performance as a typist. An intelligence test may show she fails to meet the minimum intelligence quotient necessary for this task. A psychomotor evaluation may show she lacks the necessary digital aptitudes. Personality tests may suggest that she has a psychological makeup that responds badly to a routine office situation. But before accepting a simplistic solution, it is necessary to exclude other explanations of both the initial poor performance on the typewriter and subsequent test failures. The girl may have poor physical health, concern for a sick member of the family, or lack of minimal training in the use of a typewriter.

A satisfactory test procedure must meet certain requirements of reliability, validity and discriminative capacity. If a test is *reliable*, then the response remains consistent from one week to the next, after due allowance for loss of naiveté, practice and any changes in environment. The degree of reliability is usually assessed experimentally, by splitting a test into two equal parts, and calculating the coefficient of correlation between scores for the two parts (a test/retest correlation). An acceptable level of reliability depends upon the importance of the question to be answered. More hangs upon the selection of an appropriate cosmonaut than upon the enlistment of ten *expendable* widget fixers at the neighborhood factory. Reliability can sometimes be improved by an increase in the number of questions. However, at best, the gain is proportional to the square root of the number of required responses, and if a test is unduly protracted there may be a loss of reliability due to fatigue and boredom. Some of the simpler tests such as Taylor's scale of Manifest Anxiety require no more than fifty responses, while others such as the Minnesota Multiphasic Personality Inventory include more than five hundred items.

Validity is concerned with such matters as content, predictive validity, concurrent validity and construct validity. Appropriate content is a fundamental, but often a very difficult requirement of test design. Is it possible to define those qualities expected of an ideal employee? What characterizes a good radiographer, a good stenographer, a good nurse or a good bus driver? One approach is to ask the prospective employer for desirable characteristics, but this immediately reveals a wide divergence of opinion. In one recent experiment, a large panel of hospital staff each listed different—and sometimes diametrically opposed—criteria that they would seek in appointing a staff radiographer. A second approach is to examine the demands of the job. Is a stenographer expected to type? In such a case, a test of typing speed is appropriate. Or is her function to give status to an executive and serve as an attraction in the front office? In the latter instance, a high-speed typist with flat chest and horn-rimmed spectacles might be a disastrous choice. Perhaps the simplest method of content validation is to compare test results with performance on the job. Theory suggests that a nurse should be neat, clean and strong, with simple mathematical ability. Does a test supposedly measuring such traits predict those students who will become good nurses? Unfortunately, such an empirical validation is necessarily one-sided, since it is impractical to assess the nursing ability of those candidates rejected at the initial screening. It is also difficult to allow for local variations in the requirements of the employing agency. Let us suppose that a sales position calls for stability, motivation and ambition. An employee meets these requirements, but yet has a rather dependent personality. He needs periodic reassurance of his success by an appropriate *father figure*. His initial supervisor meets this need, and he performs well. Later, the supervisor is replaced by a man who is harsh, authoritarian and insensitive, and the employee rapidly becomes a failure, with signs of psychological breakdown.

When large populations must be evaluated, reliance is usually slanted towards a series of multiple-choice questionnaires. Scoring is simpler and in a sense more objective than for a series of lengthy psychological interviews. However, responses reflect not the true nature of the employee, but rather his image of himself. Sometimes, a worker may give deliberately misleading answers; this can be detected by incorporation of a suitable lie scale (Eysenck, 1959). More often, the individual has little insight into his characteristics. This does not necessarily invalidate pencil-and-paper tests. It may be difficult to attach a precise psychological meaning to a given response pattern, but nevertheless, content validity can be established by showing a useful association between the specific pattern and a good (or bad) employment record.

Predictive validity is a measure of how well a test will predict future performance. Employers are interested not only in current performance, but in potential. Beyond the age of 25, patterns of interest and attitudes become more fixed, and the prediction of future behavior is easier.

Concurrent validity examines how well the scores on a specific test conform with currently accepted criteria for a particular type of employment.

Construct validity examines how far scores confirm previously formulated theories as to the type of person best suited to a given job. Although of theoretical interest, concurrent and construct validity have less direct application to the practical problems of personnel selection.

Some tests have excellent reliability and validity, but yet fail as a means of screening personnel because of a poor *discriminating capacity*. Most tests are arranged to discriminate about the population mean and thus work poorly in selecting atypical people. Let us suppose that we wish to select a faculty of university professors with intelligence quotients greater than 130. A standard test, such as the basic progressive matrices (Raven, 1956) would indicate quite clearly those individuals with an IQ greater than 100, but would be an insensitive method of distinguishing men with an IQ greater than 130. A specific and more advanced series of matrix problems is needed. It is always essential to ensure that the tests selected are of appropriate difficulty for the population to be examined. It is also important to decide whether the emphasis should be upon *speed* of response (a timed test) or *power* (ability to solve very difficult problems, given unlimited time).

STAFF RESOURCES

The dynamic and ever-changing needs of modern industry call for a continual re-evaluation of the skills of personnel relative to the demands of employment. It has been estimated that the resource value of trained personnel represents fifteen times the annual earnings of an industry—an important entry in any annual balance sheet. However, this resource can only be retained by appropriate salary adjustments and regular retraining programs. From the viewpoint of management, it is desirable to retain only functional personnel, and in some rapidly changing industries a high staff turnover may indicate effective management.

SPECIFIC TEST PROCEDURES

Irrespective of format, all test procedures must be used under carefully standarized conditions. If instructions are needed, these should either be issued in printed form, or read from a printed card in a uniform manner. The percentage of *correct* answers or behavior patterns achieved by guesswork is normally close to that predicted from the alternatives open to the subject, but if the person under test has had extensive theoretical or practical experience of psychological examinations, he may find it possible to choose advantageous alternatives.

The results of objective self-reports upon intelligence are usually presented as an intelligence quotient (the ratio of the observed mental age to the chronological age). The concept of a fixed intelligence quotient is somewhat fallacious. Rather, each individual has a specific potential for development. If a child is reared in an intellectually enriched environment, the potential is realized, but if he lives in a deprived environment, the final development falls far short of potential. At one time, it was thought that a maximum of intelligence was attained at about 16 years of age, and this belief is supported by the format of average Intelligence test scores. However, it is now suspected that given adequate motivation, many aspects of intelligence can improve throughout adult life. The impact of cerebral vascular disease depends upon the extent to which potential has been realized. A stroke may have little effect upon the mental function of a person who has not previously used his full powers of cerebration, but can be quite disastrous for the person who is already fully extended.

A second fundamental weakness of the intelligence quotient is the implicit assumption that intelligence is a single variable. Factor analyses have revealed a multiplicity of components, dominated by *general intelligence,* but embracing also such facets as mathematical ability, vocabulary, general information and reasoning ability. Some tests measure largely general intelligence. Progressive matrices (Raven, 1956) fall into this category. They require the matching of a series of abstract designs; however, even in such a test it is difficult to escape the influences of culture and preconceived attitudes (including frank aversion to problems of reasoning and elementary mathematics). Vocabulary tests are affected much more by environment, but given a good education and a literate home background there is a strong correlation between vocabulary and general intelligence. In a clerical appointment, none of these factors may be of prime significance; the need is rather for speed and accuracy in checking such items as address and telephone lists, and simple tests can be devised to evaluate this facility. In a workshop, the need is for an interest in and an understanding of mechanical problems. The best type of test for such employment would include the use of small mechanical models, but again it is possible to present equivalent paper-and-pencil items (Fig. 76).

Motivation is extremely difficult to measure by objective self-reports. Common approaches are to ask a subject to estimate his performance on a standard test prior to its commencement, or to rate the perception of effort at a fixed rate of working; the latter approach is particularly useful in assessing tolerance of physical effort (Borg, 1971). Attitudes and interests are not fundamental characteristics of an individual, but are culturally determined; nevertheless, they have an important bearing upon the suitability of the individual for a given type of employment. Typical tests, such as that of Allport *et al.* (1951) use a multiple-choice questionnaire to explore interest in theoretical matters, practical details, money, beauty and philosophical issues. A person with a strong theoretical bent and little interest in money would

be unlikely to be a success on the piece-rate production line of a
large factory, but might make a good university professor!

Personality is constitutionally determined, although certain aspects
of personality can become more obvious with a change in environment.
Subjective self-reports are usually used in measurement. Some tests,
such as the widely publicised Maudsley personality inventory (Eysenck,
1959) give only one or two crude scores that can be compared with
normal population values. Others, such as the Cattell 16-factor test,
yield many different personality scores, often expressed in STEN
(standard ten) units. The STEN score is so arranged that an average
person has a rating of 5.5, and 95 percent of the population have scores
between 4 and 7. In general, the search is for a worker with a well-
balanced personality, having all scores close to the population mean.
However, there are exceptions. Laboratory workers are most successful
if they have above average introversion, while conversely a receptionist
or a bus driver/conductor should be somewhat outgoing. Much
psychosomatic illness arises from a conflict between personality and
the demands of the job, and in the future it is likely that increasing
attention will be paid to the optimum personality type for specific
occupations.

The search for alternatives to subjective self-reports continues. Biographical data—such as age, birth order, family size and stability, size
of hometown, level of education, work history and credit rating—can
be obtained from an objective self-report. Since the facts involved
are readily verified, there is less likelihood of distortion or deliberate
falsification, and in general the information demanded is considered
less *personal* than the usual psychological questionnaire. The same
type of data can often be abstracted from archives and credit-rating
agencies, although the existence and use of such services raises fundamental issues of personal privacy.

Peer and supervisor ratings can be useful only if each of several
observers see a sufficient number of employees to control observer
bias. Reports are most effective when prepared by immediate supervisors. Due allowance must be made for what has been called the
halo effect. Each supervisor naturally wishes promotion for his staff,
and consciously or subconsciously recalls only the good characteristics
of an employee. If an *objective* marking system is used, nearly everyone
commonly receives top ratings. It is thus necessary to force the choice
of the supervisor, either by budgetary ceilings or by placing a restriction
upon the number of *A* ratings. It is often helpful to assure the evaluator
that the prospects of an employee are based upon several independent
ratings, so that the individual supervisor does not feel a total responsibility for the subsequent careers of his staff.

Peer or observer ratings can also be arranged in contrived situations such as a weekend house party. Superficially, this seems an expensive and time-consuming method of choosing staff, but it may be justified where extensive subsequent training of an employee is planned.

Ultimately, the current generation of multiple-choice questionnaires may be replaced entirely by some of these objective techniques. However, in most instances much research is still needed before observed scores can be converted to practical recommendations for business management.

CHAPTER 11

MOTIVATION OF THE WORKER

GENERAL THEORIES OF BEHAVIOR have been discussed in a previous chapter. At this point, it is proposed to relate such theories to the specific problem of motivating workers at various levels in the industrial hierarchy (Atkinson, 1957; White, 1959; Fitts and Posner, 1967). Although the prime interface to be considered is that relating the individual to a given organization, there must be appropriate interactions both between different groups within an organization, and also between the organization as a whole and the national environment (Schein, 1965; Katz and Kahn, 1966; Buckley, 1967; Weick, 1969).

PATH-GOAL THEORIES OF MOTIVATION

Modern management believes that the individual worker is driven and goal-oriented (Pernow, 1961; Warner and Havens, 1968). Like the pigeons in Skinner's experiments (page 208), he can be conditioned to perform in an appropriate manner through the provision of suitable immediate rewards.

Research investigators speak of a path-goal theory of motivation (Vroom, 1964). Account is taken of path-goal instrumentality (PGI)—the extent to which a given pattern of behavior leads to the realization of a given goal—and of the value of this goal (V) to the individual. Motivation then becomes a function of (PGI × V), summed over the various goals and objectives of the individual.

Performance is necessarily a function not only of motivation, but also of ability, and unfortunately these two factors may operate against each other. A person of marginal ability may find a task challenging and motivating, while an over-qualified person will perform badly because he is bored.

Although initially a somewhat theoretical concept, the path-goal model of industrial productivity has now had extensive validation in such varied segments of the working world as a public utility company, the nursing fraternity of a large hospital and even the higher echelons of management. Performance has been evaluated by self-rating, evaluation by peers or supervisors, and measurement of group output (page 230), while path-goal implementation has been explored by asking workers how they believe their work helps or hinders such personal goals as the gaining of more money, seniority and the respect of their peers. Such studies have demonstrated clearly that an individual is unlikely to perform well at any given industrial task if in his view the activity fails to contribute to a recognized goal, or the goal as recognized seems unimportant to him. Unfortunately for the production lines of industry, the realization of goals through a given pattern of behavior usually diminishes if the pattern is repeated frequently. Furthermore, the intelligent worker soon discovers any disparity between the rewards promised by his employers and the realities of employment. Much thought is thus necessary for successful long term motivation.

THE REWARDS OF THE WORKER

What are the goals and the rewards of the average worker? Certainly, they are more complex than the mere operation of a feeding box as in Skinner's experiments with the pigeons. Maslow (1954, 1962) has suggested that man responds to a hierarchy of needs. The most fundamental are physiological in character—requirements such as hunger, thirst, sleep and sex—each rooted in measurable physiological changes within the body. Assuming that such basic demands are met, priority is given to considerations of personal and familial safety. These needs are in turn succeeded by social factors, the yearning for love, affection and a sense of belonging. At a still higher level, we find the needs of the ego—the esteem of others, and self-respect engendered by feelings of competence and control over one's personal destiny. At the zenith of this hierarchy lies the goal of self-actualization—a sense of growth to maturity, coupled with an awareness of having realized to the full one's potential for development.

The lowest requirements of Maslow's hierarchy are prepotent, but if these needs are well-satisfied, they cease to be important, and higher items in the sequence become effective motivating stimuli. At least 90 percent of the physiological needs of the average North American worker are met, together with perhaps 70 percent of his needs for safety and 50 percent of his social needs, but there is satisfaction of no more than 30 percent of his ego requirements, and 10 percent of his needs for self-actualization.

Details of Maslow's hypothesis may be disputed. Certainly, the ordering of needs by the individual worker is strongly colored by his past experience. Culture influences both manifestations of need and acceptable modes of gratification. The sophisticated western capitalist may try to satiate his ego by the purchase of a luxury motor yacht, the eastern communist equally will aspire to leadership in a workers' committee, while the starving peasant of the *third world* will be concerned more with the physiology of survival than with problems of ego. Scientists seem particularly motivated by the self-esteem associated with recognized achievements and discoveries. The lower middle-class, white-collar worker may have sufficient regard for status symbols that he will accept a degree of starvation while living in an unnecessarily large house. Female workers have larger social needs than men, and respond better to opportunities for social interaction and expressions of concern regarding appropriate supervision of their labors. The needs of the individual can often be determined by thematic apperception techniques. Thus, the office employee may be shown a picture of a room containing a desk and a bookcase, with an old man talking to someone who is relatively young. If the employee describes this as a picture of father and son, he may have social needs, a requirement for affiliation and a father figure. However, if he sees it as the company president talking to a young executive, his need is for ego fulfillment through personal achievement.

Despite such personal vagaries, Maslow's basic philosophy is sound. It suggests that the main thrust of reward systems in western society will be in terms of ego fulfillment and self-actualization. At the executive level, this pattern is increasingly realized, but unfortunately the higher forms of reward are difficult to engender on the shop floor. In consequence, the factory worker shows an inherent tendency towards diminishing productivity, with a propensity for developing alternative reward systems through aggressive trade union activity. While satisfaction of the more lowly needs in the Maslow hierarchy are sufficient to avoid absenteeism and prevent an excessive turnover of employees, it is necessary to satisfy needs at a much higher level in order to stimulate a high level of performance, to secure an extra fifteen minutes of work when required, and to sustain a unique or innovative approach on the part of employees.

Finally, the motives of the employer can markedly influence the response to any reward system. If a worker feels that a company is genuinely seeking his personal fulfillment and happiness, output is likely to rise. On the other hand, if he suspects that rewards are offered simply to convert him into a highly productive robot, he develops a resistance to all types of incentive.

SAFETY AND SECURITY

The average worker today feels little threat to his personal safety. Adequate pay—perhaps even a guaranteed income—good hospital and medical insurance and proper provision for superannuation are now commonly the responsibility of the state. This is in marked contrast with the situation at the commencement of the industrial revolution, when the good employer cared also for the material needs of his employees. The actions of government, although necessary in our present society, have eroded an important cornerstone of good industrial relations.

In some industries, the work necessarily remains physically dangerous, but the average employee is singularly unwilling to use safety equipment such as hard hats and protective glasses—at least until a colleague is seriously injured. Perhaps the worker, conscious of protection through various insurance plans, feels too secure; perhaps also he is meeting a need for self actualization or ego fulfillment in breaking company rules. In either event, the most effective method of gaining employee cooperation is to build the safety equipment into the status system, by such devices as varying the color of hard hat with the grading of staff.

Most jobs now carry *de facto* if not *de jure* security of employment. However, if performance is flagging, it may be possible to induce anxiety in an employee—either a fear of loss of tenure, or perhaps more potent a fear of failure. The employee may be *promoted* to a post with little responsibility, and his status symbols such as a large office, carpet and secretary may be withdrawn. A vigorous young competitor may be appointed, or the employee may be excluded rather obviously from decision-making processes. Fear is commonly used to increase motivation in industry. Occasionally it may be successful, but more frequently it gives rise to wasteful interpersonal conflict or an avoidance reaction (page 209), with the development of an indrawn and ineffective staff member.

Positive rewards are much more effective (page 207), and in the context of personal safety there is scope for the employer to engender a feeling of security by clear norms of procedure, policy and reward systems; because of the greater security it offers, many people work best in a highly structured organization.

SOCIAL NEEDS OF THE WORKER

The young supervisor is often distressed by the amount of time that some employees—particularly women—devote to conversation. It is possible to avoid such *wastage* of business hours by arranging for the isolation of talkative individuals, but unfortunately this usually leads to a fall in their productivity, since their social needs are no longer met. On a production line, it seems normal for factory girls to be talking almost continuously; they apparently pay little attention to their work, but are so habituated to their task that in practice they make remarkably few errors.

The performance of male workers is also influenced by the need for group affiliation; this may express itself in acquiescence to unwise union decisions, or in an unwillingness to work harder than colleagues.

Philanthropic employers have sometimes provided a wide range of social and recreational amenities for their staff, including factory picnics, and (for the executive) membership in a local country club. Such activities may serve useful functions, but they commonly fail to improve productivity. This is partly because the linkage between path and goal is too tenuous—the worker fails to identify performance on the production line with membership in the social club; in Skinner's terms, the reward is not sufficiently immediate. Further, happiness is not necessarily equivalent to productivity, and it may well be that the type of activity provided is not a goal of the employee concerned. Alternatively, an employee may devote an excess of time and zeal to the social organizations of a company, to the detriment of more productive work.

MONEY AND EGO NEEDS

Money traditionally has been the main motivator of the industrialist. With the increasing wealth of the middle classes and promises of a guaranteed income for all, it is becoming a less effective stimulus than in earlier days. Many unions are now more interested in negotiating *fringe benefits* than in winning further large increments of salary; equally, middle-class university students search for jobs that offer fulfillment rather than income. Nevertheless, there is little doubt that with suitable manipulation, pay can still influence the performance of many workers.

In theory, money is active at most levels in Maslow's hierarchy. It can provide basic food, shelter and sex; it can offer a measure of security for ill health and old age; it can provide opportunity for ego boosting; and it can even help towards self-actualization by providing time for meditation or the means of philanthropy.

Most workers now view salary in ego terms—a larger house, a larger car or a summer cottage. If income is to be an effective reward at this level, it is plainly necessary for the salary to be both widely known and publicized as a measure of esteem. The secrecy of the average company payroll militates against such an objective, particularly as the pay of superiors tends to be underestimated, while that of subordinates is overestimated. Recognition of salary as a measure of esteem is helped if promotion is accompanied by rather obvious allocations of a larger room, a thicker carpet, a desk of more expensive wood, and a secretary of greater charm. The astute company president may even realize that such symbols are cheaper than a large increment of salary, and that the frequency of increments contributes more to the ego of an individual than the total achieved pay.

If money as such remains an important goal for an individual, it is only effective in controlling behavior when a clear relationship is established between the desired activity pattern and the reward. On the factory floor, a large but intermittent bonus (page 213) can be contingent upon gains in productivity. Likewise, the scientist in a research and development establishment may show an obstinate desire to pursue research at the expense of development, and if this behavior is unacceptable to his superiors, it must be made clear that prospects of promotion depend upon a change (or, euphemistically, a *broadening*) of his interests.

If an employee has few material wants, other techniques may be used to boost his ego. The employee may be appointed leader of a task force, given a holiday in the Bahamas, or (in eastern Europe) granted permission to attend a workers' holiday village. An increase of task difficulty may create a feeling of achievement. Among management, the ego is often boosted by an ability to manipulate others. Increasingly, the demand is for immediacy of reward, and whereas traditionally an achievement-oriented person was prepared to work for some years in order to reach his objective, the younger generation demands instant gratification. For this reason, they are interested in entering smaller companies; the ultimate rewards may also be small, but there are prospects of rising speedily to the top level of management within the chosen organization.

SELF-ACTUALIZATION

With self-actualization (Maslow, 1962), the worker passes from the external reward system of his employer to the internal reward of knowing that he has performed well. In such a situation, the need for external supervision and feedback is passed. A few fortunate individuals such as scientists and literary authors may have the type of work that lends itself to self-actualization, but unfortunately the concept is foreign to the repetitive tasks of the average factory.

Several techniques can ameliorate the lot of the production-line worker. It may be feasible to select employees of sufficiently low intelligence that they find fulfillment in tightening the number three wheel nut on a series of cars throughout a working day. Alternatively, the task may be enriched by moving away from the conveyor belt approach; the employee may perhaps tighten all the wheel nuts and test the wheels for balance, or complete some other clearly demarcated task that permits a sense of achievement. An ideal task is sufficiently difficult to permit some failures without discouraging the worker; the level of difficulty must thus take account not only of intelligence but also of personality.

Opportunities for further education and training may be provided (page 263). If the emphasis is upon self-education rather than attendance at formal lectures, this can provide a useful measure of self-actualization.

THE INDIVIDUAL AND THE ORGANIZATION

The values, needs and objectives of the individual should be appropriately matched with those of the organization (Schein, 1964). The sole interface between the average worker and his employer occurs at the level of his immediate supervisor. In Skinnerian terms, the supervisor is seen as the operator of the feedbox—the man who controls the rewards that are offered. He apparently has a major role in dispensing promotions, increases of pay and symbols of esteem (page 207).

The most effective type of supervisor has a warm, extroverted personality, with a genuine concern for the needs of those who are supervised. Such a person has the empathy to know the type of reward that a junior colleague would desire, and is able to present the available rewards in such terms. In contrast, a cold and impersonal supervisor can do no more than dispense money or punishment, failing to relate this either to performance or to the goals of the individual.

INTERGROUP RELATIONSHIPS

If an organization is to function effectively, there must be appropriate integration of the several groups represented therein. These may include office workers, scientists, engineers, shop-floor workers and sales staff, each with characteristic structures, values and goals. There is thus a need for an integrating or administrative branch of the organization, with characteristics intermediate between those of the other participating groups. To be effective, this administrative arm must not function unilaterally, but must be responsive to the outlook and demands of individual groups. Any reward or penalty system for the administrative arm should be based upon its effectiveness in serving as an integrator and in contributing to the success of the other groups.

THE ORGANIZATION AND THE ENVIRONMENT

The performance of the organization as a whole is normally regulated by market-place criteria. Those organizations that fare the best are normally those with the most appropriate structures. An organization concerned with production faces a relatively stable environment, with slow and readily recognized patterns of change. Marketing may be much less certain, and in research and development the current explosion of knowledge gives a need for very rapid change.

The degree of structure desirable in any given organization varies with the requirement for change. A production oriented organization is generally well-structured, whereas an organization concerned primarily with research and development functions best if it is unstructured, with few rules, few set channels of communication, and few orders of hierarchy. Size of necessity brings structure, and research and development is thus carried out most effectively in relatively small groups.

CHAPTER 12

THE TEACHING OF SPECIFIC SKILLS

W E HAVE ALREADY LOOKED AT CERTAIN ASPECTS of human learning (pages 174, 211). In this section, we shall apply psychological theory more specifically to the problems of teaching and training. The prime focus will be upon the transmission of industrial skills, but since school teachers and athletic coaches have great interest in this question, we shall have occasion to refer also to their experience (McGeogh, 1951).

PRACTICAL IMPORTANCE OF SKILL ACQUISITION

Techniques for the teaching of specific skills are assuming an ever-greater importance in ergonomics. At one point in human history, skilled work was the prerogative of the few, and there were many openings for the man with limited training or mental equipment as a "hewer of wood or drawer of water." However, with the passing years, the machines feared by the Luddites have taken over an increasing proportion of the functions of menial labor.

Many industrialists would claim that their automated equipment is a great success. Sometimes they have allowed salesmen of expensive toys to "blind them with science." Nevertheless, the majority of machines are less prone to sickness and carelessness, and once their initial cost has been met they make fewer outrageous demands of management than did the displaced laborers. Productivity is generally improved by automation. However, one significant and unhappy by-product of technological change is a diminishing demand for unskilled labor. Jobs are available, but only for those who can minister to machines of ever-increasing complexity. Unfortunately, human intelligence is not developing to parallel the increase in complexity of our industrial needs. Rather, it continues to conform quite obstinately to a standard, bell-shaped curve, distributed about a mean intelligence quotient of 100 (page 228). The choice of an industrialized nation thus lies between the permanent unemployment of those on the lower half of the distribution curve, and the development of more effective techniques for the teaching of industrial skills.

In purely utilitarian terms, it could be argued that industrial machinery will soon permit the more intelligent half of the community to provide for the needs of all. Some adjustment of attitudes to leisure may be needed, but until society has been restructured, there is little question that unemployment is a social evil. In addition to this humanitarian consideration, there are other more practical arguments for improved technological training. Even the intelligent and highly skilled worker may need re-education several times during his industrial career. Again, women are providing an ever-increasing proportion of the total labor force, and despite the efforts of the women's liberation movement, the majority of female employees find a frequent need to learn fresh skills consequent upon the growing mobility of their husbands' places of employment.

Traditionally, industrial training has depended upon a series of apprenticeship arrangements. The young lad watches and assists the older worker until he is judged proficient to operate in his own right. He is then certified, and can in turn begin to teach others. The views of government upon this approach to training will be discussed later (page 263), but we may note here that such a process of learning is usually slow and inefficient. Apprenticeship may last for several years, and in our current *fluid* industrial situation a worker may leave or even find his factory closed before he has a usable qualification. Furthermore, the good craftsman is commonly a poor teacher, and has great difficulty in explaining to the apprentice skills that are largely *automatic* (page 87). Finally, no single craftsman, however competent, has a perfect grasp of all the skills needed for his trade; thus, by tying the apprentice to one tutor we may be forcing him to learn an incorrect approach to some tasks. Errors, once committed to memory, are difficult to eradicate in the light of subsequent personal experience (page 248).

THE LEARNING PROCESS

Learning theory (Harlow, 1949) can be applied to a wide range of skills, extending from what seem largely *motor* tasks (for instance, the lifting of a heavy load onto a truck, and the operation of a heavily weighted pursuit meter) to problems that depend largely upon language or cognitive skills (for instance, the designing of a computer program, or the printing of an alphabet upside down).

The usual criteria of learning include speed and accuracy of performance. Depending upon the nature and quality of the product, varying combinations of speed and accuracy may be demanded of the worker (page 60). If a man is lifting boxes onto a truck, the employer will be concerned simply with the number that can be moved without damage, although the worker may find it useful to improve accuracy, moving more efficiently and at a lower energy cost. In a car factory, the emphasis is usually upon speed, but if the product is a Rolls Royce accuracy may be given greater importance. Speed is defined fairly simply as the number of units processed per hour, and this criterion leaves little scope for disagreement between management and labor. Accuracy is more contentious. It is usually treated as a discontinuous variable, and is expressed as the percentage of items meeting a predetermined inspection tolerance. In some situations (for instance, the lifting of very heavy loads) other criteria of learning must also be applied, such as the safety of the immediate operator and of other workers.

It is usual to study the learning process rather than the final acquisition of skill, since unless a task is very simple, learning continues almost indefinitely (Edholm, 1967). Crossman tested motor learning in a cigarette factory, and found that the speed of the machine operators improved for as long as two years; over this period, they had followed the machine through as many as three million operating cycles. The *learning curve*, an increase of speed and/or accuracy with repetition of a test, is very familiar to anyone who has carried out psychomotor experiments. De Jong suggested that the learning of industrial operations could be described by a formula of the type

$$t_n = t_1 \left(A + \frac{1-A}{n^k} \right)$$

where t_1 is the time required for the first cycle of the machine, t_n is the time for the nth cycle, (At_1) is the time required for operation after infinite practice, and k is an exponent governing the reduction of cycle time with practice.

Occasionally, the learning curve may show a temporary plateau, followed by a further advance. This has sometimes been attributed to a transition from one type of skill to another, but it is more likely that the learning process gains new impetus from the adoption of a specific trick or technique of operation. To quote one simple experimental example, a subject may be required to memorize six digits spoken over five seconds, and to repeat the numbers in the reversed order five to fifteen seconds later. Let us suppose that initially he attempts to reverse the numbers while he is listening. Many mistakes are made, and *learning* is slow. Suddenly, the subject realizes that there is time to memorize the number in a forward direction, leaving reversal for the moment when he must recall the information. This discovery is inevitably associated with a surge of performance.

Knowledge of the learning curve, and of the influence of design changes upon it is important ammunition to carry to the contract bargaining table. Often a small modification of a product may lead to a temporary slowing or reversal of the learning process, although ultimate productivity may be greater.

Task Structure

Any moderately complex task undergoes a progressive change of structure as learning proceeds. Consider a child learning to play what was once described in a grant application as a digitally operated pure-tone auditory stimulator. Initially, the budding musician is unfamiliar with a piano keyboard. He seeks to bring individual fingers to positions indicated by single notes on a musical score. Later, a progressive integration of response develops. The eyes learn to recognize chords rather than individual notes, and the full hand is positioned to play these chords. Visual monitoring of finger position is no longer necessary (page 87); kinesthetic information is sufficient. Subsequently, the speed of play becomes yet faster as whole phrases and sentences of music are read at a glance.

Many attempts have been made to analyze details of task structure, using such statistical techniques as principal component and factor analysis (Lawley and Maxwell, 1963). One approach examines the factors contributing to individual performance. A group of subjects carry out a battery of disparate tasks, and the variance of performance is distributed between a series of mutually independent (orthogonal) axes. Let us represent the tasks by the letters A to G. Let us further suppose that the first factor extracted by the computer analysis accounts for 32 percent of the variance in performance, and that the scores on all tasks except F are significantly correlated with this factor. We may then surmise that it corresponds with some criterion of *general performance* or willingness to do well.

The second factor accounts for a further 20 percent of variance, due mainly to the scores achieved on tasks A, C and G. Each of these three tasks involves a substantial element of vigorous physical work, so that we may surmise this second factor is serving as an index of motor power.

The third factor describes another 16 percent of variance, represented mainly by the scores for tasks B, D and F. Each of these involves elements of fine manipulation, and we may thus suggest that the third axis provides a measure of manipulative skill.

So the analysis proceeds. Further factor extraction could well identify such components of performance as cognitive skill, inter-limb coordination and speed of movement. However, as additional components are isolated, the proportion of the described variance attributable to error inevitably increases, and it becomes correspondingly difficult to attach real meaning to the factors or components that are extracted.

Having defined the basis of industrial performance in terms of a series of orthogonal variables, it is then possible to examine the need for each of these variables when carrying out specific tasks. It is also interesting to examine how task structure is modified by learning. Thus, in the example we have cited, it might emerge that experienced workers used less of factor two (brute strength) and more of factor three (manipulative skill) when carrying out task A. The proportion of the variance attributable to random factors (error) usually diminishes with learning; workers become more consistent in performance as their skill develops.

THE PHASES OF LEARNING

The Cognitive Phase

In this initial phase, the worker must be oriented with respect to his job. Even if the ultimate task has a predominantly motor character, the speed and success of the cognitive stage depends heavily upon non-motor factors such as an underlying knowledge of mechanical principles. To take an extreme example, if an African tribesman is suddenly removed from a small village to a large city and required to work in a modern factory, much patience may be needed to teach the operation of even the simplest machine. On the other hand, a man who has worked in a wide variety of factories for thirty years may need minimal instruction. Training films (page 259) and simulators (page 260) can markedly speed the learning process at this stage.

The Phase of Fixation

In the second phase of learning, the worker identifies the essential stimuli and their interrelationships, and develops appropriate responses. The car or truck driver, for instance, tunes his ear to engine rhythm, and learns appropriate adjustments of throttle (and in older models clutch and choke).

The Phase of Automation

In the final phase of learning, movements are no longer under direct conscious control. A vehicle driver with twenty years of experience makes few conscious decisions. Indeed, if he is familiar with the route he is driving, he may have surprisingly little subsequent recollection of a journey. At this stage, performance is limited largely by the speed with which the muscles can respond, and further improvements in performance are slight. Nevertheless, careful measurement shows that an absolute plateau of learning is rarely reached.

The ease with which a skill is lost or forgotten depends upon the stage of learning that has been attained (page 87). Recently developed skills quickly deteriorate in an adverse environment (page 65), but once automation is achieved, the rate of loss of skill is very slow. Adults of forty or fifty find little psychomotor difficulty in joining their children on a bicycle ride, although they may not have mounted a cycle for twenty years. Again, many car drivers were deprived of petrol (gasoline) for five years during World War II, and yet found little loss of driving skill when hostilities were over.

The protracted course of learning has an important bearing upon the optimum age of performance. Sports that require a large physical effort are performed best in the early twenties, when oxygen intake is at its zenith. However, most industrial operations do not tax either oxygen transport or muscular power too severely, and the output of a skilled middle-age worker may then surpass that of someone who is younger and less experienced. Even if some task has a substantial physical component, the greater skill of an older person may largely compensate for the age-related deterioration in physiological power.

CAPACITY TO LEARN

Individuals vary markedly in their capacity to learn technical skills. Much of this variation seems related to the fundamental ability of the worker—factors such as his intelligence, the quality of his sensory equipment, the anatomical and physiological characteristics of his motor system, and the ability to process, store and retrieve information within the brain.

Intelligence

The intelligence of the individual has an obvious bearing upon the rate of learning, particularly in the cognitive phase. The more complex the task to be mastered, the more crucial the intelligence level of the operator becomes. However, providing the brain is not overloaded, and learning proceeds through a sufficient series of stages (pages 207 and 211), involved procedures can be taught to people of quite limited intelligence. It is certainly rash to conclude from a single trial that some complex task is beyond a worker's mental capacity. Let us imagine a laboratory situation where a subject is suddenly confronted with a rather complex discriminative reaction time task. Initially, he is unable to complete the assignment within the allotted period, and becomes flustered, showing a very poor performance. However, if the task is attempted repeatedly, it gradually falls within his capacity, and may ultimately be carried out quite readily. It is not unknown for the eventual reaction time on such a task to decrease to the point where it is very close to that of a simple reaction test. The analogy to a conveyor-belt situation is obvious; again, if the initial speed is too fast, performance is correspondingly poor; but if the worker is encouraged, possibly by working on a slower assembly line, he will soon develop skill to the point where he can function without fluster on the main conveyor belt.

The type of intelligence needed for learning varies with the structure of the task (page 226). The *general intelligence* factor (page 229) is of particular value in operations calling for a high degree of cognitive skill. Other components of intelligence have relevance to specific tasks; thus, the computer programmer needs an ability for abstract, mathematical logic, while a typist trying to decipher almost illegible handwriting finds pattern recognition a distinct asset.

Sensory Equipment

The quality of an individual's sensory equipment inevitably influences the rate of learning. This is well recognized in the provision of special schools for the deaf, dumb and blind (page 293). Again, a driving trainee who has lost the sight in one eye is going to be much slower in mastering operation of a vehicle than a person with normal vision (page 159). But setting aside such gross abnormalities of the special senses, apparently healthy individuals still differ appreciably in the sensitivity of their receptor organs. The champion pistol shooter is likely to have an unusually good retinal discriminatory capacity, the opera singer will have unusual powers of auditory discrimination, the figure skater very sensitive proprioceptors, and the tea taster a particularly fine capacity for taste discrimination.

Anatomical and Physiological Equipment

The anatomical and physiological characteristics of the individual play a key role in learning certain mechanical tasks. Thus, a short person will have difficulty in becoming an expert basketball player, because he is unable to guide the ball through more than a short part of its trajectory. Conversely, a tall, long-legged individual is unlikely to develop into a good gymnast, because his center of gravity (page 169) is too highly placed.

Many industrial tasks call for a person within quite a narrow size range. Some compensation is possible by altering seat or bench height (page 190), but a substantial departure from intended size leads to an awkward posture, with attendant clumsiness, fatigue and alteration of learning patterns.

Muscular power is needed by the sprint runner and by the short-haul delivery man; in the latter situation a man who lacks the necessary strength will struggle ineffectively with the load to be carried, and will never master the correct technique of lifting (page 185). On the other hand, the endurance runner may find his performance is hampered by excessive muscular development; he requires rather a well-developed cardiorespiratory power. Industrial tasks that tax the cardiorespiratory system are rare in Western society. Examples are found in the occasional delivery tricycle and in certain mining operations.

As with short-term loads, a person who lacks the necessary physiological endowment will struggle excessively, and in consequence will fail to learn an efficient pattern of working.

Channel Capacity

The general problem of channel capacity is discussed elsewhere (page 85). In essence, the human brain functions as though a single channel is available to the learning process. The fundamental rate at which information can be passed through this channel probably varies from one person to another, and the effective rate certainly differs both in a given individual and between individuals according to the proportion of time that is allocated to the processing of irrelevant information. We have commented already on the problems that may face a car driver who processes information on skirt length at an injudicious moment. The impact upon school and university learning of a classroom where many miniskirts are raised to support notepads provides an interesting topic for further enquiry!

Storage and Retrieval

Even if the channel selector passes the information to the brain successfully, individuals still differ in their capacity to store and retrieve data. We have discussed concepts of short- and long-term memory (page 86–87). Some people have difficulty in learning a task because the information fades too quickly from their short-term store. In essence, they have the problem of a leaky bucket. Others have difficulty in retrieving information from the long-term store. In essence, they have the problem of a sticky bucket.

The functioning of the short-term store depends upon electrochemical processes that maintain reverberation of signals within a nerve network. The long-term stores probably have a more permanent structural basis. Swelling of active nerve terminals occurs during learning, and there may be an increased branching of nerve twigs at the synapse, a change that would link memory with the synthesis of new protein. Possibly, there is not only an increase in the quantity of protein, but also some subtle change in its chemical properties.

The existence of two types of memory storage has some practical value. As Peterson (1966) points out, it would be unpleasant if not disastrous to remember all of the telephone numbers one had ever had occasion to call. However, the separation of the two stores carries the penalty that occasionally wanted material is forgotten before it is committed to the permanent store.

Increasing knowledge of the chemical basis of memory has opened up the fascinating possibility of modifying the rate of learning by administration of suitable drugs. Prolonged treatment with antibiotics apparently hampers the transfer of information from the short-term to the long-term store. On the other hand, the cerebral stimulant Metrazol® (pentylenetetrazol, Leptazol®) facilitates such a movement; the latter compound is well-known as an excitatory mediator at the neural synapses, and in the present context may facilitate the synthesis of new protein at the nerve junctions.

Fortunately for university professors, chemical tuition is still experimental, and there is as yet no effective substitute for a good teacher. Effective storage and retrieval of information still depends largely upon the association of ideas, conscious or subconscious. One important role of the good teacher is to bring many diverse concepts within a broad, unifying framework (Murdóck, 1963). Sometimes, even bizarre associations are helpful to the memory. Consider the word list below:

> GUN
> ABACUS
> ZOO
> PROFESSOR
> FLEA
> NEEDLE

Memorization of such a series might take a long time, unless one noticed that alternate words rhymed with the digits 1,2,3. By creating visual images of an abacus leaning against a gun, a professor caged in a zoo, and a needle inserted into a flea, the series is soon committed to memory. Indeed by supplying personal rhymes for the bizarre imagery, the more dissonant list—ABACUS, PROFESSOR, NEEDLE—can also be learned quickly.

Some pupils have poorly developed powers of association, and for this reason find considerable difficulty in learning. When the problem involves the linking of cognition with motor activity, a specific class of *perceptual-motor handicap* is recognized. If detected at an early stage, the pupils concerned respond well to modified teaching procedures.

THE TRANSFER PROCESS

Past experience normally speeds the learning process; skills acquired on one machine can be transferred to the operation of another, similar piece of equipment. This is described as *positive transfer*. However, in some instances, past experience can hamper the learning of a new skill; *negative transfer* has then occurred. I experienced the latter phenomenon in a particularly humiliating form at the age of seven. Already, I was a competent cyclist, and was rather scornful when my young sister was given a gleaming tricycle for her birthday. Finally, I condescended to ride the machine, only to tip it over on rounding the first bend.

Where a novel response is required in a novel situation, *zero transfer* is likely.

Many athletes are cautious about engaging in alternative sports between seasons, for fear that negative transfer may develop. Let us suppose that Smith plans to engage in competitive tennis during the summer months, and makes the mistake of preparing himself by frequent winter games of badminton. Although the stimuli are similar for the two games, very different response patterns are required. Smith thus learns tennis more slowly than if he had kept away from the badminton court. *Pro-active inhibition* of learning is said to have occurred. When the next winter arrives, the unhappy Smith returns to the badminton court, only to find that a summer of tennis playing has caused a *retroactive inhibition* of his painstakingly acquired skills as a badminton player. On the industrial scene, the same type of problem can be envisaged in countries such as Canada, where men follow different occupations in the summer and winter months. Problems may also arise from changes in design. This point became obvious when a modification of typewriter keyboards was planned. The original QWERTYUIOP arrangement of lettering was chosen on mechanical grounds, and much more efficient arrangements could be suggested for electrically powered equipment. Nevertheless, the potential new designs remain unadopted because of the tremendous difficulties in transferring the skills of typists accustomed to traditional machines.

In general, interference is likely if a new response is needed to the same stimulus, but positive transfer occurs if an identical or closely similar response is required. If a training program is based on the teaching of *splinter* skills (page 255), it should thus contain similar response elements to the ultimately required pattern. Transfer is most likely to be effective if trainees understand the relationship of the splinter element to their ultimate goal, and the aim of the instructor should be to build as many cognitive bridges as possible between the individual elements that are to be taught.

Unfortunately, the direction of transfer is not always consistent. What appears a successful positive transfer may become negative in an emergency or other adverse environmental conditions. One good example of this is provided by the British driver in North America. For many years, he has learned the skills needed to operate his vehicle on the left side of the road. When he commences to drive in the United States, there is apparently a very successful transfer of accumulated experience. Nevertheless, when confronted by an emergency, there is a risk that the Britisher will swing his wheel in the wrong direction. The increasing interchange of vehicles associated with the entry of the United Kingdom into the Common Market may soon force a reappraisal of the British *keep left* policy.

ATTITUDE AND MOTIVATION

Any experienced teacher is aware of the importance of motivation to learning. The student who is bored by a lecturer's manner or is unconvinced of the relevance of the material presented is unlikely to learn. The same concept applies to physical skills. Lavery (1964) demonstrated that when a group of subjects was told they would later perform a simple tossing task, the skill was retained much better than if no instructions were forthcoming. Apparent loss of relevance easily occurs if a skill is broken down into its elements (page 255), and if this plan of instruction is adopted, it is important that the learner understands how the *splinter* skills fit into the total task.

Teaching, industrial training, athletic coaching—each is an art as much as a science. Whether instruction is taking place in the university classroom, on the factory floor or on the athletic field, the attitude of the person learning the new skill is crucial. Industrial apprenticeship and retraining programs sponsored by government (page 263) are doomed to failure if the individuals supported by the taxpayer lack the will to establish themselves in a new form of employment.

How may attitudes be changed? In some ways, a willingness to learn is an analogous desire to a willingness to become physically fit or to give up smoking. The average citizen can be provided with large quantities of *information* on any of these topics, yet a change of attitude rarely results. A negative attitude, whether to physical fitness or to future employment is unrelated to the information that is available; the problem is one of *affect*—the emotional, *gut* response to the issue. A negative affect is commonly a response to previous harm or failure. The immovably sedentary middle-aged adult may have had an unfortunate experience of injury or lack of success in athletics as a child, and in a similar way the reluctant trainee may have suffered physical injury or the humiliation of failure in his former employment. In both cases, any *predisposition to action* is effectively countered by the problem of attitude. The affect of the sedentary *slob* cannot be altered by further reading on the virtues of fitness, nor can the affect of an embittered trainee be improved by stern lectures on the qualities of honesty and hard work. *Gut* attitudes must be corrected.

Empathy is an important component of the instructional situation. Truax and Carkhuff (1967) have drawn attention to what they term the "Principle of Reciprocal Affect." It is difficult to adopt a consistently negative attitude towards a persistently warm, positive and empathetic teacher. Faced with such tuition, even the most negative of trainees may undergo a radical change of heart, to the point where he becomes responsive to a suitable input of information.

Intelligence undoubtedly modifies the outward expressions of affect, but nevertheless the willingness to learn of many surprisingly intelligent people is governed by *visceral* rather than *intellectual* or *factual* decisions.

Industrial organizations attempt to improve attitudes by inculcating a *team spirit* into trainees at an early stage of instruction. Undoubtedly, this is generally advantageous to learning and performance. However, much depends upon the needs of the individual. If he hopes to gain *from* the group affiliation rather than contribute *to* it, performance may actually deteriorate in the team situation (Cratty and Sage, 1964). In experimental group learning, these authors noted that well-established, social-type groups performed poorly, wasting much time in social chatter, whereas previously unstructured groups addressed themselves purposefully to the task in hand (Cratty and Sage, 1964; Cratty, 1967). The problem seemed in essence that the lines of communication and the leadership established by social interaction were not particularly appropriate to the learning of a new motor skill.

FEEDBACK AND REINFORCEMENT

Many of the standard techniques of reinforcement (page 207) can be used to speed the learning process. Feedback through provision of *knowledge of results* is important, and indeed essential information for a purposeful change of behavior (Fig. 77). A university student

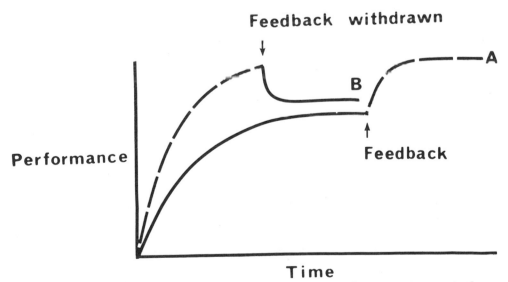

Figure 77. The influence of knowledge of results upon performance. In case A, there is a sudden jump of performance when feedback is provided. In case B, there is an equally sudden deterioration of performance when feedback is withdrawn.

may be blissfully unaware that he is learning from an out-dated textbook or is spending too little time upon the learning process until he obtains a mark of 35 in his midterm test. Likewise, the new employee at a factory may not realize he is passing too many defective widgets on an inspection line until an experienced supervisor points out the faulty items to him. The feedback not only corrects errors, but also serves to reinforce previously conditioned behavior. Returning to the students, one may find *positive reinforcement:* "Because I am learning, I will earn another beautiful A +." Another may find *negative reinforcement:* "Unless I learn more effectively, I will get another lousy C−." Although learning can occur without such reinforcement, nevertheless experience shows that the feedback of marks speeds the process of instruction.

While some individuals find positive reinforcement from the satisfactory completion of an assignment, others learn best if their efforts are cut short at any one session; a *need tension* then builds up, an overpowering desire to master the particular skill or technique (Ziegarnik, 1927).

The choice of feedback depends upon the characteristics of the individual and his attitude towards learning (page 233). Knowledge of results provides a yardstick of progress, and is a powerful stimulus to a goal-oriented personality. On the other hand, it may have almost no impact upon a ghetto Negro (Fitts and Posner, 1967). If performance is to be used as the reinforcing stimulus, it is most effective when the individual has a 50 percent chance of achieving the established goal (Atkinson, 1957). This is unlikely if a trainee is urged to compare his output with that of a mythological worker-hero (as in some communist states). The most effective challenge is a small increment upon the individual's personal performance (White, 1959). The objection to self-pacing is that little incentive is given to the *lazy* person; however, most people seem to pace themselves as learning proceeds, in the sense that aspirations are increased by success (Locke and Bryan, 1966). The alternative of striving to emulate the worker-hero implies that all reinforcement is withheld until the task has been mastered; learning is unlikely to proceed smoothly under such circumstances (Skinner, 1953).

The nature and amount of reward provided in any schedule of reinforcement has more effect upon performance than upon learning. This implies that a simple observation of current performance does not always reveal the full extent of learning. Let us suppose that a worker has been employed at a fixed hourly rate for some years. One week, he is shifted to piecework. There is an immediate and dramatic gain of output, not because of new learning, but rather because the extent of existing learning is revealed. The opposite phenomenon is seen in the worker who has learned very rapidly under a piecework system, and suddenly finds this incentive withdrawn.

Psychological theory suggests that a reinforcing stimulus is most effective if applied immediately (page 212). However, in some learning situations, an immediate knowledge of results is impossible. Examinations, for instance, require a substantial period to mark by hand, and sometimes even longer to mark by computer. Nevertheless, delayed feedback can be quite effective, particularly if care is taken to emphasize its relevance to the initial learning experience.

OPTIMUM PATTERNS OF LEARNING

Practice of a given skill may be either continuous or discontinuous. In general, distributed practice provides a more effective use of available training time than sustained or massed practice (Fig. 78). The larger the motor component of an activity, the greater the discrepancy between massed and distributed practice. If no rest pauses are allowed, fatigue inevitably leads to poorly coordinated, uneconomical movements and even frank errors of performance. As a consequence of frequent repetition, these errors become incorporated into the individual's interpretation of the task, and are subsequently difficult to eradicate.

Over a prolonged training program, the difference between distributed and massed practice becomes smaller, since both groups have a respite during the hours of leisure and sleep. Such intervals give an individual opportunity to recover from boredom and fatigue, physical and psychological. The mind also *reminisces*, going back over the task to review errors of technique and possible new approaches. Thus, when massed practice is resumed after a night's rest, the learning curve may rise to approach that for distributed practice (Fig. 78).

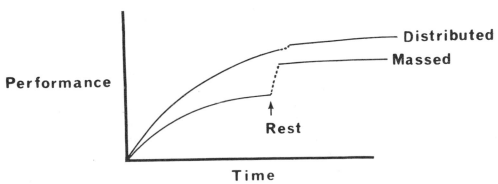

Figure 78. A comparison of learning curves for distributed and massed practice. Note that the difference between the two approaches is largely dissipated by rest (reminiscence).

Reminiscence is rather analogous to the mental practice that Cratty (1967) has urged in the teaching of motor skills. Excessive mental practice is unlikely to help the acquisition of a skill, but some opportunity to ponder over the task is preferable to relentless physical activity.

Reinforcement can occur during rest periods. However, if injudiciously applied, it may reinforce not learning but rather the tendency to forget already inherent in massed practice. Such *negative learning* is not simply a matter of muscular fatigue. It can be transferred from one hand to the other, and can even arise from watching others perform the same type of activity. Nevertheless, its main basis is probably incorrect performance secondary to fatigue.

Immediately following a rest period, performance may be markedly impaired relative to a previous *plateau* of learning. The psychomotor component of *warm-up* is well recognized by sportsmen. Recent practice of an athletic skill provides much more effective preparation for a contest than other methods of attaining comparable body warming. If the phenomenon is formally investigated, it is found that the loss of performance becomes smaller during successive rest intervals, and the rate of return to the initial learning curve also increases (Fig. 79).

ORGANIZATION OF THE LEARNING PROCESS

The key to learning many tasks, both mechanical and mental, is the development of an appropriate plan of attack. Consider a man who is required to memorize a series of nonsense syllables of the type BLX, SRY, TZU . . . . If ten such syllables are presented, learning may occupy from three to twenty trials in different individuals. On questioning, a variety of learning techniques will be found. One man may have grouped the syllables into five sets of six letters. Another may have placed a sexual connotation upon each of the nonsense

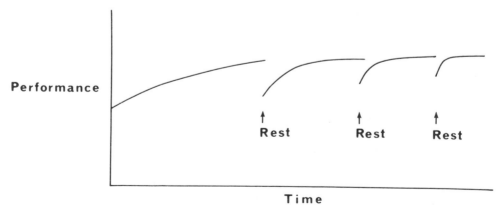

Figure 79. The initial decrement of performance following a rest period.

phrases. A third may have sought to establish a visual image of the series; the use of visual imagery is particularly common in children. Mnemonics are a favorite trick with some students. Thus, the initial letters of the essential amino acids are recalled as "Very Many Hairy Little Pigs Live In The Torrid Argentine." Memory is here helped by rhythm and by some similarities of pattern (for instance, between the familiar word Argentine, and the less familiar arginine).

When material from a book is being memorized, the best plan is to read the entire chapter or even the entire volume, and then to break it down into parts of a size that can be processed by the brain. It is difficult for the short-term memory to handle more than seven items at a time (page 86; Posner, 1963; Melton, 1963; Postman, 1964). Thus, when teaching the alphabet to small children, it is commonly presented as a rhythmic song of the type

ABC, DEFG
HIJK, LM
NOPQ, RSTU
VW, XYZ

In the same way, an adult confronted with a binary computer printout such as 10100011001110 (page 331) must break this information into groupings that can be handled by the brain, either

$$10, 1000, 1100, 1110$$

or

$$10, 10, 00, \underline{} 11, 00, 11, 10$$

The need to split a motor skill into its elements depends upon its complexity. If the task lies within an individual's capacity, then the fastest approach to learning is a practice of the whole (Cratty, 1967). However, the short-term memory is easily overwhelmed by a complex motor problem. The optimum approach then seems to divide the task into several elements, with the objective of subsequent recombination. Formal psychological evidence on the value of splinter-skill teaching is limited, but the procedure has been used in military training, on the factory floor, and in the correction of faulty athletic techniques. Standard work-study methods (page 175) and movement analysis pinpoint the essential components of a task. Training is concentrated on the most difficult elements, and when these have been mastered, the overall task is reassembled. Sometimes, assembly can proceed in a discontinuous fashion, but owing to the need for transfer of the partial skills (page 248), shaping is most rapidly achieved by a gradual metamorphosis of the task (page 211). There is much to commend the idea of placing the short, imperceptible steps of such a learning process under the control of the learner (as in programmed learning, Kay *et al.*, 1968). Senior teachers then concentrate their experience upon the design of learning systems, and the functions of the personal tutor become (1) to focus attention upon desirable elements of current performance, (2) to add new elements of the required skill as soon as practicable, and (3) to minimize the risks inherent in an exploratory approach to learning (Bilodeau and Bilodeau, 1961).

Belbin and his associates (1957) applied the task-splintering approach to teaching the craft of textile repair. Unfinished cloth commonly has missing threads that destroy its value; the missing threads are replaced by a small group of highly skilled workers. Traditional apprenticeship schemes were rather ineffective in teaching this craft. Only a few workers seemed capable of identifying the different weaves of cloth and of visualizing the appropriate manner to insert the missing threads. The new training plan proceeded through three stages:

1. the trainees learned the different weaves of cloth from large scale models,
2. they were given opportunity to identify and correct errors in such models, and
3. the scale of the models was progressively reduced to that of standard bolts of cloth.

Employees trained by the new approach were found to work faster, more accurately and more consistently than regular apprentices, and furthermore the required training time was much reduced.

In the textile factory, it was relatively easy to visualize the necessary skills, but in many forms of employment the perceptual clues are subtle and poorly documented. A wine taster knows a good wine, but would find it extremely hard to set down the criteria of judgment that he is applying. The same is true of the tea blender. However, with persistence, the clues used by the skilled worker can be identified. Thus, the *notes* emitted by vacuum cleaner motors are used to diagnose faults of assembly, and tape recordings or even a frequency display of undesirable noises might be a useful teaching device for a sound-oriented skill. Again, in cheese factories, reliance is placed upon texture, and standard rubber cheeses of approved firmness can be used to instruct the trainee (Edholm, 1967).

THE EMOTIONS AND LEARNING

We have already commented upon affect and its influence upon the learning curve (page 250). However, the negative effect of excessive anxiety deserves further emphasis. As with ultimate performance (page 215), there is an inverted U-shaped relationship between the level of arousal and the rate of learning. The source of anxiety may be the risk of personal injury, but equally there may be a fear of damaging some expensive piece of machinery.

Anxiety levels are particularly high during the early stages of mastering a complex skill. The individual concerned may have a substantial personal stake in the outcome of the learning experience; completion of a probationary period, promotion or summary dismissal may be consequences of various levels of performance. The clever instructor manipulates arousal levels to his advantage. If the trainee seems tense and anxious, all instructions are given in a quiet, reassuring, deliberate and nonthreatening manner. Sometimes the nervous apprentice is helped by applying *positive* labels to his feelings—"excitement," "desire to do well," "anticipation"—or the appropriate use of light humor. If further arousal is judged necessary, the manner changes to include ridicule, threats and the adoption of a formal, demanding manner. Excessive use of arousing techniques is generally undesirable, and may lead to avoidance reactions (page 210).

The interaction of emotions and learning varies according to the personality, age and expectations of the individual. Let us suppose that two middle aged men decide to take up skiing. Both are passed on the slopes by a host of young and athletic-looking skiers, and both fall ignominiously in the snow. One of the two men had some pretensions as an athlete when he was younger, and is humiliated by his apparent lack of success. The other was never a winner at sports. He thus laughs at his misfortune, and moves to a more gentle slope. Here he learns to ski quite happily, and eventually becomes reasonably proficient.

Interest is currently focussed on methods of reducing anxiety in order to improve both learning and ultimate performance. Sometimes, no more than simple reassurance is needed. Students attending an examination are told that slight anxiety is common to the group, and that such anxiety can improve performance by boosting synaptic transmission and retrieval; reassurance of this type is apparently sufficient to produce a measurable increase in examination scores. Jacobson (1967) has suggested the value of more specific relaxation exercises. He has reasoned that a person cannot remain tense and deliberately relax his muscles. Relaxation can be encouraged by allowing the subject to watch electrical records of muscle activity. Once the technique has been mastered, stressful stimuli are presented simultaneously with relaxation exercises, and the subject is *desensitized*. Golf is one sport where performance readily deteriorates with excessive muscular tension, and a number of professional golfers have claimed benefit from relaxation techniques. In other activities calling for unusual strength, performance may be boosted by a modest increase of tension.

Relaxation is of value in the board room, and indeed a reduction of blood pressure may spare the lives of executives trained at considerable cost to their employees.

Hyperactivity is a common obstacle to learning at all ages, but especially in children. Previously, the search has been for a suitable tranquilizer, or a drug that would activate the inhibitory circuits of the brain. However, the use of relaxation techniques may prove a more satisfactory method of treating such children.

THE INFLUENCE OF ENVIRONMENT UPON LEARNING

Academics make much of the cultural environment that precedes learning. As with anxiety, the problem is one of creating optimal arousal. The student must be presented with adequate stimuli, but not to the point where he becomes distracted, overloaded or anxious.

A comfortable environment usually facilitates learning, but a warm fire and a soft chair may send the student to sleep. If an adverse environment is presented suddenly, the deterioration of performance is larger in a recent trainee than in a worker who is thoroughly familiar with his job. The latter finds the task so simple that under normal conditions he has a wide margin of spare capacity (page 87); this he calls upon in his adversity.

The challenge of environment has some relevance to design. Let us suppose that aircraft A is more difficult to fly than its competitor, aircraft B. The makers of aircraft A explain that with a somewhat longer course of training, pilots can reach an accepted standard of efficiency. However, they fail to point out the smaller margin of spare capacity. In an emergency situation, aircraft B remains within the competence of the average pilot, but aircraft A becomes an extremely dangerous machine.

TRAINABILITY

Industry would welcome some simple yardstick that would predict the likely success of entrants to a training program. However, in view of the many variables that influence the learning curve, it is hardly surprising that no practical method is yet available.

The problem is familiar enough to those concerned with university selection. Here, criteria of *general intelligence* and attitudes to learning at school have some theoretical relevance, but offer little practical advantage over such simpler tests as "Does father know the Dean?"

or "Can the boy play football?" In industry, the problem is made more complex by the absence of any *general motor skill;* most mechanical aptitudes are highly task specific. Further, current performance bears no necessary relationship to ultimate skill. Let us suppose we wish to select a series of trainees for driver instruction, those with reasonable immediate competence can be picked out using a vehicle simulator, or better by having them drive a truck around a test range. But such as assessment leaves unanswered the crucial question: which men will improve with training, and which will remain only moderately competent?

TEACHING DEVICES

Nowhere is there a greater need for ergonomics than in education. Professors who criticize the construction industry for using fifty-year-old methods themselves persist with approaches hallowed by five hundred years of university teaching. Some economists have warned that if the current trend of diminishing productivity in schools and universities remains unchecked, the costs of education will soon outstrip the gross national product (Shephard, 1972).

Educational Films

There is certainly little justification for the professor who solemnly stands before a class and reads a set of notes he prepared some twenty years previously. If there is any need for such a stilted performance, it could be met more economically by a film or a television screen cassette.

However, the majority of educational films have as yet offered no serious challenge to the good lecturer. Films are expensive to prepare, and thus must be used in a number of centers for a number of years. Relevance to a specific course is inevitably lost, and in a rapidly changing world the dated nature of the presentation becomes painfully obvious. The author recalls one film on aviation medicine, shown in 1969, that revealed gleaming "Super-Constellation" aircraft and stewardesses proudly outfitted in the *new look* of the early 1950's. The speed and depth of a film presentation may not suit the background of a specific class. The good lecturer is sensitive to this problem, and adjusts his talk accordingly; however, there is no such feedback between the cinematograph and the student, and if the film topic is unpopular, the machine may continue to grind away, blissfully unaware that the occupants of the lecture theater have either departed, or (if attendance is compulsory) are engaged in the grossest forms of horseplay. Some of these problems can be overcome if individual students are provided with personal projection booths. They can then select films of interest to them, and in the case of filmstrip material they can regulate the speed of projection. The script accompanying a filmstrip could also be programmed to allow for individual differences in knowledge of a subject, although this is not usually done in practice.

Despite such strictures and the recent public reaction against purchase of expensive and ineffective gadgets, there is probably a place for the film and television in a university. Events that cannot be seen around a crowded demonstration table can be projected onto a large screen. Distant parts of the world can be visualized. Difficult concepts can be explained by animation. The need for staffing of routine introductory lectures can be reduced. Highly specialized graduate topics can be taught on a province-wide, or even a nation-wide basis. And such tuition can proceed not merely over the leisurely ten to twelve, two to four timing of many professors, but on a twenty-four hour basis, thus making much better utilization of expensive university facilities.

Programmed Learning

The majority of readers will have encountered a number of programmed textbooks. A few are well designed, but many take a large quantity of paper to convey remarkably few facts. The unfortunate student has to buy not only a full account of material that may interest him, but also sections devoted to unraveling problems that never bothered him or were resolved some months previously. The remedy here seems to reserve programmed texts to limited editions, borrowed by the student while he is learning; they are an inappropriate permanent repository of facts.

The possibility of programmed filmstrips has been noted above. A variety of more sophisticated teaching devices are in the experimental stage. Some permit the student to enter into a rather stilted dialogue with a teleprinter or a slide projector. In this way, he can begin to learn skills such as diagnosis. The first slide describes the appearance of a patient, and perhaps invites the student to ask one of five questions. The answers are provided on the next five slides, each leading to an array of five further possible questions. Eventually, the student reaches a diagnosis and prescribes a treatment. The appropriateness of the diagnosis and treatment are rated by the teleprinter, together with comments on the time taken to reach the diagnosis, and the expense to the patient of any investigations that are proposed. In this way, logical and economical schemes of diagnosis may be developed (page 353).

Use of Simulators

In view of the increasing complexity of modern machinery, the ever-increasing costs of repairs, and personal considerations such as employee safety and customer satisfaction, increasing use is being made of simulators not only in medicine but in many branches of commerce and industry.

One example is the manikins used to teach artificial respiration, cardiac massage and venipuncture to medical students and nurses (Shephard, 1966). In aviation, the *Link Trainer* is well known as a device that can reproduce many features of normal and emergency take-off, flight and landing, including movement, noise and vibration. Even more sophisticated simulators have prepared space explorers for their task, while on a humbler scale many airline passengers have escaped spilt soup and coffee through the cabin mock-ups used in the training of stewardesses. Driver-training centers now use quite elaborate vehicle simulators, and further supplement such experience by tuition on a driving range before the vehicle and instructor are exposed to the hazards of real traffic. In commerce and in the armed services, many techniques of battle are now learned through *war games* (page 348).

The value of these various types of simulator depends largely upon the ability of the designer to identify critical skills and to incorporate them into his model. The respiratory manikins used in teaching first aid can be quite glamorous, yet give the rescuer an unrealistic impression of airway resistance and lung compliance in a drowning man (Shephard, 1966). Thus the first aid worker is taught to blow too vigorously or too gently. The skin texture is not comparable with real life, and such important details as sealing the lips around the victim's mouth and pinching his nose are badly learned. The forces needed to manipulate the jaw and head are poorly indicated. And finally, the normal feedback of breath sounds and chest movement is distorted or lacking. Such defects may be acceptable if no real-life alternative is available, although there is as yet no clear evidence that first aid workers trained on a manikin are particularly successful in transferring this skill to the resuscitation of a casualty. Certainly, the weakness of existing models presents a continuing challenge to design.

The vehicle-driving simulator apparently cuts down the teaching time for naive trainees, and can give more experienced drivers useful instruction in the handling of emergencies such as a tire blow-out or brake failure in heavy traffic. Specific errors of technique, such as failure to use turn signals or excessive braking can also be identified and corrected without danger to the pupil or other road users. Nevertheless, there is a considerable problem of transfer (page 248) between the simulator and a vehicle, since such normal channels of feedback as engine noise, movement of the car relative to the road and vehicle sway are lacking.

Most simulators can be criticized on the grounds that their feedback of information is incorrect. Normal elements are lacking, but other aspects are exaggerated. Thus, in the University of Toronto medical teaching machine, the normal appearance of the patient (which must be interpreted by the clinician) is reduced to a series of didactic statements of facial color, breathing rate and the like. Other aspects of the machine—its relentless forced questioning of the student and the final cost/benefit analysis of treatment—provide more feedback than the student will encounter in normal practice.

TEACHING OLDER WORKERS

Most western nations have an aging society, so that maintenance of productivity in the older worker has considerable economic importance. In the future, most employees will need retraining on several occasions prior to retirement. Machines are changing continually, and not only in the factory. Trams (streetcars) have been replaced by trolley buses, then by petrol buses, and finally by diesel buses; quite a proportion of the drivers have remained with transit authorities throughout these changes. Propeller-driven aircraft have been replaced by turbo-prop machines, and these in turn have given place to jetliners, super-sonic and vertical take-off planes. Ferries are being replaced by hover-craft. Winding lanes with a 30 mph speed limit are being replaced by 70 mph expressways, and the elderly driver may be just adjusting to this situation when he finds the speed limit signs are changed from 70 to 105. And on the domestic scene, the geyser and mangle of the laundry room have been replaced by spin dryers and heaters. Man must adapt continually to a bewildering array of new machines at work, in travel and at home.

The ideas of an older person become rigid, and short-term memory is often impaired. It is less clear whether these changes represent true physiological aging or merely disuse; certainly, memory is better preserved in those old people who have occasion to use it, and loss of orientation is often rapid under the passive conditions of the average nursing home. Problems of an aging memory make verbal instruction unreliable as a teaching method. Older workers also lack both the visual acuity and the patience to read closely-printed instructional handbooks, and the best approach to learning seems a course of appropriately graded on-site activity (as in Belbin's Weavers, page 256).

Many North American firms are anxious to fire and reluctant to hire older workers. Nevertheless, the older age groups have some positive attributes, including a wealth of experience and in many respects a great emotional stability. Given the proper setting, an older person can continue to give useful service at least to the time of normal retirement.

THE ROLE OF GOVERNMENT

Currently, government is taking a long and hard look at the processes of academic learning. However, the present section will consider more specifically the role of government in industrial training, including both formal apprenticeship and other methods for the development of skilled craftsmen (Warren, 1968). The concerns of government are for a balanced, occupationally mobile work force, adequate for present and future needs, with at the same time opportunity for all to make their potential contribution to society.

In the past, the need for trained workers was met fairly adequately by formal or informal apprenticeship ("sitting with Nellie") and—particularly in North America—by immigration. However, the appropriateness of this approach is being questioned increasingly in our fast-changing technological society.

Apprenticeship

The problems inherent in traditional apprenticeship are similar in most Western nations, but will be illustrated by specific reference to Canadian experience.

The first apprenticeship legislation was introduced in 1928. The rate of entry to specified trades was regulated, and appropriate scales of payment for apprentices were defined. Systematic courses of instruction were prescribed, and the ratio of trainees to established journeymen was laid down. Additional trades were later listed in the interests of public safety. Motor vehicle mechanics were licensed in 1936, and subsequently hairdressers also were required to undergo certification. There are currently nine trades in Ontario where work is permissible only as an apprentice or as a duly certified journeyman. Contravention of regulations is punishable by a fine of up to $1000.

A detailed review of apprenticeship arrangements was begun in 1964 at the urging of both labor and management. In general, the building trades favored retention of apprenticeship, and indeed were exerting pressure towards the compulsory certification of electricians, plumbers, sheet metal workers, steam fitters, and air conditioning and refrigeration engineers. On the other hand, industrial craftsmen affiliated with the Congress of Industrial Organizations were opposed to rigid classification. They pointed out that there was no possible definition of "trade" other than the operational criterion "a group of skills over which some body had established jurisdiction." Partly because of the slow pace of building technology, there was little difficulty in classifying construction workers, but many of the newer industries' workers used parts of traditional skills, and in any particular factory occupational groupings were determined largely by technological history.

The interim report of a governmental *General Advisory Committee on Industrial Training* appeared in 1966. This concluded that compulsory certification "makes no recognition of the fact that the skill mix will vary in industry. . . . it may inhibit the evolution of effective work patterns. Developing technology requires more flexibility than that permitted by the rigid definition of trades by regulation." Perhaps to appease the construction unions, a scheme of *voluntary* certification was adopted for certain trades, and already there are signs that such *voluntary* certification is becoming a mandatory condition of employment.

The final report of the governmental committee listed several specific objections to the current system of apprenticeship, as follows:

1. Its rigidity limits its usefulness in industries where occupations have evolved outside traditional craft demarcations.

2. A simple, two-stage assessment of proficiency (apprentice and journeyman) is inappropriate to modern industrial organization, where many levels of specialization are required. Thus, in the training of motor mechanics, three varieties of apprentice have traditionally been recognized (body repairers, fuel system technicians and electricians). This has some relevance to the corner garage. However, with the increasing complexity of vehicles and the emergence of large dealerships, a much greater degree of specialization is commonly needed; at least eleven distinct types of motor mechanic can now be distinguished.

3. No basis is provided for the recognition of workers with more (or less) skill than the certified journeyman. This presents a particular difficulty in cities such as Toronto where there is a large immigrant population. Some of the immigrants have considerable craft experience from their home country, and yet are ineligible to work in Canada. A motor mechanic from Italy, for example, can obtain a provisional certificate for up to one year, but at the end of this time he must pass the formal examination for his trade. The structure of the examination is usually foreign to the cultural background of the immigrant; he may be still rather defective in his English, and thoroughly unprepared to undertake a multiple-choice examination. Some tests are conducted under the pathetic conditions of a middle-aged man trying to use his young child as an interpreter, and it is hardly surprising that attempts to bribe examiners are common.

The barriers to full utilization of an immigrant's skill are numerous. He must discover how to obtain a notarized statement before a provisional license can be issued. Then he must obtain both a municipal license and union membership, paying substantial fees. Finally, he is faced with the certification examination. After many attempts, this hurdle is crossed, only to find that employers refuse to hire a man who lacks *Canadian* experience. Inevitably, many well-qualified immigrants find themselves on welfare rolls or in minimum wage-rate employment.

4. Occupational mobility is restricted. We have noted the frequent changes of trade that will be required of future generations of workers. Under existing legislation, each change of trade involves a lengthy second apprenticeship, even though a large percentage of the required skills may already be familiar to the worker. Further, current schemes take no account of the possibility of lateral and/or upward promotion of a man within a large industrial organization.

5. Apprenticeship is often coupled with restrictive legislation. Limits are set on apprentice/journeyman ratios, and lengthy time-serving periods are prescribed without examining the real times needed to acquire the skills and knowledge of a trade. Thus, a tenth-grade education or its equivalent is now a requirement of certification. Until 1964, eighth grade was considered adequate. No account is taken of the relevance of literacy to the trade in question, nor is allowance made for wide regional differences in what constitutes a tenth-grade education. A large proportion of entrants do not complete their apprenticeship because they fail to gain the necessary experience. A man working in a garage that simply repairs transmission systems (or installs windshields or mufflers) has no hope of covering the requirements for certification, and a five-year training program is plainly inappropriate for a worker who will remain in such employment.

Alternative Training Schemes

If apprenticeship is to be rejected, what may be set in its place? One possibility is the *spectrum principle,* currently advocated in the United Kingdom. Such a scheme would involve accurate definition of various jobs in terms of their work functions, the development of appropriate curricula to provide the skills essential to such functions and the setting of a training period related to curriculum needs rather than some arbitrary concept of time-serving.

This type of approach was endorsed by the Canadian manufacturers' association in a brief to the Select Committee on manpower training as early as 1963. Their recommendation was that training should be organized in the form of blocks, some being knowledge common to a number of trades, and others highly specialized skills. They suggested that by identifying a hierarchy of blocks, including basic training, production, office and field skills a theoretical framework could be drawn up that would permit the individual to take training in an orderly way, moving laterally as often as necessary in response to the changing needs of industry, and moving vertically as far as individual ambition, capacity, effort and opportunities would allow. Such a scheme would permit a much freer interchange between (for instance) journeyman and technologist grades, and could well eliminate the current situation where well-qualified technicians have to register as second-year apprentices in order to perform some relatively simple functions within a factory. It would also encourage the disappearance of traditional craft titles such as electrician; in any event, these have become so blurred in the industrial setting that they no longer have clear meaning. The necessary analysis of occupational skills could well prove helpful, not only in the design of training programs, but also in the more difficult task of selecting trainees who will ultimately attain a high level of performance. Further, it may become clear which blocks of instruction are best given in the classroom and which in the factory. One block may be of a theoretical nature; for instance, a lathe operator may need to know various symbols, tolerances and the effects of temperature on his metal. Another block may involve the mechanical skills of using the lathe; here, the factory equipment is an essential ingredient of practice, and instruction is best provided within industry. Such blocks of knowledge can continue to be added throughout a working career. Accreditation of the classroom components could proceed along traditional lines, while the blocks of knowledge accumulated within industry could be evaluated by testing the time required for a trainee to bring a product to some objective criterion of tolerance.

Although some pilot projects are underway, the full restructuring of industrial training will take a long time to complete. In the meantime, there is an urgent requirement for workers with highly specialized training in many industries such as machine sewing, textiles, aircraft assembly, foundries, machine shops, welding, furniture manufacture, food processing and leather cutting. It is not feasible for the public education system to bear the entire burden of keeping the labor force in these various industries *au fait* with modern technology. Hence, there is an increasing trend towards short-term, in-plant courses, supplemented where necessary by formal classroom instruction. Initially, these programs were established on the basis of a *cost-sharing* partnership between government and industry, and in some industries the net result was that the government subsidized low wages. In the light of such experience, the government decided to recompense employers on the basis of numbers graduating from a given program, thus leaving the manufacturer the responsibility of screening entrants and absorbing the cost of any *drop-outs*. As currently administered, the instructors may be provided by government, or existing staff may be given further training so that they can serve as instructors.

Such courses have the important advantage that they are custom-designed to the needs of a specific firm. Training is based on the equipment that will ultimately be used, and problems of *transfer* of skills (page 248) are minimal. Given a sympathetic employer, the results are excellent. An in-house course was arranged for the Toronto Transit Commission to retain the track grinders who were displaced when the streetcars were sold. The men attended a three-year course, running two hours a night, two days a week, and at the end of this time men as old as sixty were able to assume the new responsibility of maintaining escalators on the subway system.

The main disadvantage to the worker is the specificity of experience that is given. In the event that the factory should close, his training may be poorly suited to employment elsewhere. However, this difficulty should largely disappear if a modular or block system of certification can be introduced.

CHAPTER 13

CIRCADIAN RHYTHMS AND
IRREGULAR HOURS OF WORK

TRADITIONALLY, MAN HAS OPERATED on a closely defined and consistent time schedule. Over the past 150 years of industrial activity, the hours of duty have been reduced progressively from twelve or fourteen to a more modest seven or eight per day, but sleep has been held to a rather constant night-time allowance of eight hours.

Such an ordering of life is unfortunately becoming increasingly difficult to sustain in modern post-industrial society. Many large chemical refineries and smelting operations have a prolonged *warm-up* phase, and it is uneconomic to run such plants on other than a twenty-four hour basis. Other facilities such as large digital computers have a negligible warm-up time, but because of a high capital cost they can only be justified when used both day and night. Urban growth increases the demand for fire and police protection, ambulance and other emergency services on a regular twenty-four hour schedule. And as more and more of the populace operate around the clock, there is a growing demand for nocturnal service workers, particularly in transport and catering. Employees are driven increasingly to accept shift work as a normal basis of operation.

The upper echelons of society also face problems of irregular hours, although often in a different guise. The physician has traditionally experienced weeks of hospital duty with curtailed and frequently interrupted periods of sleep. More recently, the great increase of intercontinental air travel has exposed substantial numbers of aircrew, politicians, scientists and salesmen to large and rapid changes of latitude, with enforced alterations not only in personal schedules of wakefulness, but also in environmental clues as to the true time of day. Exploration of the arctic and the antarctic, the continental shelf, and outer space (Halberg *et al.*, 1970) equally creates problems of altered environmental clues. In the space situation, unless a *normal* length of day is imposed by mission-control, each astronaut tends toward his inherent free rhythm, and interpersonal conflicts may arise from the enforced companionship of a short-cycling early riser (*lark*) and a long-cycling late riser (owl).

Such changes of work pattern present important problems to the ergonomist. Do abnormal periods of wakefulness affect human

268

efficiency? If so, what is the speed of adaptation, and should this determine the optimum number of night shifts? For how long following
an intercontinental flight is the decision-making process of a politician
disturbed? For how long is he restricted in his capacity to interact
favorably with other politicians? Does the repeated crossing of time
zones have a cumulative effect upon the efficiency of aircrew? And
how is work tolerance altered by unaccustomed exposure to twenty-four
hours of daylight or night?

NORMAL SLEEP

Our discussion is conveniently initiated by a brief review of normal
sleeping patterns: the duration and types of sleep, and the period
required for arousal.

Duration of Sleep

The normal duration of sleep can be studied with reasonable accuracy
by having subjects record on a card the time of going to bed and
their estimate of the period that elapsed before they fell asleep (Lewis
and Masterson, 1957; Tune, 1968). More precise information on the
extent and type of sleep is obtained by observation of the subject,
together with recordings of respiration, electroencephalogram and eye
movements.

The average person sleeps for eight hours out of every twenty-four,
but there is considerable interpersonal variation. Masterson (1965)
found one of ten tradesmen who claimed a daily average of only 5.9
hours sleep throughout a sixty-day study. Such individual differences
may be related to corresponding variations in the inherent periodicity
of circadian rhythms.

Medical personnel have traditionally taken very short periods of
sleep, particularly while on duty in surgical departments; in such circumstances, less than five hours of sleep per night may be obtained
throughout a three-month appointment. During a subsequent vacation,
there is reversion to the normal eight-hour sleep pattern, but with
the possible exception of the first night off-duty there is surprisingly
little attempt to catch up the accumulated deficit. The sleep of medical
students is also somewhat restricted during term time; Masterson noted
an average of 7.4 hours, with reversion to no more than 7.8 hours
while on vacation. Nurses, contrary to popular sentiment, enjoyed a
substantial 8.1 hours of sleep per day, and problems of sleep deprivation
arose only in those with an overextended social calendar.

The absence of any serious attempt to catch up arrears of sleep shows
the substantial adaptability of the body, and suggests that the normal
eight-hour period is conditioned as much by social factors as by any

true physiological need. This view is supported by observation of arctic communities; here, too, social constraints normally produce eight hours of sleep per day, but there is increasing departure from this pattern at seasons of the year when diurnal clues of daylight intensity are removed.

Types of Sleep

Over the last decade, it has been increasingly recognized that not all sleep has the same quality. Various stages may be distinguished on the basis of electroencephalograph waveform and other physiological properties, but the most important distinction is between forebrain and hindbrain sleep (Dement and Kleitman, 1957; Dement, 1960). The former requires the mediation of the cerebral cortex, and is characterized by a progressive muscular relaxation, with slowing of the heart rate, regular breathing, and falling oxygen consumption. Hindbrain sleep has an electroencephalogram similar to that of a subject who is awake, with bursts of alpha rhythm; the muscles are completely relaxed, the pulse and respiration are less regular, and there are intense visual dreams accompanied by rapid eye movements (REM) and sometimes penile erections. A typical sleep pattern is illustrated in Figure 80. During the first two thirds of the night, there are perhaps three incursions into relatively deep forebrain sleep, but little REM sleep; this is concentrated mainly in the period when the depth of forebrain sleep is diminishing. Arousal from forebrain sleep becomes progres-

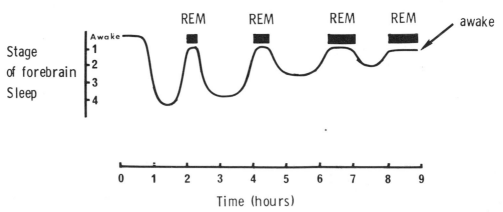

Figure 80. The typical course of an eight-hour period of sleep (based on studies of Dement and Kleitman, 1957). The stage of forebrain sleep can be judged from the electroencephalogram. While the subject is aroused, the waveform is desynchronized. As he relaxes, the slow alpha rhythm appears, and when he becomes drowsy this is supplanted by small, high-frequency waves. Moderately deep sleep (Stage 2) is shown by *spindles* of 14 cm/sec waves. In Stage 3, these are accompanied by large, slow delta waves, and in Stage 4 the delta waves are the sole form of rhythmic activity.

sively more difficult as its depth is increased. However, an alarm is least effective during REM sleep; at this stage, the noise merely becomes incorporated into a very vivid dream. Older people spend more time in light forebrain sleep (Stage 1) and less in deep sleep (Stage 4). They are thus readily woken by outside disturbances, particularly if on shift work. However, if they remain in bed, relaxed, they soon fall asleep again, and obtain a satisfactory total sleep period.

REM sleep seems a biological necessity, and if a person is selectively deprived of this type of sleep, more of it appears on subsequent nights. This has operational implications. A worker who is forced to sleep intermittently, or who uses a sedative at an unusual time of day may achieve the anticipated eight-hour allowance of sleep and yet develop fatigue because he has been deprived of REM sleep.

Speed of Arousal

Many workers on night call are allowed to sleep by their telephones. It is thus important to know the speed of arousal when they are awoken. Langdon and Hartman (1961) tested skilled performance in the period two to twelve minutes after awakening at twelve midnight and 3 to 4 AM. Initially, the speed of performance was 25 percent slower than in the daytime, but improvement was quite rapid, and within seven minutes performance had apparently stabilized at the normal level for that particular time of night (efficiency would of course have been low at 3 AM even if the worker had not been allowed to fall asleep, since arousal shows a diurnal rhythm).

SLEEP DEPRIVATION

As with many forms of psychological stress, the subjective impression is that sleep deprivation causes a substantial impairment of performance. Nevertheless, it is hard to demonstrate this by formal psychological tests (Wilkinson, 1965). This is partly because there are wide individual differences in sensitivity to sleep deprivation; anxious, neurotic and extraverted people, and those with a low level of intelligence seem more susceptible than stable and intelligent individuals (Cappon and Banks, 1960; Corcoran, 1964). Also, the normal diurnal rhythm of wakefulness persists for several days, even if there is complete sleep deprivation; thus, a person will usually be more sleepy and less efficient if tested in the early morning than if evaluated during standard laboratory hours. Much depends also upon the nature of the task. Some, such as a serial choice reaction time or the crossing-out of letter e's are affected by short periods of sleeplessness. With others, such as typical pursuit meter tasks, impaired performance is seen during the second night, while simple tests of mathematics, learning and communication remain unimpaired until the third or fourth night. It would seem that if an impairment of performance is to be demonstrated, the test must be sufficiently difficult to challenge the *reserve capacity* of the brain, and yet not so demanding as itself to restore the arousal of the subject. The most vulnerable tasks are repetitive and uninterest-ing operations, and performance often can be increased quite dramati-cally by providing knowledge of results; the subject is still able to appreciate short-term goals, but has less interest in long-term perfor-mance. Complex tasks may resist quite severe sleep deprivation because of the interest that they create; on the other hand, if two tasks are of equal interest, the impairment is greater for the one that is the more complex. Furthermore, the stimulating effect of a difficult test disappears if the novelty of both the test and sleep deprivation are lost through frequent repetition of the experiment.

The most common form of psychomotor disturbance is a periodic lapse of vigilance while the general level of performance is well main-tained. This has been attributed to a brief period of sleep, lasting perhaps one to two seconds. In the laboratory situation, a reaction time task would show a normal median value, but an unusual *tail* of occasional poor responses. The apparent normality of the worker between lapses makes this a particularly dangerous form of impairment in any job where continuous vigilance is needed (vehicle operation, radar control and the like). If the task is self-paced, there may be no loss of accuracy, but there is an annoying slowness; the worker pauses each time he senses a *micro-sleep* is about to occur. However, if the task is very familiar and automatic in nature, there is no pause, and an error occurs; this is the usual effect of sleeplessness on conveyor belt inspection. If the pace is forced, errors are particularly likely with an unfamiliar task, for instance a pilot who is landing at an airfield that he has not visited previously. The decrement of psychomotor func-tion shows a progressive development during the period without sleep, the loss of efficiency being superimposed upon normal diurnal changes of psychomotor performance.

Arousing stimuli (page 215), including noise, excessive warmth, phys-ical activity and drugs of the amphetamine class (Laties, 1961) reduce the effects of lack of sleep. On the other hand, temperate warmth and large doses of alcohol exert a potentiating effect.

Given adequate incentive, the body can develop sufficient arousal to compensate for as long as 100 hours of sleeplessness. However, the correct performance of a difficult task imposes a much greater stress upon the body; this is revealed by the various physiological indices of arousal, such as muscle tension, heart and respiration rate, blood pressure, skin conductance and galvanic skin response (page

215). Biochemical evidence of strain is more conflicting. Luby *et al.* (1962) have shown increased blood levels of adenyl phosphates, particularly ATP, but also ADP and AMP; on the other hand, reported changes of adrenocortical activity (as monitored by blood glutathione levels, eosinophil count and 17-ketosteroid excretion) have been very variable. In an attempt to reconcile conflicting reports, Wilkinson (1965) suggested that adrenocortical activity was increased in the early stages of sleep deprivation, particularly if the subject tried to overcome his drowsiness, but that in the later stages activity was normal or depressed.

Drowsiness, with yawning, itching eyes, and slowed reactions becomes most intense between 3 and 6 AM. On the second night without sleep, drowsiness is worse and there is an intense desire to close the eyelids, sometimes with an associated diplopia (double vision). There are also disturbances of behavior. Irritability is covert if not manifest, and the attitude becomes increasingly negative (Laties, 1961). There is little attempt to persist with demanding work, particularly if group effort is needed (Murray *et al.*, 1959). Pain thresholds are reduced, but this probably reflects poor moltivation rather than increased arousal (Wilkinson, 1965). The mean body temperature falls, perhaps as a consequence of diminished physical movement and muscle tone, but there is no agreement as to whether the diurnal rhythm of body temperature is increased or diminished.

Disorders of perception appear after about 100 hours without sleep. The individual is in a drowsy state, and increasingly loses touch with reality. Illusions of shape, size, movement, color and texture are first appreciated as such, but later become frank hallucinations which the subject accepts as real. There is often temporal disorientation, ranging from an apparent slowness in the passage of time to an unshakable belief in an erroneous concept of time. There may also be cognitive disorganization, at first no more than some slowness of thought, but progressing to rambling and incoherent speech. (Morris *et al.*, 1960). Behavioral changes are often a caricature of normal personality; problems are more marked in neurotic individuals, and in those with a history of psychosis there is a real danger that sleep deprivation may lead to a recurrence of the abnormality.

The effects of grossly curtailed sleep have been studied less extensively. However, they are of great importance to all who remain on night call, such as hospital surgeons, military patrols and control room operators. Wilkinson (1969) found that vigilance was impaired if a sleep debt of five to six hours was accumulated; it apparently made little difference whether this was brought about by one night with two hours or less of sleep, or by two nights each with five hours or less of sleep. Deprivation of this order is a hazard of intercontinental travel, as aircrew and passengers attempt to sleep in an unfamiliar hotel room, perhaps overheated and with an excess of street and guest noise. If the person who is resting remains in bed relaxed, sleep will usually soon return, but if he becomes restless and paces his room, a serious deficit may be incurred. If the total number of hours sleep is reduced, the main deficit occurs in the REM component; after several nights of sleep curtailment, there is some compensation in that REM sleep begins to appear earlier during the night.

Recovery following sleep deprivation is rapid; indeed, if one night of extended sleep is allowed, it is difficult to show any subsequent decrement of psychomotor performance. Much naturally depends upon the extent to which the normal sleep/wakefulness rhythm has been lost and the time needed for its re-establishment.

CIRCADIAN RHYTHMS

A number of biological variables show a twenty-four-hour periodicity (Bünning, 1964). Thus the body temperature is low in the early morning, and climbs by about 1° F to reach a plateau near noon. Urine output is much smaller at night than in the daytime. Similar rhythms are apparent for wakefulness, respiratory and circulatory variables, the cellular composition of the blood, and steroid secretion. It may naturally be questioned how far such rhythms are truly endogenous phenomena, and how far they represent a secondary response to socially conditioned changes of physical activity or other extraneous factors.

Rhythms such as the menstrual cycle illustrate the existence of biological clocks. Mills (1966) has specified the criteria of an endogenous rhythm as follows:

1. persistence in the absence of external time clues such as clocks, lighting, meal and sample times,
2. persistence despite a phase shift in external clues (as when a man is displaced by an aircraft through four or more time zones),
3. persistence despite false time clues (such as the use of clocks operating on a 21- or 27-hour cycle),
4. persistence under constant environmental conditions, including the absence of temperature and light fluctuations (as in cave exploration).

Some rhythms, particularly those shown by cardiovascular and respiratory variables (blood pressure, pulse rate, regional blood flow, respiratory rate, sensitivity of the respiratory center), are apparently secondary manifestations of cyclical changes in wakefulness and body temperature. Others, such as the volume of urine flow and the level of arousal will adapt to a phase change over the course of several days, but some variables such as the excretion of potassium ions (Lobban, 1965) meet all of Mills criteria of a true circadian rhythm. The controlling *clocks* are probably located in the limbic region of the brain, and lesions of the hypothalamus can disturb rhythms of both temperature and arousal. One classical example is the *Pickwickian syndrome*, exemplified by Sam Weller, the fat and sleepy boy in the *Pickwick Papers*.

Lobban (1965) has emphasized that a protracted period is necessary for full adaptation of all biological clocks to a new time schedule. Permanent arctic residents show little evidence of renal rhythms, but in visitors normal excretory patterns can persist throughout two months of arctic summer. If subjects are constrained to a 21- or 27-hour day, adaptations of body temperature occur quite quickly and water excretion also adjusts over several days, but a twenty-four hour pattern of potassium ion excretion is still dominant after eight days of life on the new regimen. Nevertheless, given sufficient time, even potassium excretion is modified; one group showing such a change are arctic miners who remain on a night shift throughout an entire winter.

Aschoff (1965) completely isolated men from normal time clues while making various distortions of the normal light/darkness cycle. A sudden transition to a constant level of illumination was associated with severe disruptions of biological rhythms and poor performance of fine sensory and motor tasks. After some three days of adaptation to the regimen, individuals began to develop characteristic personal cycles, varying in length from twenty-four to twenty-six hours. Sleep/wakefulness rhythms remained relatively stable over much longer periods of isolation (six months and more), although the ability to judge the passage of time was progressively lost. On leaving the experimental chamber, the subjects reverted to a normal twenty-four-hour cycle over the course of three or more days. On the basis of these observations, Aschoff suggested that many biological functions had an inherent periodicity of twenty-four to twenty-six hours, and that such rhythms could be *tuned* to oscillate in phase with environmental clues having a twenty-four hour periodicity.

SHIFT WORK

From the ergonomic point of view, recurrent and untimely cycles of wakefulness, increased appetite and increased urinary output can wreak havoc with the attempts of a shift worker to sleep (Colquhoun *et al.*, 1968). His adaptation to the abnormal routine is slowed not only by biological problems, but also by persistent time clues (particularly intensities of light, temperature and noise that are inappro-

priate for sleep); the home environment of the shift worker deserves
more attention than it commonly receives. During waking hours, circadian rhythms can modify the results of simple fitness tests such as the
Åstrand prediction of aerobic power (Klein *et al.*, 1968), an important
consideration in the large-scale evaluation of industrial workers.
Temperature rhythms modify the responses to extremes of heat and
cold, while circulatory and respiratory rhythms alter the hazards presented by various toxic materials. However, in the average working
situation, the most important changes are in psychomotor efficiency and
vigilance; Kleitman (1939) has shown that the performance of many
tasks (e.g. simple mathematics, encoding and reaction-type tests) is
greatest around noon and lowest at night (Colquhoun, Blake and
Edwards, 1968). Thus, until adaptation of the diurnal rhythm has occurred, the shift worker (and also the intercontinental traveler) will be
at a disadvantage in any skilled operation, whether it be production line
inspection, radar monitoring, or the drive on the airport expressway
(Alluisi and Chiles, 1967).

For many workers, the social and psychological problems of shift
work are more important than physiological and psychomotor disturbances; night duty leads to a marked diminution in social contacts,
and if the wife is working a day shift there may be little normal home
life. When long periods of night work are undertaken, even hobbies,
sports and other interests must be modified. The choice of routine
seems between a short and a very long period of night work. If the
duration is two or at most three days, the worker will remain below
peak efficiency, but will not transpose his various rhythms to the night
phase; adaptation to the subsequent day shift will thus be relatively
easy. With a week of night work, more complete adaptation occurs,
but the man is then inefficient on returning to the day shift. One popular
and apparently fairly effective short-cycling shift arrangement involves
a four-day sequence of morning, evening and night shifts, followed
by a twenty-four hour rest period.

INTERCONTINENTAL TRAVEL

The problems of the intercontinental pilot have been recognized
for many years (Post and Getty, 1931; Strughold, 1952). Difficulty arises
when four or more time zones are crossed in rapid succession.
Desynchronization of the wakefulness cycle impairs the decision-
making process, and normal efficiency may not be fully restored for
five or more days. Some authors (for instance Hauty and Adams, 1965)
have found little difference between east/west and west/east flights;
others (for example Halberg *et al.*, 1967; Siegel *et al.*, 1969) report

more difficulty following westward flights. Presumably, the response depends not only upon the direction of flight but on the time of day at which it is undertaken; from the passenger's point of view, a west/east transatlantic flight usually involves not only the crossing of time zones, but the loss of a night's sleep. Repeated desynchronization can lead to cumulative fatigue, due mainly to loss of sleep.

In the near future, the problems of aircrew may be greatly reduced by supersonic aircraft that can make a double intercontinental journey in the course of a day. When the crew of subsonic aircraft are required to make long journeys, it is necessary to place arbitrary restrictions upon their flying hours; for instance, they may alternate ten-hour watches to a maximum of fifty hours. A careful check is kept upon the wakefulness cycles of individual crew members, and where possible critical operations such as landing and take-off are scheduled to avoid periods of impaired vigilance. During an extended mission of perhaps two days' duration, the crew are able to take progressively less sleep while off duty (Atkinson *et al.*, 1970); this is presumably because the time at which they are trying to sleep becomes more abnormal for them. Depending upon the extent of sleep disturbance, up to five days may be needed to restore normal patterns of sleep, and in recognition of cumulative sleep deprivation, the International Civil Aviation Organization has proposed calculating a postflight rest period as follows:

$$\text{Rest period} \atop (\text{1/10 of a day}) \quad = \frac{\text{Travel time (hr)}}{2} + (\text{Time zones} - 4) + \frac{\text{departure}}{\text{coeff.}} + \frac{\text{arrival}}{\text{coeff.}}$$

Departure and arrival coefficients are shown in Table V. In applying the ICAO formula, the calculated rest period is rounded to the next highest half-day, but rest stops of less than half a day are not scheduled unless the journey involves overnight flying. The need for a rest allowance is well shown by the data of Hartman (1971). Although the aircrew that he studied slept regularly for seven to eight hours per session during their flight, this was presumably sleep of inadequate quality;

TABLE V. ARRIVAL AND DEPARTURE COEFFICIENTS USED IN COMPUTING ICAO REST PERIODS. (BASED ON ATKINSON *ET AL.*, 1970; SEE TEXT)

Local Time	Departure Coefficient	Arrival Coefficient
8 AM to 12 PM	0	4
12 PM to 6 PM	1	2
6 PM to 10 PM	3	0
10 PM to 1 AM	4	1
1 AM to 8 AM	3	3

the average durations of sleep for the first three days postflight were respectively 9.9, 9.2 and 8.9 hours, and on the first day after flying 32 percent of the aircrew slept for twelve hours or more.

The businessman can usefully minimize postflight desynchronization by pretuning his biological rhythms for several days; he can, for instance, go to bed one to two hours later on each of three to five days preceding a westward journey. Alternatively, he must allow an equal postflight period of adaptation. The majority of mild sedatives such as Seconal® (secobarbital) are not particularly helpful in permitting sleep at unusual hours; they may actually inhibit REM sleep (page 270), and are liable to leave a persistent drowsiness through much of the following day. Flurazepan hydrochloride (Roche) is now reported to induce sleep without loss of the REM phase (Siegel *et al.*, 1969); however, informal sedation by mild exercise, a warm bath and a comfortable environment seems a preferable approach in most circumstances.

PART IV

PRACTICAL APPLICATIONS OF ERGONOMICS

CHAPTER 14

THE AGING WORKER

CERTAIN PROBLEMS OF THE AGING worker have already been considered. However, the topic is of sufficient importance to merit fuller discussion. In some European countries, 25 percent or more of the population are already over the age of 65. Canada, by virtue of continued immigration, still has a relatively young population. Nevertheless, the proportion of old people seems destined to increase throughout the world, as birth control measures become more widely applied and zero growth concepts gain in popularity. Contrary to public opinion, medical science has not as yet contributed materially to the total of elderly people; over the past half-century, the life expectancy of a 60-year-old man has increased by less than one year. However, research on the chronic and degenerative diseases is increasing, and it may well be that in the near future medical discoveries will further increase the problems associated with an aging population (Sauvy, 1970).

The necessity for an older person to remain in productive employment is debatable. In industrialized societies, automation is yielding rapid gains of productivity. The means of production is also evolving rapidly, and in strictly economic terms it may be cheaper to consign the outdated worker to a life of leisure than to embark upon a lengthy and expensive retraining program (page 262). Retirement ages have fallen progressively, from 70 to 68 and now to 65 for most men, and to 60 for many women. However, this leaves a terminal age span of at least ten years in the male and as much as fifteen to twenty years in the female; there is thus a challenge to devise satisfying and creative activity that will fill the post-employment void. Some enlightened companies are initiating preretirement programs to suggest uses for the unaccustomed leisure time.

PHYSIOLOGICAL SEQUELAE OF AGING

Aerobic Power

The role of aerobic power as a determinant of daily work tolerance is discussed elsewhere (page 55). There is a steady decrement of aerobic power between the ages of 25 and 65 years; expressing data relative to body weight, readings for the male worker may drop from 40 to

50 ml/kg min to 25 to 30 ml/kg min, with proportionately similar losses in the female (Shephard, 1969a). Individual links in the oxygen transport chain are each weakened by 25 percent or more over the adult span. Lung volumes are reduced, and although there is not much change in the stroke volume of the heart, the maximum heart rate decreases from around 195/min to 160/min or less. The capillarity of the muscles is also reduced, and at exhaustion the blood lactate concentration (60 mg/100 ml) is only about half of the figure found in a younger person. On the other hand, a given submaximum workload produces a similar oxygen consumption and heart rate in both young and older workers.

The permissible daily workload of the older person is still not agreed. If a 40 to 50 percent ceiling is imposed, as in younger workers (page 55), then many older men and the majority of older women will be severely restricted in what they can undertake (Shephard, 1969b). An energy expenditure of 4.8 kilocalories/min is equivalent to 50 percent of aerobic power in the average 65-year-old man, while the corresponding figure for a 60-year-old woman is no more than 3.7 kilocalories/min. Furthermore, 50 percent of all workers fall below these average readings; 95 percent confidence limits are such that one 65-year-old man in twenty would be restricted to a daily work rate of 3.2 kilocalories/min, and one 60-year-old woman in twenty would be restricted to 2.3 kilocalories/min. Such limits apply to healthy individuals; even more marked restrictions would be necessary for those who were grossly obese or had disease of the cardiorespiratory system.

In practice, many older men find it necessary to work at tasks that would cost a young person 4.8 kilocalories/min or more. Because of limited technical knowledge, the older person is often relegated to what is physically the most demanding work in a factory. Nevertheless, fatigue is not widespread. Does this mean that the older person can operate for eight hours at more than 50 percent of his aerobic power? This seems inherently unlikely. It is more probable that the energy cost of a given task is reduced in the older worker, due to the accumulated physical skill of the employee, and the acceptance of a slower rate of working.

If the aerobic power of an older person is becoming marginal for his work, replacement can be delayed by several measures aimed at increasing personal fitness. These include:

1. treatment of any anemia and cardiorespiratory or other intercurrent disease,
2. specific rehabilitation, including muscle-building exercises and the administration of anabolic steroids following industrial injury,

3. encouragement of weight reduction (potential for 15% gain of relative aerobic power),
4. encouragement of regular recreational activity (potential for 10% to 20% gain of aerobic power),
5. encouragement to stop smoking (potential for 5% to 10% gain of aerobic power, with additional diversion of blood flow from chest to leg muscles).

If an older person can be persuaded to adopt all of these measures, the total effect upon the performance of a task that was previously of marginal severity can be quite striking. In the event that anaerobic work is still occurring, the task can often be brought within the compass of an elderly worker by a suitably extended allowance of rest pauses.

Muscular Strength

A decrease in muscle strength obviously restricts the potential for heavy lifting and other forceful manual tasks; it may also enhance the tendency for an accumulation of lactate in sustained rhythmic activity, thus leading to more rapid fatigue (page 53). However, the most important practical consequence of muscle weakness is an increased risk of injury. The reduction in physical bulk of muscle leaves the bones and joints more exposed to contact trauma, and less force is available to support and protect the system in an emergency; the problem is compounded by more ready development of fatigue and slower compensatory reactions to internal (slipping, loss of balance) and external (collision) stress. The back, knee and ankle are all vulnerable to injury, and if degenerative changes are occurring in the intervertebral discs, there should be some reduction in the load that is carried. Weakening of the abdominal muscles also reveals any congenital predisposition to the development of a femoral or inguinal hernia.

Formal studies of muscle strength (Asmussen, 1964; Shephard, 1969) show that the male worker reaches his peak in the mid-twenties, but there is little loss of strength until the final decade of work. There is a substantial loss of muscle protein between the ages of 55 and 65, this change being shown by both a decrease in overall body weight, and also a fall in maximum isometric muscle force. Probable explanations include a decrease in leisure activity as the worker's family leaves home, and a diminished output of anabolic hormones.

Muscle wasting is particularly liable to occur if the worker is immobilized following industrial injury, and in such circumstances he may benefit from a formal program of muscle building (Fried and

Shephard, 1969), including the possible administration of anabolic steroids.

Vision

The first practical sign of aging for many workers is the development of longsightedness. This is due to an increasing rigidity of the lens structure, with consequent problems of accommodation when performing close work. The nearest point of clear focus recedes from 12.5 cm at age 30 to 18 cm at 40, 50 cm at 50, and 100 cm at 70. The task of the worker can be helped by the use of larger type, an increase of space between the eyes and the work surface, and the provision of suitable spectacles. However, complaints of visual fatigue during sustained close work become more likely with age (page 75).

There is a progressive decrease in visual acuity with respect to both monochromatic and polychromatic light. This is due to scattering at opacities in the cornea and lens, possibly with some true reduction in the sensitivity of retinal receptors. From the ergonomic point of view, there are several practical consequences:

1. Anomalies of color vision become more frequent. The worker who as a young man had marginal color recognition deteriorates to the point where he is unable to distinguish a red from a green signal.
2. Dark adaptation is slow and rather incomplete. MacFarland (1963) suggests that there is a 240-fold difference of visual acuity between a young and an old person when both have fully adapted for night vision. The older worker is thus unsuited to any task that calls for careful search procedures in dim or night light,
3. Glare sensitivity is much greater in an older person. MacFarland notes that in the presence of glare the older person requires a 50-fold increase of target illumination relative to a young person. The implication is that night driving presents a serious hazard to the older worker.

Hearing

Some deafness is a frequent accompaniment of aging. As with industrial hearing problems (page 128), the first loss is usually in response to high frequency sounds (4000 to 5000 c/sec). Indeed, if the worker has been exposed to a noisy environment for long periods, it is difficult to separate the effects of aging and noise exposure. There is much interindividual variation, and some workers retain good auditory acuity until retirement. Nevertheless, there appears to be an age-related deterioration in hearing that affects a substantial proportion of people independently of any industrial noise exposure.

Central Nervous System

Almost all functions of the central nervous system show some deterioration with age. Failure of short-term memory is a very common complaint, and has been attributed in part to lack of practice (page 85). Nevertheless, formal studies show that the memory span of the average older person is reduced from seven to six digits (Birren *et al.*, 1963). The input selector of the cerebral computer (page 84) also works less efficiently, so that simple psychomotor tasks such as card sorting are performed more slowly than in a younger person (Rabbitt, 1964).

The older executive places increasing reliance upon memo pads, a careful diary and a good appointments secretary. The older employee responds best to familiar and well-ordered procedures, reacting badly to a change of stereotype (page 27) such as an alteration in the mode of operation of a switch or lever.

The pace of change in industry is such that retraining of the older person is commonly necessary (page 262). Learning proceeds more slowly in those who are older, and for this reason they may find advantages in a self-paced teaching device. Because of the slow response of the input selector, it is important for an instructor to maintain an orderly sequence of presentation and avoid switching from one topic to another. Often, the response to tuition is helped by techniques that demand active participation, rather than passive attention to lectures or a series of verbal instructions that are soon forgotten (Belbin and Down, 1965).

Reaction times are slowed by at least 20 percent, more for complex tasks. There is some slowing of both nerve conduction and muscle movement, often supplemented by arthritis, but the main cause of delayed reactions is an increase in the time needed for signal identification and decision making. More time is required to respond to any external instructions, and if vision is still good, printed material becomes a more effective method of communication than complicated verbal directives. Hesitancy and excessive caution do not present any serious problem if the worker is allowed to function at his own speed, but paced work becomes increasingly burdensome. Slowness can be annoying and even dangerous in a group situation. This is particularly true on fast motor routes, where the failure of an elderly driver to clear a turn in the normal time is a frequent cause of accidents.

Tasks that call for manipulative skill are performed slowly, but the older worker may compensate for this with his greater patience and accuracy. The speed and accuracy demanded by a given job (page 60) should be assessed carefully, and where possible the age of employees should be adjusted accordingly.

In jobs that demand innovative skills and creativity, the experience of the older person must be set against an increasing rigidity of attitude and a difficulty in establishing new concept patterns within the cerebral cortex. There is increasing irritability and paranoia, the employee feeling—sometimes with good reason—that a company is planning his demotion or elimination. At the same time, motivation may falter. By the age of 40 or 45, personnel have commonly achieved their goals of house purchase and establishment of a family, and have also reached either the top of their employment ladder or at least their *level of incompetence*. There is thus a strong temptation to *coast* from this point to retirement, avoiding mistakes by not making any hard or crucial decisions.

In scientific investigation, the optimum age of employment may be particularly low. Lehman (1953) suggested that the major discoveries of psychology were made by investigators in their early thirties; however, in defense of the older scientist it should be stressed that major discoveries are the exception rather than the rule at all ages, and that the advancement of knowledge depends also upon the competent exploitation of new ideas, a task best carried out in the well-run laboratory of the experienced investigator.

AGE AND EMPLOYMENT

Accidents

The previous section has noted a number of reasons why older people might have an increased risk of accident and injury: weakening of muscles and bone; slowing of reactions; confusion by changing stereotypes; failure of hearing, sight and color vision; and involvement with others less patient than themselves.

Nevertheless, the majority of statistics show relatively few accidents among elderly employees. This can be explained in part by a weakness in the statistics; injury rates are commonly expressed for an overall occupational class such as mining or foundry work, whereas tasks within a given industry show a clear-cut age distribution. Griew (1964) demonstrated that if the trade of "aircraft worker" was broken down into specific categories such as "electrician," "miller" and "grinder," a number of trades showed an increased incidence of accidents among the elderly. Further, there is some self-selection, only the more fit workers continuing in hazardous employment until old age; those who are accident prone are eliminated either by management or by personal selection of less demanding work. However, it is also likely that the greater caution and experience of the older worker make an important contribution to accident prevention, and current statistics do not sup-

port risk of injury as a general reason for refusing employment to
an older person.

Disease

A large proportion of older workers have some form of chronic disease (Brown and Shephard, 1967), so that a precise definition of
"normality" becomes difficult. Varicosities frequently afflict older
women, and through a peripheral loss of blood volume reduce working
capacity (Carlsten, 1972). Some elevation of resting blood pressure
(hypertension) is common in both sexes, and the increase of blood
pressure with moderate work is greater in the elderly; this leads in
turn to an increased cardiac workload, with a greater chance of effort
angina and sudden death (Kavanagh and Shephard, 1963). The risk
of sudden death rises progressively over the age of 40, and is a particularly important consideration in such occupations as airline pilot
and train driver; the annual medical examination for such groups should
include an exercise electrocardiogram. In men, chronic nonspecific
lung disease (bronchitis and emphysema) is very likely, particularly
if the employee is a heavy smoker and has worked in a dusty industry;
breathlessness from this condition can restrict physical effort, and
absence from work with recurrent chest infections becomes likely.
Diabetes, infections of the gallbladder, and renal problems are all
more likely in the elderly.

Those who employ the aged worker must thus anticipate occasional
episodes of major disease. However, this is not a contraindication to
continued employment of the affected person. If treated kindly, the
average employee is responsive to such care, and rewards his firm
by work of premium quality. Further, it must be emphasized that the
major difficulty on most production lines is not the predictable absence
of a genuinely sick older person, but the repeated minor and irresponsi-
ble absenteeism of younger staff members.

Working Conditions

Perhaps because the vascular system is less responsive to changing
thermal stimuli, older people require a more precise regulation of
environmental conditions. The preferred temperature (page 67) is
somewhat higher than for a young person, as would be anticipated
from the decrease of basal metabolic rate with age. Nutrient requirements decrease in parallel with metabolism. On the other hand, the
elderly are less tolerant of an added external heat load, and the summer
of the midwestern United States is sufficient to cause the demise of
a number of older people.

Overall Recommendation

It is difficult to make an overall recommendation on the employment of the elderly. Due consideration must be given to the demands of the job, the age distribution of the available labor force, and the physiological and psychological age of the prospective employee. It is a fact of everyday experience that some men are past useful employment at 60 or 65, while there are other outstanding specimens whose mental and/or physical powers are well preserved at the age of 80 or even 85. Uniform legislation that takes account only of chronological age does less than justice to either of these groups.

CHAPTER 15

ERGONOMICS OF THE HANDICAPPED

SCOPE OF THE PROBLEM

THE HANDICAPPED FORM A SUBSTANTIAL segment of the potential working population in most nations. In this section, we shall look at some of the ergonomic problems that confront the handicapped in industry and the home (Asmussen and Poulsen, 1966), with special emphasis upon the blind. The amputee is discussed in Chapter 8.

About 100,000 U.S. citizens, mainly of employable age, receive federal aid for economic blindness, while some 500,000 others are rated as "totally disabled." The proportion with partial disability is obviously much larger. In the United Kingdom, a register is kept of 650,000 persons who are substantially handicapped in obtaining or holding employment on account of injury, disease or congenital deformity; all substantial employers (20 or more personnel) are required to accept a standard percentage of disabled employees, and in certain industries well-suited to the handicapped a special quota is imposed.

The causes of disability are varied. Congenital problems include spastic paralysis (Corcoran *et al.*, 1970), deformities (including the thalidomide victims; Pascoe, 1971), the deaf, blind and dumb (Cartmel and Banister, 1968; Cumming *et al.*, 1971) and those with mental deficiencies (Hayden, 1962). The young adult may be stricken by poliomyelitis or spinal injury secondary to an accident of war, traffic or industry. Older workers may be afflicted by multiple sclerosis, arthritis, cerebral vascular disease, amputations secondary to atherosclerosis and diabetes, or loss of special senses. Often, disabilities are multiple, and occupational training or rehabilitation may be made doubly difficult by extreme age, cultural deprivation or true mental deficiency.

The task of the therapist is not only to develop muscle power and increase the range of movement about specific joints, but also to teach a new approach to life. If the disability is of sudden onset, morale may be severely shattered, and in this context the Olympic Games for Disabled Persons makes a most valuable contribution.

CONDITIONS OTHER THAN BLINDNESS

Many interesting challenges are presented to the ergonomist by the disabled person. If a limb has been amputated, can an effective prosthetic replacement be devised? Will the proposed design permit an adequate flow of information between the user and his prosthesis (page 197)? If the power of a limb has been weakened, either by amputation or by muscle wasting, can suitable tests be devised to establish the potential for brief isometric effort (as when operating a car) and for sustained rhythmic activity (as when operating a wheelchair)? Is it possible to teach a worker new skills that fall within the competence of his restricted powers, or alternatively can the common tasks of home and employment be redesigned to accommodate his limitations (Poulsen, 1963)? In this connection, it is vital to have reliable figures (Durnin and Passmore, 1967; Asmussen and Poulsen, 1963) for the steady caloric cost, the peak forces and the range of movements required in everyday tasks. Minor modifications may enable the handicapped to compete very effectively with the average worker; better lighting may eliminate a need for stooping, an adjustable height of chair or working surface may avoid a need for stretching and reaching, and a better control of environmental temperature may reduce the discomfort of arthritis. Sometimes, it is possible to help the disabled by the addition of power controls; this approach has been exploited most widely in vehicle design, but obviously could be extended to many industrial processes. A large proportion of the suggested ergonomic measures—better lighting, better seating, environmental control and power tools—seem likely to benefit not only the handicapped but all classes of worker. The concept of employing the disabled in normal factories is relatively new; at one time, the emphasis was upon segregation in *sheltered workshops,* but this has the obvious psychological disadvantage of branding a man as a cripple throughout his working life. In fact, given proper encouragement, psychological support, and small adaptations to the place of employment, many "cripples" are better motivated and more productive than ostensibly healthy members of the labor force.

Methods for the transport of the disabled require some thought. In a young person, either crutches or an artificial limb can give a very useful range of mobility, but in older patients the aerobic power following amputation may be so low as to make these unrealistic forms of therapy (Kavanagh and Shephard, 1973b). Such patients, together with the victims of spinal paralysis, must often resign themselves to a life based upon a wheelchair. Should this be arm propelled to maintain fitness, or battery-driven to avoid fatigue of overworked arms (Voigt and Bahn, 1969)? Should it be rugged enough for use over rough ground, or a lightweight collapsible vehicle that can readily be packed into a car? If the patient can operate a car, what type of design is most appropriate (Asmussen *et al.,* 1964; Bogh and Poulsen, 1967)?

In some nations, the emphasis has been upon modification of a standard power-equipped vehicle to allow hand operation of brakes and throttle; this carries little stigmata of abnormality, and permits the disabled person to make journeys with his family. In the United Kingdom, the emphasis has been upon the alternative of a single-passenger vehicle subsidized by the Department of Health (Isherwood, 1970). Presumably in the interests of safety, there is but one seat, and the carriage of passengers is illegal; thus in any emergency, the driver is forced to fend for himself. The engine is small and atypical, so that the average garage has neither the spare parts nor the expertise to make a repair A form of bar steering carries the throttle and brake, and throws a high proportion of the workload upon the left hand. Storage of a wheelchair is difficult, particularly for a disabled person. The vehicle is unstable, and lacks the weight for effective traction on an icy or slippery road surface. Finally, there are all the psychological problems associated with isolation from family and friends on entering a vehicle so obviously designed for a permanent cripple. Like so many current items of equipment, the British invalid car is essentially a *nonsystem* (page 24) and there is much to commend modification of a more normal vehicle. Whether the patient drives or not, by suitable arrangement of seat heights it is often possible for him to wriggle from a wheelchair onto a relatively standard car bench. If he cannot leave his chair, then it is necessary to provide a vehicle with sufficient vertical height to accommodate both the invalid and his chair.

Building design needs a substantially increased allowance of floor-space to permit the maneuvering of wheelchairs; elevators, corridors and doorways must all be of greater than average width, and stairs must be replaced by gently sloping ramps (Leschly *et al.*, 1960). Electrical switches and sockets must be set at wheelchair height, and doors and windows must be operated by remote control. Bathroom furniture must be redesigned, with alterations in the height of basin, bath and lavatory, easily operated taps and provision of handrails or hoists. Other changes in furniture pattern may also be helpful, such as the purchase of cabinets with hopper-type drawers. It is naturally easiest to arrange such special facilities in a home for the disabled, and in some European countries beautifully designed buildings have been constructed along these lines; however, they give rise to all the psychological problems of segregation, and although it may be more difficult to arrange, it is preferable to keep the handicapped person within a home used also by normal healthy people.

The leisure needs of the handicapped are equally specific. Churches with steep outside staircases, streets with six-inch curbs, cinemas with narrow aisles, and underground lavatories with small cubicles all seem

designed to keep the cripple at home. However, he has perhaps greater
need of recreation than the average worker. In recent years, voluntary
associations of the handicapped have become quite aggressive in
demanding appropriate facilities, and increasingly airports, theaters
and other public facilities are including in their design suitable space,
access ramps and toilets for the disabled. Listings are now available
of hotels, restaurants and cinemas that make adequate provision for
the crippled.

BLINDNESS

The legal definition of blindness in the United States and Canada
is a visual acuity (page 149) of less than 20/200 after correction of refractive errors; this implies that a patient is unable to read the large letter
E at the head of an eye chart, although he still may have a useful
appreciation of both large objects and colors. The number of people
with a defect of this order is larger than indicated by the previously
quoted economic assistance figures, some 2 per 1000 of the U.S.
population, and as many as 40 per 1000 in middle-eastern countries;
to this total must be added the vast group—perhaps 5 percent of the
U.S. population—who have lost the sight of one eye.

A large proportion of blindness is preventable, as can be appreciated
from the list of principal causes:

cataract (often operable)
glaucoma (treatable if detected at an early stage)
hypertension (treatable in many cases)
diabetes (treatable)
retrolental fibroplasia (caused by administration of an excessive percentage of oxygen to premature infants)
rubella (may cause a combination of blindness, deafness and congenital heart disease if it occurs during pregnancy; preventable by
vaccination)
injury (responsible for a surprisingly low 8% to 9% of all cases of
blindness; again often preventable by corneal transplant or retinal attachment operations).

Having regard to etiology, it is not surprising that more than 60
percent of all blind persons are over the age of 65. Only 12 to 15
percent of those registered are employable; success is more likely if
the patient is young, and less likely if he suffers from multiple handicaps. It is important from the psychological point of view that a blind
person seeking employment should have at least average productivity,
and he should be told if he is not achieving what is expected of him.

Nothing is more denervating than a feeling that an appointment has been made on a compassionate or charitable basis. At the same time, a blind person must be counseled that his earning potential will be below average for his intellect, and that his chances of advancement will be less than for a normally sighted person.

Immediate Adjustment and Training

Where blindness is of sudden onset, an initial orientation course is of value; this varies from 3 to 24 weeks in different centers, and in some instances classes are open to more long-standing cases of blindness who have failed to make an adequate adjustment to their condition.

A prime emphasis is upon conquering the fears of the blind, such as loss of employment and consequent reduction of status within the family, problems of movement and difficulty in interpersonal communication. Orientation and mobility must be developed through the training of other sensory receptors and particularly the use of a long cane when walking. Manual dexterity must be improved within the limitations of the patient and new hobbies such as basketry, woodwork and metalcraft must be learned to occupy leisure hours. Techniques of speaking, particularly the art of looking naturally at an unseen audience must be taught, and where intelligence permits, training should be given in the use of a braille typewriter. Safe forms of physical recreation such as swimming and dancing must be emphasized, and social workers should lead frank group discussions of the problems of return to the working world. Throughout this period, the patient must be observed closely, and an assessment made of his capacity for independence both at work and at home.

Towards the end of such a course, vocational counseling is given. This includes extensive testing of aptitudes and temperament. Employment opportunities for the blind now range far beyond the traditional basketweaving in a sheltered workshop. Professional life as a lawyer, economist, teacher or social worker is quite practical for those of sufficient intelligence. Technical level workers can find employment as computer programmers, transcription typists, and x-ray film processors. As with the ordinary populace, the majority of the blind must seek employment in industry, and it is now well established that mechanical equipment is not hazardous to a blind person if he has received appropriate training; in fact, the accident rate is usually *below* the average for normally sighted workers. The instructor must initially describe danger points on the equipment to be operated, stressing the identification of individual parts by shape, weight, texture and size. An assembly operation that involves small and easily lost components is unsuitable for a blind person. As with other types of disability, small modifications to the working area may be helpful. Many legally blind people have some vestigial appreciation of form and color, and can be helped by a marked color contrast between the machine and the factory wall. Bench stops to assist in the correct manual alignment of parts are also useful.

Government is now providing increasing aid to vocational training programs for the blind, both through normal community channels (as in the computer programming course for the blind at the University of Manitoba) and through specific institutions (such as the Canadian National Institute for the Blind). The C.N.I.B. offers workshop training to improve both manual dexterity and tolerance of a normal working day, business experience through its catering department, and dictaphone typing courses for business girls. The latter are unable to use shorthand, and must in consequence become more efficient than the average secretary at transcription work.

Communication with the Blind

Information flow in the blind is largely dependent upon the special senses of hearing and touch. Equipment such as watches and other instruments can be adapted for touch reading. The main limitation arises when reading books and correspondence.

The traditional solution, widely available for more than a hundred years, has been the use of braille. This codes letters and syllables in terms of six dots, arranged in two vertical columns. A writer operates an embossed zinc plate with six keys, *playing* each letter or syllable like a chord on the piano. Computers have now been programmed to provide a braille readout, but considerable difficulties arise from the structure of the language, such as the use of headings and abbreviations, and only about 70 percent of material can be transcribed directly from a computer tape; at the present time, a braille typewriter is cheaper and more accurate.

Braille translations of classnotes and even textbooks can be provided for university students, but problems arise from the cost of preparing what are extremely limited edition materials and also from the physical bulk of the finished article. Despite developments of microbraille, a *Pocket Oxford Dictionary* occupies a sixteen-inch cube!

A number of alternative reading devices are in the experimental stage. These use a group of photocells to scan each letter and emit either vibrations that are transmitted to the fingers, or a series of auditory tones; with the latter type of coding, it is possible to introduce the principle of compatibility. For instance, the letter W can be represented by four tones (descending, ascending (short), descending (short) and ascending). Unfortunately, none of these devices as yet operate sufficiently rapidly to provide a worthwhile means of communication. Typical vibration systems run at no more than sixteen to twenty words per minute, and auditory tone systems do not exceed forty to fifty words per minute. Speed will only be increased as the machines can be *taught* normal reading patterns that involve instantaneous recognition of whole words and phrases; unfortunately, the logic of such techniques is so involved that it has defied useful coding.

For recreational reading, the most practical solution is the use of long-playing tape cassettes; if books are read by professional actors with verve and understanding, they provide most satisfactory leisure entertainment. A single cassette will run for as long as twelve hours, providing material from one to two books.

The student may elect to take a braille notepad or typewriter into the lecture hall, but usually is defeated by the bulk of his notes. The same problem of excessive bulk arises if whole lectures are taped, and the most effective plan seems for the blind person to sit in a quiet corner and whisper a summary of salient points into a small tape recorder as the lecture proceeds.

CHAPTER 16

PROBLEMS OF UNDERDEVELOPED NATIONS

IN THE HIGHLY DEVELOPED COUNTRIES of western Europe and North America, the task of the ergonomist is changing, and in certain areas is actually diminishing. We know how to produce all that we can conveniently consume, usually at a minimum physiological cost, and the *purpose* of our economic system is currently being redefined in terms of more humanistic goals such as the personal fulfillment of the worker.

The majority of *underdeveloped* nations have not yet reached this point in evolution. A lack of mechanical equipment still forces men to labor to the limit of their physiological powers, often hampered by an adverse climate, undernutrition and disease. The balance between energy expenditure and the rewards of primitive hunting and agriculture is precarious—famine is never far distant (Kemp, 1971). And at the same time, the advancement of *civilization* is frighteningly rapid. All the psychological problems encountered by an older worker in a developed country are seen in a very acute form.

ENERGY EXPENDITURES

One of the hypotheses put forward by investigators associated with the International Biological Programme was that primitive peoples still dependent upon hunting and/or marginal agriculture would be more fit than their white counterparts as a consequence of a greater expenditure of physical energy in daily life. How true is this stereotype of the primitive nomad and villager?

Much depends upon the terrain. Is the hunter crossing firm and flat grassland, climbing steep ridges, or wading through deep snow? How close is the game to the settlement? Can food be obtained within a few hours' journey, or is it necessary to carry overnight equipment (Lee, 1969)? Is a vehicle such as a sledge or boat used, or must a heavy carcass be carried long distances upon the back? Do the family accompany the nomad in the search for food? If a settled life is chosen, is the peasant hoeing a rich alluvial soil, or tending a hard and stony outcrop?

The work undertaken by some primitive peoples would undoubtedly be classed as heavy by normal industrial standards. Data for Canadian

Eskimos living in the eastern Arctic are shown in Table VI; the majority of these readings are based on oxygen consumption as measured by a portable respirometer, but one (chewing skins) is of necessity based on heart rate measurement. The caloric cost of this particular item seems overestimated, probably because a relatively small muscle group is active.

The community that we have studied (Igloolik, 69°N) is currently undergoing rapid transition. Prior to 1960, the people lived a nomadic life, ranging widely over the hunting grounds of Baffin Island. Over the last ten years, they have migrated to a fixed settlement of some five hundred people, with prefabricated housing, schools and churches. A considerable amount of heavy equipment has been brought to the village, including power cranes and a variety of tracked vehicles. The Eskimos have purchased snowmobiles and power boats, a nursing station has been established, and in 1969 an airstrip was built, bringing the village within a five-hour journey of Montreal. Despite a superficial sophistication, community life contains many items of activity found in older ergonomic surveys, but no longer typical of life in southern cities (Godin and Shephard, 1973). There is as yet no piped water supply, and in consequence much effort is spent upon the carriage of water in the summer and equivalent quantities of ice in the winter. There are no garbage disposal machines and no sewage pipes, so that both domestic waste and human excreta must be collected in large bags and transferred to tracked vehicles for removal from the village. Many of the women walk about the community and perform their household chores while carrying a baby in the traditional *yappa*. The household appliances of the south are not universally available; clothes are often washed by hand, and bread or bannock must be made rather than purchased at the store. Some homes have a fair amount of furniture, but many tasks are still performed while sitting or kneeling upon the floor. Vehicle driving is a much more energetic pursuit than in the south, particularly when a tracked vehicle is traveling fast over rough ground; furthermore, any necessary field repairs must be carried out personally rather than through a garage mechanic. The hunting grounds for seal, walrus, caribou and char are accessible from Igloolik, and even those who have accepted the white man's role in such pursuits as grocery clerk, carpenter, painter and electrician still devote a significant proportion of leisure time to the search for game.

At least 20 percent of our sample of Eskimo men accepted hunting as their primary occupation. Because of the physical difficulties associated with following and monitoring subjects, activity data for this group is more sketchy. However, it was possible to supplement relatively infrequent measurements of oxygen consumption by ventila-

TABLE VI. ENERGY COST OF ESKIMO ACTIVITIES. ALL VALUES STANDARDIZED TO VALUE APPROPRIATE FOF A 65-kg MAN OR 55-kg WOMAN. DIRECT MEASUREMENTS OF OXYGEN CONSUMPTION EXCEPT WHERE MARKED BY * (RESPIRATORY MINUTE VOLUME ESTIMATE) OR † (HEART RATE ESTIMATE).

MEN

Sedentary 2 kilocalories/min 8.4 × 10³ J/min	*Light* 2.0–3.3 kilocalories/min 8.4–13.8 × 10³ J/min	*Moderate* 3.3–5.4 kilocalories/min 13.8–22.6 × 10³ J/min	*Heavy* 5.4 kilocalories/min 22.6 × 10³ J/min
Driving dog team 1.4	Office clerk 2.2	Soapstone carving 3.4	Igloo building 5.4
Standing at floe edge 1.6	Watching seal hole 2.2	Snowmobile driving (Skiddoo®), no load 3.4	Checking nets by canoe 5.7–7.6*
Snowmobile passenger 1.9	Tractor driving (Caterpillar®) 2.3	Snowmobile driving (Bombardier®) 3.5	Seal hunt 5.8*
	Snowmobile driving heavy load one passenger 2.5	Carpentry 3.5	Snowmobile delivery 5.9*
	repairs 2.7	Cement mixing 3.6	Loading meat 6.4*
	Light garage work 2.9	Sled passenger 3.6	Skinning seal 6.5*
	Electrician 2.6	Oil delivery 3.8	Garbage collection 6.5–7.8
	Grocery clerk 2.7	Outside painting (scaffolding) 4.0	Walrus hauling 6.7*
	Making knives 2.9	Hauling nets 4.0	Digging ice hole 6.8*
	Boat repairs 3.0	Garage work 4.2	Loading sledge 6.9*
	Water distribution 3.0	Janitorial work 4.2	Feeding dogs 7.2
	Painting indoors 3.2	Snowmobile repairs (Bombardier®) 4.3	Walrus skinning 10.2*
		Warehouse work 4.3	
		Cutting dog meat 4.3	
		Tractor driving (Case®) 4.5*	
		Generator maintenance 4.8	
		Aircraft unloading 5.0	
		Ice distribution 5.0–5.3	

WOMEN

Sedentary	*Light*	*Moderate*	*Heavy*
Dishwashing 1.7	Washing clothes 2.3	Walking (baby in Yappa) 3.8	Chewing skins 5.3†
Snowmobile passenger 1.8	Housework 2.4		
Sewing 2.0	Scraping furs 2.7–3.1		
Soapstone carving 2.0	Making bannock 2.9		
	Washing floors 2.9		
	Walking 3.0		

Based on material presented by the author (with G. Godin) at the Symposium on Polar Human Biology. (Cambridge, 1972).

tion readings and diaries kept by our observers. The majority of Eskimo hunts had rather long periods of relative inactivity—the snowmobile trip to Baffin Island, the boat voyage to the walrus breeding grounds, the relentless watching of the seal hole. But interspersed with such light work, there were episodes of quite vigorous labor—the strong paddling of a canoe in a sudden squall, the hauling of a boat onto the shore or a walrus onto an ice floe, the stripping of skins, the throwing of 20- to 30-kg hunks of meat into the hold of a large whaling vessel, the carrying of 50- to 70-kg caribou carcasses over several miles of hilly territory, and the building of an igloo to provide shelter for the night. Intermittent but intense activity is likely to develop cardiorespiratory fitness (Shephard, 1973c) and our studies have confirmed a large maximum oxygen intake in those villagers who were making frequent hunting trips (Rode and Shephard, 1973). The aerobic power of the 20- to 30-year-old hunters was 36 percent greater than that of a comparable white population, and the leg extension strength was also 50 percent greater (although on account of a short leg length, the Eskimo's advantage in torque was only 13% to 18%).

Because few of the villagers were completely sedentary, even those employed by the government, making souvenirs, or living largely upon welfare had a greater aerobic power than residents of large North American cities.

Our findings are somewhat at variance with previous studies of primitive populations (Cumming, 1966). The reported aerobic power of groups such as Eskimos, Arctic Indians, Bantu and Kalahari bushmen has ranged from 41 to 50 ml/kg min—very normal figures for young city dwellers. One other population known to have an above-average aerobic power are the nomadic Lapps (53 ml/kg min); this group travels long distances over rough terrain in the course of tending their reindeer herds. Some investigators have argued that in communities with a low average aerobic power, even the hunters are not really active. They spend no more than one or two days per week in the search for game, and compensate for this activity by an excess of lethargy on intervening days. This may be true of idyllic tropical villages such as Easter Island. But in Igloolik, hunting is not an intermittent or seasonal adventure. Throughout the eastern Arctic (Kemp, 1971), the serious hunter works almost continuously, with only brief periods of respite in the village. Sometimes, he will travel for twenty-four hours at a stretch, spending as much as 4000 kilocalories of energy—a heavy metabolic loading for what are basically small people. The women of Igloolik also do not lack for physical activity. International agencies set a 15 to 20-kg limit upon the recommended load for women workers (I.L.O., 1964). The Eskimo housewife fulfills her duties while carrying a baby of one to three years' age upon her back, sometimes working under the primitive conditions of a field camp, and denied most of the power equipment used by the city householder.

There is no strong evidence that other primitive populations have an easier life than the Eskimo. In many communities, acculturation is less advanced, and mechanical equipment is less readily available than in the Arctic. The Kalahari bushmen may walk as far as twelve miles per day in search of Mongono nuts (Lee, 1969), while many native women will carry both children on their backs and large burdens upon their heads. Possible explanations of the previously reported low values for aerobic power include:

1. the use of small and unrepresentative samples of a community,
2. the inclusion of older subjects, and
3. the existence of malnutrition and disease.

ADVERSE CLIMATE

The primitive worker has two climatic disadvantages. Firstly, his natural habitat may be characterized by extremes of climate—the bitter cold of the Canadian arctic, or the steaming heat of a South American jungle. Secondly, because of limited technology, he may be exposed to adverse conditions more frequently than a white person would be (page 66).

This is well borne out by data from recent scientific expeditions to the Antarctic. Because of skillful use of prefabricated housing, tracked vehicles and aircraft, coupled with modern clothing, few of the explorers have been seriously exposed to ambient conditions. The Eskimo, equally, has a well-developed cold technology; caribou clothing has excellent insulating properties, and windbreaks and temporary shelters are constructed from soft snow with remarkable speed and efficiency. Nevertheless, there is no question that the Eskimo hunter is still exposed to cold. Sometimes he must travel many miles perched high on a heavily laden sledge, and at other times he must remove his mittens and work for an hour or more repairing the ice-cold metalwork of his snowmobile. There are several practical consequences of the harsh environment:

1. Since hard physical work is performed intermittently, the insulation of clothing must be adjustable to accommodate variations of activity.
2. Unless the energy expenditure required for occupational activity is sufficient to maintain body temperature, metabolism must be supplemented by occasional bursts of deliberate voluntary effort. Thus, when on a long sledge journey, the Eskimo will boost

his limited heat production by running briskly over the ice whenever he becomes unduly cold.

3. The basic energy cost of outdoor activities is increased by heavy clothing, a greater muscle viscosity, and possibly a trend towards anaerobic metabolism. If shivering develops, this further increases energy expenditures.

4. The Eskimo becomes acclimatized to the cold, both over a single winter, and also through a lifetime of exposure. In particular, he is able to maintain manual dexterity when the fingers of an average city dweller would be numb and useless. Such adaptations are being lost with the transition to a settlement culture, and currently the Igloolik village council is faced with pleas for a school bus on the grounds that children have difficulty in walking the length of the village in cold weather.

In a hot climate, activity is limited by the reverse problem of heat dissipation. The heat loss of a worker is largely dependent upon the evaporation of sweat, and even if all of the secreted sweat were evaporated (1 to 2 liters per hour), energy expenditures would be restricted to about 10 kilocalories/min (page 66). In dry heat, there is no problem in performing quite heavy work, but in humid heat much of the sweat rolls to the ground, and it is difficult to sustain more than moderate effort.

UNDERNUTRITION

Mild undernutrition is not necessarily a disadvantage to the laborer; body fat is depleted, and aerobic power, expressed as milliliters of oxygen transported per kilogram minute, is proportionately increased. The cost of activities that involve the raising and lowering of body weight is also diminished (page 307), and in a hot climate the elimination of body heat proceeds more readily. The main disadvantages are some reduction of glycogen stores (affecting endurance in protracted effort) and (in the arctic) the negative adaptation of diminished insulation.

Gross dietary deficiencies impede normal growth and development. At least ten amino acids are essential to protein formation, and six of these must be supplied to adults if wasting of muscles is not to occur. Thus, serious food shortages may produce a population with little flesh, seriously lacking in muscular strength and endurance.

Most primitive peoples are thin relative to white, city dwellers, and some are frankly malnourished. One of the simplest methods of assessment is a simple height/weight table. This approach was used by Dreyer (1920) at a time when malnutrition was still a major concern of Western civilization; body weight is still regarded as a very useful method to screen the working capacity of miners recruited in central Africa (Wyndham *et al.*, 1963).

In primitive civilizations where 80 or more percent of the adult population are engaged in growing or hunting food, it takes but a slight alteration of climate or of hunting conditions for famine to arise. Older villagers at Igloolik are able to recall episodes of starvation before the coming of the white man. Now, any shortage of game can be made good by the use of welfare checks at the village grocery store. The current problem in the Canadian Arctic—as in many developing communities—is not hunger, but a wise choice of foods. Excessive purchase of refined carbohydrates is leading to premature dental decay, and in those families that fail to budget wisely protein intake may be inadequate. This situation is exacerbated by the trend towards snowmobile hunting. The snowmobile ranges further afield than the traditional dogsled, and thus encourages the hunter to collect furs at the expense of meat.

The young adult Eskimo, like most tribesmen, has little subcutaneous fat. The average skinfold thickness of our sample was only 6 mm (Rode and Shephard, 1973a), compared with 12 to 16 mm in a city dweller (Shephard, 1969a). If 4 mm of the fold is attributed to a double layer of skin, then the Eskimo body is covered by a 1 mm layer of fat, compared with a 4 to 6 mm layer in our urban sample.

DISEASE

Chronic infection may lead to a debilitating general weakness that prevents the effective performance of heavy physical work. Further, specific diseases of the bloodstream and the respiratory tract may strike at individual links in the chain of conductances carrying oxygen from the atmosphere to the working tissues, thereby limiting cardiorespiratory endurance.

Until recently, tuberculosis was rife in northern settlements. This commonly restricted the bellows function of the chest, leading to a poor matching of ventilation with pulmonary blood flow, and often an impaired diffusing capacity in the lungs. Crowded homes and village meeting halls still give rise to a high incidence of other chronic respiratory infections, but the pulmonary function of the Eskimo is now *better* than that of the white, city dweller (Rode and Shephard, 1973b). Anemia remains common in tropical countries. Malaria, bilharzia, ankylostomiasis and a variety of intestinal parasites share responsibility for this condition. Red cell production of the unfortunate native may be ten times normal and yet he is unable to maintain enough red blood pigment for the effective carriage of oxygen.

In most instances, the diseases concerned could readily be cured in a white community, but remain to reduce the productivity of natives who lack even elementary medical facilities. The problem of health care delivery is being rapidly rectified in most countries. State nurses are now attached to Canadian Eskimo and Indian settlements, and with the building of airstrips, more difficult diagnostic and therapeutic problems are readily resolved in the large hospitals of southern Ontario and Quebec.

However, other problems are increasing with acculturation. As in some of the emerging nations, the disintegration of the traditional social framework in larger Eskimo and Indian settlements is accompanied by a high incidence of alcoholism. There may be associated obesity, and (because of poor protein intake) myocardial degeneration.

PSYCHOLOGICAL PROBLEMS

Extinction of game and wild crops has destroyed the natural habitat of many primitive peoples. Ironically, increased nursing care has hastened the destructive process by initiating a vast upsurge of population (Godin and Shephard, 1973). A region that provided subsistence hunting for 250 Eskimos is useless for feeding 500 or 1000.

The transition from a nomadic Paleolithic existence to a modern urban-type culture has proceeded with breathtaking rapidity. In ten years, the Arctic has seen bone arrowheads and dogsleds replaced by high-powered rifles, tracked vehicles and aircraft; the housewife also has moved from tent and igloo to a three-bedroom bungalow, and children have jumped from rudimentary learning to a twelve-grade educational system.

It is hardly surprising that problems have arisen in introducing such people to western patterns of life. Soap was a new concept to many Eskimo housewives until the arrival of a dedicated public health nurse. Time was an equally unaccustomed constraint. One can imagine the practical difficulties of running an airline when the ground crew suddenly decide that the weather looks auspicious for a week of hunting. Maintenance of equipment is a major problem; simple preventive procedures that would be second-nature to a southern mechanic just do not occur to the Eskimo. The same story is repeated in many parts of Africa and Asia, valuable equipment being destroyed by lack of elementary precautions such as routine lubrication. On the other hand, many Eskimos are very adept at improvising after disaster has struck, and can repair a damaged snowmobile under circumstances that would be fatal to a white person.

The overall reaction to white culture depends upon tribal characteristics and the degree to which the native peoples have been exploited.

Some are glad to see the relief of hunger, the arrival of education and other manifestations of *progress*, but many labor under an oppressive feeling of inferiority. The ultimate disposition of such peoples is a continuing concern for many governments. Numbers are currently incompatible with traditional modes of survival. Must an ever increasing population be condemned to subsistence on government handouts? Can local crafts and tourism be developed to the point where the economy will become self-sustaining? Or must such people be subjected to the even greater turmoil of incorporation into a large metropolis?

CHAPTER 17

ANTHROPOMETRY IN THE SERVICE OF DESIGN

PHYSICAL ANTHROPOMETRY DESCRIBES the dimensions of the body and its various components under static conditions. Such data is a necessary preliminary to movement analysis (page 173). The information is also relevant to descriptions of body type, the standardization of biological data, calculations of the mechanical efficiency of effort, and the design of equipment, controls, furniture and clothing.

BODY TYPE OF THE WORKER

Descriptive Approaches

Several distinctive body types have been recognized since classical times. Initially, the different body forms were related to personality and susceptibility to disease. More recently, it has been suggested that what has been termed "somatotyping" may help in the selection of both athletes and those undertaking vigorous industrial work. Certainly, a long and thin body type is an advantage in running and jumping, while a heavy, thick-set build is an asset to the footballer. Again, a muscular worker is more likely to be a success in furniture removal than is a lean and poorly muscled individual.

While the overall shape of man has not changed greatly over the last 2000 years, the names allocated to the various body types have shown a continual evolution. In the classical era, Hippocrates described the short, thick-set "habitus apoplecticus," red-faced, jovial and forceful, but liable to death from apoplexy, and the long, thin "habitus phthisicus," more inward looking, and liable to die of phthisis (tuberculosis). Halle (1797) noted four types: the fat "abdominal" man; the strong "muscular" man; the long, slender-chested "thoracic" man; and the large-headed "cephalic" man. Kretschmer and Enke (1936) distinguished the round and compact form of the "pyknic" from the muscular "athletic" build and from the long, thin and asthenic "leptosome." More recently, Sheldon (1940) classified his population into fat "endomorphs," muscular "mesomorphs," and thin "ectomorphs." All of these authors were describing essentially the same phenomenon. Most of us can recognize a typical member of any one of the three basic groups, but unfortunately the majority of the population occupy an intermediate position. Sheldon estimated that 7 percent of a population could be classified as pyknic, 12 percent as athletic, and 9 percent

305

as leptosomal, leaving a vast 72 percent that shared features of two if not all three body types.

In attempting to overcome this problem, Sheldon suggested a semiquantitative assessment. A series of measurements were to be made on standard photographs taken from the front, side and rear of the individual. All readings were to be expressed as ratios to standing height. Subjects with a broad face were to be regarded as mesomorphs or endomorphs rather than ectomorphs. If the chest/height ratios were larger than the waist/height ratios, a person was mesomorphic, whereas if the reverse was true he was endomorphic. Each individual was to be rated for each of his three basic characteristics on a seven-point scale (seven signifying a maximum effect). Thus a man might be classified as type 642—strongly endomorphic (6), slightly mesomorphic (4), and with little evidence of ectomorphy (2).

Sheldon's approach remains somewhat subjective. Further, it is by no means clearly established that the scales used are linear, or that the three factors identified are orthogonal (statistically unrelated to each other); indeed, there is some evidence that the young mesomorph degenerates into an older endomorph. In essence, Sheldon's rather elaborate system does no more than specify such characteristics as muscle size, percentage body fat and bone dimensions. It is thus preferable to use more direct measurements of the variables of interest (Dupertuis and Emmanel, 1956).

Muscle Mass and Body Strength

From some points of view, muscle strength is more important than muscle bulk. Thus, the length of lever arm required on a given manual control will be governed by the average force that a given group of muscles can exert for a particular angulation of a joint (page 171). Further, industrial training is not necessarily associated with a gain in muscle bulk; as a worker becomes more skillful in performing a strenuous physical task, he learns to distribute his efforts over a larger total number of muscle fibers, so that the maximum force may be increased without gain in muscle size.

Muscle force is measured quite readily by dynamometers and cable tensiometers. However, the results obtained depend upon motivation and acquired skills, and force measurements are not particularly well-suited to either initial assessment or specific evaluation following sickness or compensable injury.

When dealing with poorly nourished peoples in underdeveloped countries, the tolerance for arduous work is closely related to muscle mass and thus body weight (Wyndham, 1966; page 302). The physician often uses simple measurements of muscle girth to assess wasting following industrial injury. This is a rather unsatisfactory procedure. A large girth may reflect well-formed muscles, but it may also be attributable to an excess of subcutaneous fat or heavy bone structure. Furthermore, muscle wasting may be accompanied by an almost equal gain of subcutaneous fat. The girth technique can be much improved if it is supplemented by readings of bone size and skinfold thickness (Jones and Pearson, 1969).

Alternatively, soft tissue radiographs of the thigh can be taken in the postcroanterior and lateral planes, with measurement of the muscle shadow at a standard distance above the knee joint. Technical questions such as the optimum positioning of the limb, the influence of variations in muscle tone, and dilution of the muscle shadow by intramuscular fat and edema fluid have yet to be fully resolved, but nevertheless this approach offers a simple and relatively incontravertible measure of muscularity.

A third method is to sit the worker within a whole body radioactivity counter. Such a machine detects emissions from the naturally occurring potassium isotope^{40}K. Since most of the body potassium is intracellular, radioactive emissions are proportional to muscularity. Naturally, this type of apparatus is expensive and available only to a few specialized laboratories of industrial physiology.

Body Fat

The industrial importance of body fat lies in (1) its insulating properties, (2) its influence upon the energy cost of standard activities, and (3) its impact upon life expectancy.

Heat loss from the body core is restricted by the relatively avascular layer of subcutaneous fat. To the worker in a cold environment—the arctic, the continental shelf or a refrigerated food store—a certain degree of obesity may be an advantage (pages 66–70). However, it is a definite disadvantage when work must be performed in a hot climate such as a bakery, a steel furnace, a deep mine or the engine room of a ship.

An excess of body fat increases the energy cost of most industrial tasks. The influence of body weight upon oxygen consumption has sometimes been expressed in the form

$$\dot{V}_{O_2} = A + B(W)^n$$

where $\dot{V}_{O_2}$ is the steady state oxygen cost, W is the body weight, A and B are constants, and n is an exponent of 0.75 to 1.0 (Brown, 1966). Practical application of this equation has not been particularly helpful, for two main reasons: (1) resting energy expenditure is not distinguished from the cost of work. Many industrial activities add but little to the resting oxygen consumption, and the energy cost of labor is over-shadowed by a much larger resting metabolism. (2) In order to solve the equations on small calculating machines, the unproven assumption has often been made that n = 1.

Recent studies (Godin and Shephard, 1973) have separated oxygen consumption into three components, as follows:

$$\dot{V}_{O_2} = A + B(W)^{0.75} + C(W)^n$$

A is here an error term, B is a constant relating resting metabolism to body weight, and C is a constant relating the cost of industrial effort to body weight. For sedentary and clerical work, the exponent to the third term is small and statistically insignificant. Low exponents (n = 0.1 to 0.2) are also found when carrying out light arm work (such as operating a machine press), but values close to unity occur when the body weight must be raised and lowered (as when lifting heavy boxes).

At the white-collar level, the adverse effects of obesity upon life expectancy are a significant economic consideration for the individual employer. From the clinical point of view, obesity is often regarded as a pathological, all or none condition in which the body weight exceeds the prescribed norm by a fixed margin. However, the percentage of body fat in an industrial population shows a continuous distribution from the very thin to the very fat worker. It is not unknown for a reduction of body weight to be specified as a condition of employment when hiring senior personnel or arranging their insurance and medical benefits. In the armed services, also, promotion may be contingent upon meeting minimum standards of leanness. Airlines usually specify height/weight ratios for stewardesses, although here emphasis is upon an aesthetic purpose rather than any relationship between body weight and health or efficiency of work.

Five simple indices of body fat may be derived from measurements of height (H) and weight (W):

1. the weight/height ratio (W/H)
2. Quetelet's Index (W/H^2)
3. the ponderal index (H/W$^{1/3}$)
4. excess weight relative to actuarial standards
5. **Tuxford's index** $\left(\dfrac{W}{H} \times \dfrac{336-m}{270}\right.$ in boys; $\dfrac{W}{H} \times \dfrac{308-m}{235}$ in girls, where m is the age in months).

The first three measures are sometimes used for industrial surveys. On theoretical grounds, Quetelet's index or the ponderal index might be thought better than the simple weight/height ratio, but in practice all three indices show rather similar coefficients of correlation with independent estimates of body fat (Shephard *et al.*, 1971). Tuxford's index is applied primarily to children. The present author prefers to use excess weight as a simple criterion of fatness. In 1959, the Society of Actuaries proposed *ideal* body weights for men and women of various heights, subdividing allowable ranges into figures for "small," medium" and "large" framed individuals. *Ideal* figures were those for the 25-year-old North American insured population, and in many industrial situations advantage might be gained from a somewhat lower body weight. Frame classification is arbitrary, and is often colored by the obesity of the judge. The present author thus ignores frame size, and rates the individual relative to the average *ideal* weight for a person of the same height (Table VII).

Since height and weight can be measured with considerable accuracy, the various simple measures of body fat all give a fair indication of the obesity of a community or work force. However, they may do less than justice to the individual employee, since a large body weight may reflect either fat or muscle. The present author overcomes this difficulty by coupling excess weight with ancillary data on skinfold thickness. A second useful approach is to keep an annual record of an employee's weight. If there is a gain of thirty pounds between the ages of 25 and 40, this is more likely to be fat than muscle!

Much of the body fat is subcutaneous and the thickness of a double fold of skin and fat gives a good impression of the overall body fat content. Some observers measure as many as ten skinfolds, but as

TABLE VII. AVERAGE *IDEAL* WEIGHTS OF INDIVIDUALS IN RELATION TO HEIGHT. DATA OF SOCIETY OF ACTUARIES (1959) AS MODIFIED BY PRESENT AUTHOR (1972).

| Height (no shoes) | Ideal weight (wearing indoor clothing) | |
| | Men | Women |
Inches	Pounds	Pounds
58	—	107
59	—	110
60	—	113
61	124	116
62	127	120
63	130	123
64	133	128
65	137	132
66	141	136
67	145	140
68	149	144
69	153	148
70	158	152
71	162	—
72	167	—
73	171	—
74	176	—
75	181	—

with most examples of data collection, the accuracy of body fat predictions increases approximately as the square root of the number of readings that are taken. Three skinfolds (Weiner and Lourie, 1969) were proposed for use in the International Biological Programme (triceps, subscapular and suprailiac). These were chosen on anthropological grounds, and it is still debatable whether they are the best three folds to monitor changes in body fat with aging and obesity. Nevertheless, in many contexts, they give as much information as the measurement of six to eight folds (Shephard *et al.*, 1969). As with any production line task, some feedback of results is necessary to ensure accurate skinfold measurements; indeed, unless a reference patient is examined periodically, field workers can easily develop interobserver errors of 25 percent or more.

Soft tissue radiographs provide a more sophisticated estimate of body fat. Other procedures are open to the physiologist, including underwater weighing, measurements of body volume by gas displacement, and ingestion of materials such as tritiated water that are readily distributed throughout lean tissues. However, these methods are too complicated for large-scale use.

STANDARDIZATION OF BIOLOGICAL DATA

Many of the biological variables of interest to the ergonomist —aerobic power, anaerobic power, working capacity, muscle strength and so on—vary with the size of the worker. Should standardization be attempted, or is it sufficient to direct attention to the absolute readings?

Physiologists have traditionally expressed many of their findings as a ratio to body surface area. The latter is approximated by the formula

$$S = W^{0.425} \times H^{0.725} \times 71.84.$$

DuBois (1927) reached this conclusion many years ago through tedious experiments in which subjects were wrapped in long strips of paper. He prepared a nomogram for carrying out his calculations, and it is now a simple matter to program the same formula for a small computer (Shephard, 1970a). DuBois was interested in metabolism and body heat loss, and his choice of body surface area as a basis of standardization was obviously appropriate (although it can be argued that a linear regression of resting metabolism upon body surface area is preferable to a simple ratio). Other bodily functions such as lung volumes seem more closely related to sitting or standing height (Shephard, 1970); indeed, if a broad range of sizes including children are to be covered

(Rode and Shephard, 1973b), the data may best be described by H_n (where n = 2.7 to 3.0). Work-related functions such as aerobic power and muscle strength logically seem related to lean body mass. However, the measurement of lean body mass is quite complex, and such data are therefore standardized by expressing results as a simple ratio to total body weight.

It may be objected that a weight ratio penalizes the fat person. The classic reply to such an objection has been that it is a fair penalty. The body weight must be carried about and adds to the cost of industrial activity. As we have seen above, this is true of tasks that involve lifting the entire body. If only a single body segment is active, the penalty is smaller, but the gross oxygen cost may still vary somewhat with body weight, because the heavy person has a greater resting metabolism than someone who is lighter.

Muscle force has traditionally been measured with a dynamometer or a tensiometer calibrated in kilograms. If it is intended to determine the crushing force of a doorman's handgrip, a dynamometer reading may be adequate. However, if the problem is that of turning a heavy lever, it is necessary to take account, firstly of the torque that can be generated about a given joint, and then the leverage that the limb can apply to a specific control. A long arm gives a person speed, but robs him of power. The ectomorph (page 305) who attempts to manipulate a heavy control has several disadvantages, including poor musculature, dissipation of the available force by his long limbs, and poor balance due to a high center of gravity.

MECHANICAL EFFICIENCY AND TASK DESIGN

The design of a machine can be evaluated from its efficiency in carrying out its intended purpose. Similarly, in a man/machine system that calls for a substantial expenditure of physical energy, one criterion of task design may be the mechanical efficiency of the human operator.

Human efficiency may be expressed as a gross or a net value. The gross figure relates the overall energy cost to the work performed, while the net figure makes allowance for energy usage under resting or basal conditions. Energy expenditure can be measured by direct calorimetry, but this involves living and working in a special chamber; under industrial conditions, it is estimated indirectly, from observation of the worker, records of food consumption, or measurements of oxygen intake. Food studies work quite well on closed communities (ships or military patrols), while portable respirometers can be taken onto the shop floor. One liter of oxygen intake is equivalent to between 4.7 and 5.1 kilocalories of energy usage, depending upon the nature of the fuel (fat or carbohydrate).

The work that is performed may be largely external. The simplest example of this is seen in the laboratory, when a man pedals a bicycle ergometer fitted with a heavy friction belt. Under such circumstances, the mechanical load can be calculated rather precisely from the frictional force and the number of flywheel revolutions per minute. The equivalent machine for the working man is a delivery tricycle. Much of the work is again attributable to friction, here occurring between the wheels and the road surface; the load can be calculated from the distance traversed (meters) and the coefficient of sliding friction between the wheels and the road surface (0.005 to 0.030 kg per kg of loaded tricycle). Depending upon the speed of the delivery boy, additional work is carried out against wind resistance and any contrary wind force.

In other situations, much of the total work is performed against body weight. The author can recall building sites in the north London suburbs where laborers would spend their days climbing ladders with a small load of bricks supported over one shoulder. Let us suppose that the bricks weighed 20 kg, that the body weight of the laborer was 80 kg, and the height of climb was two floors (5 meters). The total mechanical work achieved by each ascent was 500 kg-m, and assuming a 20 percent efficiency of ascent, the cost to the body was 2500 kg-m. The man then returned to ground level to fetch more bricks. In physical terms, he dissipated his entire store of 400 kg-m of work while making a carefully controlled descent of the ladder. This cost the body a further 500 kg-m of energy, a quarter of the price of lifting the body. The purpose of the exercise was presumably to lift bricks, and the useful contribution to this purpose was a mere 100 kg-m. The true efficiency was thus

$$\left(\frac{100}{2500 + 500}\right) \times 100, \text{ or } 3.33\%.$$

A system should function more efficiently than this, and if many bricks are to be lifted it may be acceptable to throw them to a higher level, or if their unit price demands careful handling to install a manual or power-operated lift.

When a man is walking on level ground, no external work is performed. Although it may be difficult to convince a mailman of this point, the potential energy of the body at the end of a walk is the same as it was initially. However, the mechanical efficiency is obviously not zero, and it can be calculated if account is taken of the movement of individual body segments. As with the builder's laborer, energy is dissipated in lifting and subsequently controlling the descent of body parts (page 195). If movement is rapid, appreciable energy losses may occur in acceleration and subsequent deceleration of the body (Margaria, 1971). The work performed can be estimated if the centers of gravity of individual body segments are defined, and their displacements followed by serial photographs. As with the cyclist, a runner also loses appreciable energy through wind resistance and friction at the ground surface.

It is not always practical to measure the mechanical work performed in industry. Fortunately, it may not be essential to have this information. If the weekly food consumption diminishes, or the steady state oxygen cost of an activity is reduced, it is highly probable that a more effective working plan has been devised. The main virtue of assessing absolute efficiency is to test the potential for improvements in task design. The inherent biochemical reactions (the breakdown of glycogen to CO_2 and water, the conversion of adenosine diphosphate to adenosine triphosphate, and the coupling of actin and myosin filaments within the muscle) have an overall efficiency of about 25 percent (di Prampero, 1971). Thus, if we estimate the mechanical efficiency of an industrial operation at 2 percent, it is worth giving thought to its improvement. However, if its efficiency is already 23 percent (as in the bicycle ergometer), no big gains can be expected. Applying this reasoning to the building laborer, we could spend much time teaching him to climb ladders, and might thus raise his efficiency from 3.33 to 3.5 percent; however, the main source of inefficiency is the carriage of body weight up and down the ladder, and greater gains of performance are likely from eliminating this mode of working.

DESIGN APPLICATIONS

Basic Design Data

Data on the height and weight of the body as a whole, together with the dimensions of individual body segments has now been collected on many different populations. Multiple tailors undoubtedly store such statistics, and a number of military reports acknowledge the value of this information to the clothing quartermaster (Morant and Gilson, 1945). Vital statistics also find ready application to the design of crew space, consoles and control panels.

The designer may predict the ideal size and shape of a seat or the layout of a console simply from accumulated anatomical and physiological statistics for the population in question. Alternatively, personnel may be asked to complete subjective ratings of equipment, and films may be taken under normal operating conditions. Commonly, a combination of these various approaches is necessary before putting a new design into general service.

The need for rigid conformity to minimum dimensions depends upon purpose. Consider a vehicle. The main purpose of a private car may be social status rather than economic transport, and comfort and appearance then take precedence over the dictates of body dimensions. On the other hand, in a military vehicle (whether an aircraft, space capsule, tank, submarine or destroyer) there is a constant search for a minimum of space compatible with working efficiency. Working space is pruned to the point where a minimum distance separates the limbs from control panels, three- and four-deck bunks make the maximum use of sleeping accommodation, and even hatches are held to the smallest possible size compatible with the escape of crew members.

Many items are designed for the *average* man. Thus, a work counter for a standing task is set at the optimum height for an average member of the work force. This is an economical solution, but may be fatiguing for both short and tall workers. The best solution, seen in some modern typewriter tables, is an adjustable bench height. Alternatively, the height can be set somewhat above the optimum, with removable floor platforms available to smaller workers.

Other designs take account of population extremes. An escape hatch has little value if it will not accomodate a large worker. A control pedal is equally useless if a small person either cannot reach it, or cannot exert the necessary pressure. Unfortunately, the shape of the distribution curve is such that it is both difficult and wasteful to accommodate everyone. In civilian applications, it may suffice to meet the needs of 95 percent of the population (that is, ± two standard deviations from the population mean); for example, a 7 to 8 cm adjustment of car seat position may be allowed, to bring the foot controls within reach of 95 percent of purchasers. In the military context, the need for such adjustments can be minimized by eliminating recruits of extreme body type. Aircrew, in particular, must fall within specified upper and lower size limits.

A further general consideration is the impact of environment upon space requirements. The pilot of a military aircraft may be quite tense, and thus will be sitting erect. He will need a very different type of seat from someone relaxing in a cinema or in the comfort of his own home. In the latter situation, a man will lounge untidily and need a seat at least two inches longer than that of the pilot. Social conditioning also plays a role, and presumably for this reason the seat length preferred by women increases much less under relaxed conditions.

Height

Standing height is less equivocal than most biological variables. It can be determined very simply with a rule and a rigid T-square.

However, because of its simplicity, the measurement if often made on large samples of a population, and quality control of data is then important. In some surveys, observers have been too impatient to remove their subjects' shoes, and there is now no way to determine whether heel height was half an inch or four inches.

The current height of the British male population may be gauged from careful measurements of a large sample of airmen completed by Morant and Gilson (1945). The mean height of this group was 172 ± 6 cm (68½ ± 2½ inches). Unfortunately, a single careful survey will not suffice for all time, as the average height of most populations is continually increasing (Rode and Shephard, 1973c). The problem of predicting the future load upon a system is well exemplified by the architecture and furniture of Europe. North American visitors soon interpret the phrase "crowned heads of Europe" in terms of painful encounters with doorways and bedsteads designed for medieval man. Over the past century, the size of prepubertal boys has increased by 1.0 to 1.5 cm per decade, faster in those from working class homes than in those with professional parents. Further, puberty and its associated growth spurt have advanced by about 0.35 years per decade (Harrison *et al.*, 1964). Such historical trends are of interest to those concerned with the equipment of schools (Oxford, 1969) and the production of children's clothing. Occasionally, the trend to taller children reaches public consciousness through a need to adjust height standards for children's fares on public transit. Since girls reach puberty earlier than boys, they also have an earlier growth spurt, and for a brief period are the taller sex; it may thus be the daughter of the family who becomes involved in an altercation with a bus driver regarding the payment of adult fares.

Height remains relatively constant over much of the adult span, but tends to diminish in later life. This is partly psycho-cultural—the adoption of a less vigorous and aggressive posture—but also reflects organic changes such as compression of the intervertebral discs and increased curvature of the spine.

Regional Dimensions

The size of individual body segments can be predicted from standing height, but because of differences in body type, the predicted values are not particularly accurate (Barter, 1957). In children, there is the added complication that individual body segments grow at different rates. In general, the hand matures faster than the forearm, and the forearm faster than the upper arm, while a similar ordering of growth is found for the lower limb (Harrison *et al.*, 1964). The majority of segmental measurements (sitting height, leg length, thigh length, chest, waist, seat, thigh and neck girths) are fairly reproducible, interobserver discrepancies amounting to less than 5 percent. However, wrist girth, arm length and shoulder breadth are difficult to measure consistently unless special anthropometric tools are available; if a simple tape measure is used, observers vary markedly in the tautness with which this is held.

Most regional measurements of size conform fairly closely to a normal distribution curve, unless there has been deliberate selection upon the part of the employer or employee. The foreshortened aircrew height curve has already been noted. Another interesting example arises in the London Transport bus system. Here, drivers have a systematically larger abdominal girth than the conductors at the time of initial tailoring measurements (Morris, 1966). Presumably, this is partly self-selection, the more sedentary and obese Londoners seeking the post of driver. However, employer selection may also play some role, since the operators of public service vehicles are rigidly screened, and in some cities no more than one of forty applicants is accepted.

The majority of girth measurements increase over the span of adult life. Although there is a slight loss of muscle, there is usually a substantial gain of body fat. Obesity commonly develops between the ages of 30 and 40 in men, and after the menopause in women. Overall dimensions may diminish after the age of 60; the wasting of muscle accelerates, and there is not necessarily any further deposition of fat.

The virtue of extensive regional anthropometric measurements is debatable. The International Biological Programme specified a *basic* list of twenty-one readings, with an additional seventeen recommended measurements (Weiner and Lourie, 1969). Such careful documentation may be justified when attempting to identify the biological background of a central African tribesman, particularly if the logistic costs of an expedition are high. However, in the ergonomic context, it is often more useful to concentrate upon a few relevant items. Thus, a chair manufacturer may wish to know crown-rump length, thigh length, and lower leg length (Jones, 1969; B.S.I., 1965), while a shoe manufacturer will be interested in body weight and detailed figures for foot and ankle dimensions. In any event, there is a fair intercorrelation of the various dimensions, and this inevitably limits the new information contributed by additional data. The majority of girth measurements are related quite closely to body weight, while the length of individual body segments bears a substantial relationship to standing height.

Weight

Weight, like height, seems a simple variable to measure, but in practice population measurements may show large systematic errors. Not only are inadequate checks kept upon the performance of scales, but *corrections* ranging from 1 to 5 kg are applied for the supposed weight of clothing worn by the subjects. Unfortunately, details of such corrections are commonly omitted from the published results

The weight of children has been advancing over the last century, due to increases in adult size, earlier maturation, and (in recent years) overnutrition. In adults, weight parallels girth, increasing between the ages of 30 and 40 in men, and after the menopause in women.

One simple ergonomic application of body weight is in the calculation of payloads for aircraft. On normal scheduled routes, a considerable safety margin is allowed, but on bush routes where a mixture of passengers and cargo is carried—sometimes with fuel for the return journey—the anticipated weight of thirty well-clothed male passengers, for example, must be predicted rather closely. Similar calculations are needed for elevators; often, these are driven by relatively small motors and have rather limited braking systems. Springing for chairs could also usefully take account of body weight, although a comfortable design is more often reached empirically from a series of subjective trials (Shackel *et al.*, 1969).

The relevance of body weight to nutritional status (page 301) and to the survival of executives (page 309) has already been noted.

THE WORKING AREA

When assessing the compatibility of the human operator with his assigned working area, it is necessary to consider not only his static dimensions, but also the potential range of limb movements, passive and voluntary, and the force that can be exerted in different postures (Woodson and Conover, 1964; Rohmert, 1971). Further, the lines of vision must be acceptable; if accurate readings of a control panel are required, parallax should be avoided, and if careful steering of either a vehicle or a crane is necessary, the space surrounding the driver should be freely visible without awkward changes of posture. The sloping hood (bonnet) of the modern car has reduced the forward blind area, although in some models a lowering of seat height (to reduce wind resistance) has cancelled much of this advantage; there is still a blind zone of at least 4 meters in front of the average car (Jones, 1969). Occasional awkward movements to reach a control or to obtain necessary visibility may be an acceptable design compromise, but if frequent movements do not lie within range of the normal operating position, bad posture will lead to cumulative fatigue over the working day (Beswick, 1966; page 54).

Let us consider first the working area of a sedentary person. An executive is seated at a massive and status-rewarding desk (page 236). The normal usable work area is that reached by a sweep of the forearm, with the upper arm hanging at the side (Fig. 81). In an average man, this will comprise a semicircle with a radius of about 40 cm (McCormick, 1957). Unfortunately, much of this potential reach is lost in the 15 to 18 cm that separate both the chair and the elbow joint from the edge of the desk. At the cost of supporting a larger fraction of the total body weight, it is possible to use the full arm. This will traverse a semicircle with a 60 cm radius. However, even if we assume that the executive is ambidexterous and need grasp no more than the bottom two inches of a ten-inch file cover, it is obvious that much of the desk surface has no purpose other than to serve as a status symbol. The usable area can be much extended by provision of a rotating or castor-mounted chair, and by an L-shaping of the desk, as in some modern typing bays.

The optimum height of the working surface varies with the task and the pressure that must be exerted. Desks where writing is undertaken are best arranged at a height of about 74 cm, with a distance of 15 to 18 cm between the chair and the desk. Where the desk is attached to the seat, as in many university auditoriums, a slight (10 to 15°) backward tilt of the writing surface is an advantage; most writers also appreciate a side support for the arms (Grandjean, 1969).

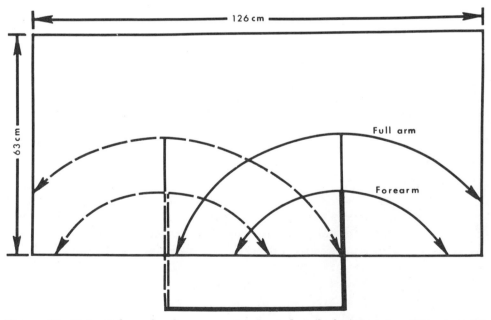

Figure 81. Potential reach of an executive seated at desk measuring 126 cm × 63 cm (based on data of McCormick, 1957).

Dining tables tend to 2 to 3 cm higher than writing surfaces, to minimize spillage of liquid foods, and if a surface must serve for both writing and dining, the comfort of the writer can be increased by an adjustable chair, or more simply by provision of a firm supplementary cushion. If a typewriter is to be operated, the bulk of the machine is such that an advantage is gained from lowering the working surface by at least three inches, rather more for a manual than for an electrically operated machine.

Because a proportion of the body weight can be supported through the arms, a low desk is less tiring for the seated worker than a high one. Thus, if in doubt, it is best to err in the direction of too low a desk. Where the entire working day is spent in the sitting position, attention should be paid to the problem of individual variations in sitting height. The best remedy is an adjustable seat or desk, but this lacks the cheapness and durability sought by most schoolboards. Oxford (1969) has suggested providing six different sizes of school chairs, with an appropriate mix of large and small sizes for a given grade; some choice of chair size is also desirable if the same design of desk is to be used by men and women.

One expression for the comfort of the seated worker is provided by the sum of heel to roof space and toe to back space (Jones, 1969). In order to accomodate 95 percent of the population and allow two inches of head room, the *sitting envelope index* thus derived should amount to 253 cm for a car driver, 260 cm for a front passenger, and 240 cm for a rear passenger. In fact, most European cars provide rather less space than this, just over 230 cm for the driver, about 245 cm for the front passenger, and 215 to 230 cm for the rear passengers. If two inches of head room is assumed, lecture theater seating also provides about 230 cm.

Some industrial operations such as a small manual telephone switchboard or a mail sorting operation present the seated worker with a vertically oriented working surface. Here, the possible range of reach is usually governed by problems of near vision (McCormick, 1957). If the head of the worker is kept at a minimum distance of 50 cm from the panel, the average operator can work within two overlapping circles, each with a diameter of a little under 90 cm, giving a total lateral reach of 110 cm. If the panel is brought to within 25 cm of the face, the reach is extended by some 25 cm in each direction, to a total of some 160 cm laterally and 140 cm vertically; however, such close work is fatiguing to the eyes (page 75). Controls that require frequent adjustment should lie between elbow and shoulder height. The seated operator can develop maximum force on a hand control at elbow level. Pulling strength in a forward/backward direction is greater than pushing; thus a lever such as an emergency brake is designed to pull on. However, in the lateral direction, a push is more effective than a pull. Horizontal movements of the hand can be performed more rapidly than vertical movements. The steering wheel of a car should thus be turned by a push from a hand held near the top of the wheel, the opposite hand serving mainly as a brake when the movement has been completed. Flexion of the arm is carried out faster than extension (Woodson and Conover, 1964; Medford, 1951).

Cycling is an interesting form of seated activity, since very strenuous and sustained work is required of the leg muscles. Many long-distance cyclists elect to use a high seat with drop handlebars. The prime purpose of this arrangement is not to minimize wind resistance, but rather to produce sufficient flexion of the hip joint ($> 45°$) that the gluteal muscles of the rump contribute to the cycling task.

The optimum bench height for a standing task depends upon the relative importance of physical effort and of careful visual inspection of the finished product (Brown, 1972). Light inspection and sorting tasks are performed best at about 106 cm, some 7 to 8 cm below elbow level. A task that needs extensive arm movement and application of body weight (for instance, hand planing) is performed better at a height of 90 to 93 cm. However, even if fairly heavy manual work is required, too low a bench will lead to fatigue and an unnecessary dissipation of energy in stooping (Jorgensen, 1970). Low-level work can be performed more economically by kneeling or crouching; work above head level, such as painting or hammering throws a fatiguing strain upon small muscle groups, and the circulatory response to a given oxygen cost is thus exaggerated (Åstrand, 1971).

CHAIR DESIGN

Chairs have a wide range of legitimate purposes (Shackel *et al.*, 1969):
1. long-term sitting without arm support (in reception and conference rooms, libraries, theaters and concert halls),
2. long-term use with arm support (office and desk use).
3. short-term use with arm support (eating a meal),
4. driving a vehicle, and
5. carrying out physical work upon a flat surface.

The optimum design of chair will vary with its intended purpose, although for reasons of economy, there may be a search for a common design to meet the needs of desk, table and conference hall. In choosing dimensions, the average height of the intended population should be noted, together with the likely proportions of men and women users.

Unfortunately, expert ergonomists with previous experience of rating seem unable to make a valid assessment of comfort based upon their personal use of a chair; presumably, the individual's impressions are unduly influenced by his own body shape. The best available method of assessment is thus to make specimens of the chair in question available to a segment of the population, and to request ratings on a twenty-point scale that ranges from complete relaxation through restlessness to stiffness and unbearable pain. This simple assessment is supplemented by more specific questions. Is the seat so high that it presses upon the thighs, or so low that the thighs are largely clear of the chair? Is the seat so long that it presses into the popliteal fossa, or so short that the buttocks overhang the rear? Is the seat so narrow that movement is restricted, or so wide that it is difficult for the subject to slide out? Is the backward inclination of the seat too slight, so that one slides off, or too great, so that one becomes wedged in the angle? Is the backrest so high that the small of the back is unsupported, or so low that the middle of the back is unsupported? Are the sides of the chair excessively curved, clamping the body? Is there adequate clearance for the feet and calves, and so on?

The overall comfort rating diminishes with time, faster with some chairs than with others. The deterioration of comfort is slower with arm support than without. It is thus important to test a chair over an adequate period and under the intended conditions of use. The most common ultimate complaint is of discomfort localized to the buttocks, and some guidance to bad design can be obtained from pressure recordings in this region. Figures should not exceed 60 gm/cm² for short-term use and 30 gm/cm² for long-term use (Rebiffé, 1969), although many current designs give readings of 90 gm/cm² under the ischial tuberosities.

Film studies of restlessness are not well correlated with comfort (Floyd and Roberts, 1958). Presumably, the frequency of postural change is strongly conditioned by cultural factors. Children seated on impossibly hard chairs are warned not to fidget, and girls are taught to sit with their legs neatly crossed even at the expense of a temporary paralysis of their sciatic nerves.

The British Standards Institution (1965) recommended the following dimensions for a general purpose chair (Fig. 82):

Seat height   17 inches (43 cm)
Seat depth   14 to 18½ inches (35.5 to 47 cm)
Seat width   16 inches (40.5 cm)
Seat slope   0° to 5°
Back slope   95° to 105°
Back height   13 to 18½ inches (33 to 47 cm)
Back pad   > 4 inches (10 cm) (> 5 inches, 12.5 cm if convex).

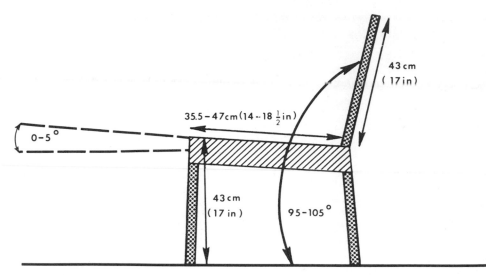

Figure 82. Dimensions of a comfortable general purpose chair (after British Standards Institute, 1965).

Variations about this specification depend upon the anticipated posture of the user. With a relaxed pose, the length of the seat must be increased, and a greater backward tilt is also preferred (Fig. 83). Men generally prefer lounging, while women continue to sit upright even when relaxed. People with recurrent backache (notalgia) may benefit from a more convex loin welt, and a more concave back (Grandjean, 1969).

Esthetic considerations may enter into the final choice of design, but such factors have been given too much weight in the past. Color

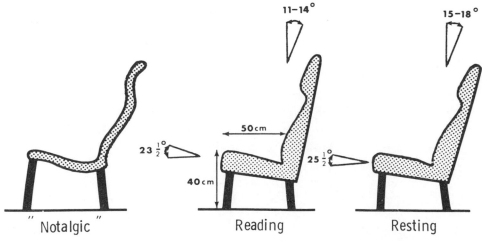

Figure 83. Optimum designs of chair for reading and relaxation, with modification for notalgic individual (recurrent backache) (after Grandjean, 1969).

should be chosen on the basis of need to increase or to diminish arousal (page 215). If the chairs are to be moved frequently, attention must be paid to ease of stacking and noise produced by movement. Tubular frame chairs should be packed with sound-absorbing foam rubber, and tipped with rubber or soft polyethylene rather than hard nylon.

Group Seating

The spacing of seats is of some importance to designers of buses, aircraft, theaters and auditoriums (McCormick, 1957). In all of these situations, the objective is a relaxed posture, and a backward tilt is used to reduce the load on the body muscles. In a theater, a relatively short backrest may suffice, but in a vehicle a longer backrest minimizes head movements (and thus motion sickness, page 138) at the expense of some increase in vibration. If the user of a seat is watching a play, he may be content with 53 cm of elbow room; experience has shown this will satisfy 95 percent of people. However, if it is necessary to eat a meal (as will be the case on a transoceanic aircraft), comfort may call for a rather larger space allowance. Much depends upon previous conditioning of the passenger; a person accustomed to traveling first class on planes and trains will feel severely restricted by the smaller space allowances of economy class. The knees commonly project 20 to 23 cm in front of the seat edge. Thus, given a 50 to 51 cm seat width and some padding of the back support, a minimum allowance of 80 cm is needed per row of seating. More space is required at exit windows of an aircraft, and an experienced traveler soon learns the seat numbers that carry this bonus. In a theater, the pitch of the floor must be calculated with some precision. The eye to target line of the viewer should be at least 13 cm above the head immediately in front. By staggering the arrangement of seats, the next head can be set at a distance of two rows (160 cm), giving a pitch of about 5° to the orchestra stalls (Fig. 84). Once the floor of the auditorium exceeds stage height, the audience begins to look downward, and the pitch of the seating must increase progressively. In older European theatres, the uppermost balcony has dangerously steep steps. Thus, if site size permits, it is better to enlarge the ground floor, and to use at most one balcony, setting this well back from the stage.

Cab Seating and Design

Both space requirements and seat design for a cab or cockpit must be integrated with the various functions that are to be performed. Design must facilitate purpose. The functions of a vehicle operator include visual monitoring of the road and/or instrument panel, regulation of speed and (in most vehicles) steering. Escape must also be easy in an emergency. The average car ignores many anthropometric

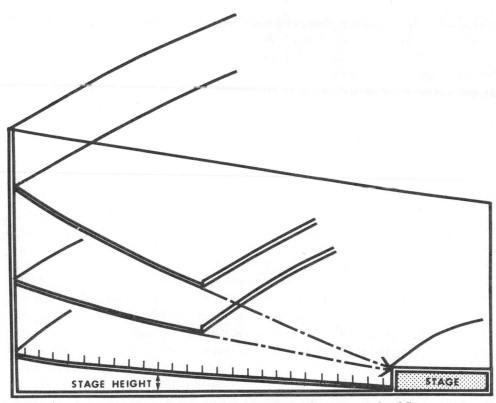

STAGE HEIGHT

STAGE

Figure 84. Arrangement of seating in a theater, to illustrate pitch of floor.

characteristics of the driver, and this faulty layout makes for difficulty in operation and an increased frequency of accidents. Thus, the brake pedal may be placed dangerously close to the accelerator, permitting the foot to slip from one to the other or catch underneath, particularly if the pedals are worn or the driver is wearing wet and clumsy boots. In some English cars, the position of the accelerator pedal places a severe load upon the iliopsoas muscle, and the fatigue of a long drive has given rise to symptoms that can be mistaken for appendicitis (Beswick, 1966). Prolonged sitting in such occupations as taxi driving may lead to a shortening of the hamstring muscles, with fatigue on standing erect. Drivers with long legs often find that the steering wheel does not provide sufficient knee space, particularly when applying the footbrake; a steering column gearshift further restricts knee movement. In some older cars, it was necessary to lean forward to operate the hand brake. This placed the head in a very vulnerable position as an emergency developed; thus, if a handbrake is used, it is now placed at the driver's side.

The seat height is much lower than that recommended for other applications (about 22 cm in a modern car) because of the desire to minimize wind resistance. A forward/rear seat adjustment of 75 mm permits comfortable operation of foot pedals by both long and short individuals (page 314).

Seating for Shop Assistants

Most countries have legislation requiring provision of seats for shop assistants, and it is possible to increase the percentage of useful time on the shop floor by arranging that customers can be served from a sitting position. In a bank, it is desirable that the hand height should match that of the customer, so that a tall stool or raised platform is needed by the assistant. In supermarket operations, the hand level of the assistant must be set at a convenient height for handling common heavy purchases.

DESIGN OF STAIRS AND LADDERS

The choice between the use of ramp, stairs or ladder depends upon the incline, the nature of the load to be raised, and (if outdoors) the likely weather conditions.

If wheelchairs or items of equipment mounted on castors are to be moved to a higher level, a ramp provides an acceptable alternative to a lift (elevator), particularly if the incline is no greater than 20° and the total climb is no more than a few feet. In some climates, outdoor ramps may require heating strips to prevent icing.

Stairs provide the most efficient method of ascent where the slope has an incline of 20 to 50 degrees. Laboratory physiologists have accumulated a fair amount of information on stair climbing from the performance of standard exercise tests (Shephard, 1969a and 1972a). Nevertheless, this data refers to a more vigorous rate of ascent than is common in daily life. The rise of the individual step can range widely from 23 to 46 cm with little change in the mechanical efficiency of the climber. However, 46 cm steps are uncomfortable even for the average male worker, and lead to a very ungainly pattern of climbing in children and others with short legs. Some subjective evaluations have set the preferred height as low as 13 to 20 cm. Much depends upon the height that must be climbed and the time available for ascent. A person with average coordination can climb at 120 paces per minute with little risk of falling. At faster rates, the natural frequency of the limb is approached (page 123); efficiency decreases, and there is a serious chance of tripping. Thus, a 13-cm step holds a 70-kg man to a work rate of $13 \times 70 \times 120 = 1090$ kg-m/min. This is more than adequate for a long climb or one that must be repeated throughout the working day, but can be unsatisfactory for a young person who suddenly decides to sprint up two floors. Nevertheless, it is possible to approach the

potential limit of anaerobic power on such a staircase by running
upwards two stairs at a time. The minimum size of tread is determined
by the shoe length of the largest likely user. A man may wear a 30-cm
shoe, and when descending a staircase his equilibrium becomes unstable if more than 10 cm of shoe overhang the front of the stair tread.
In order to avoid striking the riser, a small safety margin must be
allowed, giving a minimum tread-width of about 24 cm.

Very steep stairs or companionways are used in small boats to
economize upon floor space. However, these are dangerous to negotiate
in rough weather. A second solution to a restricted space, found in
the Boeing 747 airliner, but also familiar to the medieval builders
of castles and churches, is the spiral staircase. The minimum width
of such a staircase must be about 60 cm. The innermost foot is then
traversing a circle of some 23 cm radius, and if this foot is allowed
a 24 cm tread, a maximum of six steps can be accomodated per circuit.
In this distance, a height of at least 250 cm must be climbed in order
to provide minimal headroom and accomodate the structure of the
staircase. It is hardly surprising that the average spiral staircase has
a combination of steep steps and a narrow tread. Instability is more
likely to develop during descent than during ascent. Because of the
problem of treadwidth at the center of the staircase, it is undesirable
that tourists should try to pass one another on castle stairs. However,
where an alternative exit route cannot be arranged, the descending
party should use the outer part of the staircase.

For very confined spaces and inclines greater than 50°, the best
solution seems a ladder. Individual rungs may vary in spacing from
18 cm to 40 cm, although most adults prefer a 30-cm spacing. The
width of a portable ladder is held to 25 to 30 cm in order to conserve
weight, but a slightly greater width may be advantageous for a permanent installation. The efficiency of climbing is about 26 percent for
a leg-over-leg approach, and 21 percent if both feet are placed on
each rung; the cost of descent is between a quarter and a third of
the cost of ascent (Kamon, 1970).

CHAPTER 18

MAN AND THE COMPUTER

COMPUTERS HAVE BEEN DISCUSSED at many points in this book, ranging from respiratory models (page 100) and systems analysis (page 16) to the provision of a simple analogy for the functioning of the human brain (page 85). We have noted that while the immediate goal of the ergonomist is to match the man to the machine, in many humdrum tasks his ultimate objective is to *close the loop,* replacing man by some form of computer (page 30). In an increasing number of instances such as the control of traffic flow (page 367), this objective is close to realization, and the spectre of automation is creating redundancies and a need for retraining in many industries (page 262). It thus seems appropriate to look briefly at the possible functions of digital, analogue and hybrid computers; to examine salient features of design in an ergonomic context; to assign respective areas of competence to the three types of machine; and to illustrate a few current and potential applications of computer technology (Peterson, 1962; Chance *et al.,* 1962; Sippl, 1967; Spicer, 1968).

FUNCTIONS OF COMPUTERS

At least eight distinct spheres of function may be described:

1. *Equation solving.* The mathematician or scientist may demand the solution of complex algebraic, differential, polynomial and matrix equations. The first two can be handled by digital or analogue computers, but the second two require large digital computers.

2. *Process simulation.* Both analogue and digital computers can be programmed to simulate a physical or a biological system and to test the validity of current hypotheses describing its behavior. Specialized electromechanical analogues can be used to instruct vehicle operators (page 261) and aircraft pilots (pages 261 and 346).

3. *Process control.* An analogue or hybrid computer can carry out *on-line* computations of data (page 112), and the information thus processed can be used as a *feedback* to modify the rate of an on-going physical or biological process.

4. *Data storage.* The large digital computer can rapidly store and yield vast quantities of data, thereby offering a possible solution to problems created by the *knowledge explosion.* It may also be programmed to *learn,* and to make a variety of logical decisions on the basis of its acquired knowledge.

5. *Data handling and synthesis.* The large digital computer can greatly speed the statistical analysis of voluminous data, seeking functional relationships and intercorrelations between a wide range of experimental variables.

6. *Construction of tables.* A large digital computer can print mathematical, physical and biological data accurately and in great detail.

7. *Translating and indexing.* There is an increasing potential to utilize the large memory capacity of modern computers to provide translation, indexing and cataloguing services.

8. *Design and instruction.* Both analogue and digital computers have a role in helping the processes of design and instruction.

DIGITAL COMPUTERS

The essential concept of the digital computer is that information (page 9) is handled in numerical form. There is a capacity (1) to store data and instructions, (2) to perform a sequence of arithmetical operations, and (3) to transport information from one part of the machine to another. Occasionally, it is convenient that the input or output appear as letters (as in machines that will circulate individualized charitable appeals or election propaganda), but in such cases a translating device must be interposed between the computer proper and the input or output.

Computing machinery has a long history. The oldest digital device is the *abacus*. This was apparently introduced into Babylon in the sixth or seventh century BC as a development of the Phoenician *abak*, a sand-strewn writing surface. Loose counters of bone, metal or glass were placed on a ruled table, or (in portable versions) were arranged on wires or strings. In accordance with local concepts of mathematics, the Mediterranean abacus carried up to eighteen beads per string, while the Roman system had but five.

A second early form of computer was the *pendulum clock*. Time was converted from analogue to an appropriate digital value by fixed interval sampling (the period of the pendulum). The rocking of the pendulum arm allowed the drive wheel to advance by one tooth, and this small displacement was amplified by suitable gearing, displaying angular movements of n times the displacement (hours), 60 n (minutes) and 3600 n (seconds) on the dial.

The use of hand-operated *desk calculators* was first proposed by such famous mathematicians as Leibnitz and Pascal. Commercially produced instruments became common in the latter part of the nineteenth century. *Punch-card* machines were devised by Hollerith in 1930 as a means of sorting and listing data. A *first generation com-*

puter was developed by Eckert in 1946. This was capable of carrying out 100 simple arithmetical operations per second, a thousand-fold increase relative to the usual speed of a desk calculator. The memory size was 2000 *words*. Over 18,000 radio tubes were used in the machine, giving a heat production of 150 kilowatts! There were thus considerable problems in stabilizing electrical and thermal conditions for the apparatus. *Second generation computers* such as the I.B.M. 7094 began to appear in the late 1950's. These had an operating speed one hundred times faster than that of Eckert's machine, with a memory for up to 200,000 words. By adoption of transistorized circuitry, heat production was reduced to negligible proportions. *Third generation computers* such as the I.B.M. 360 series first became available in 1965. They are characterized by modularity, with a wide range of optional input and output devices such as three dimensional viewers and pens. The speed of operation is further increased to the point where a hundred or more investigators can each carry out a thousand operations per second on a time-sharing basis. Multiprogramming helps ensure use of an optimum combination of operating instructions. Versatility is provided by plug-in components to carry out specific mathematical operations, and storage capacity is enormously increased by developments of microelectronics.

The degree of interaction between the human operator and a digital computer varies with its complexity, but irrespective of the degree

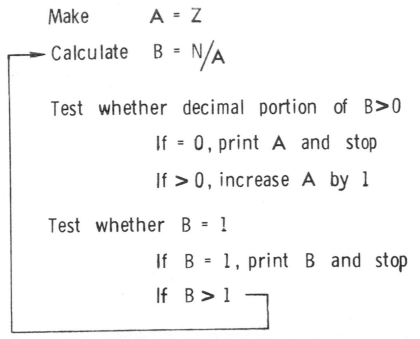

Figure 85. Example of a simple algorithm to determine whether N is a prime number.

of involvement and interaction, there is an inescapable requirement to describe the required process in the logical stepwise format of an *algorithm*. Thus, let us suppose that we wish to determine whether N is a prime number. The necessary operations may be written as shown in Figure 85. With a simple desk calculator, the human operator must enter each number manually, and after manipulating appropriate controls the answers must be transcribed into a notebook. However, with the largest electronic computers, machinery can be used on a twenty-four-hour basis, even in the absence of the operator.

A typical digital computer layout is illustrated in Figure 86. Basic components include input and output devices, a logic unit and a storage register. The *input* of information has traditionally been from punch cards. However, this is partly an accident of history in that some of the larger manufacturers of computers were originally concerned with the production of card-sorting equipment. The reading of punch cards requires an electromechanical device that is inefficient relative to current internal computing speeds. To avoid a bottleneck at the input, data is conveniently transferred to a magnetic tape or magnetic disc store. Other possibilities for the input are a typewriter (thus permitting *conversation* with the computer) and various forms of electronic chart reader. The *output* is normally through a high-speed typewriter,

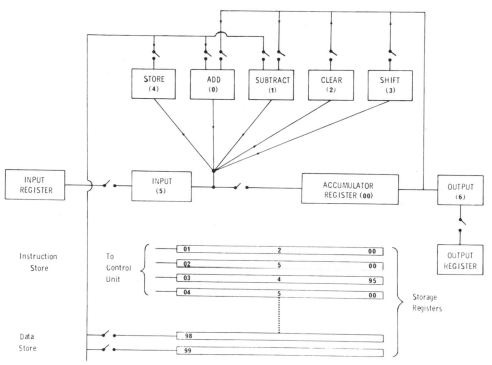

Figure 86. Basic layout of digital computer.

but processed information can also be transferred to magnetic tape, punch cards, tape or chart printers, or an oscilloscope display.

Storage methods are in a phase of rapid change. Many current machines use a magnetic core. Information (on/off; yes/no; true/false; 1/0) is represented by the direction of numerous small magnetic fields (clockwise/counterclockwise), arranged in the form of an interconnected lattice. In some smaller computers, the same function is served by magnetic tape mounted on a rotating drum. In both types of memory, the store is broken into *words,* each word consisting of 1 to 4 *bytes* (1 byte = 8 pieces of binary information, or bits; see page 9). The location of a given word is identified by a matrix-type *address* (for example, 41 = fourth column, first row of memory lattice). The binary notation is chosen mainly because an on/off circuit is simpler to design and more reliable than a satisfactory ten-stage switch. Nevertheless, there are also theoretical advantages to binary logic. Thus if the number 16 is represented as 10000, there is a need for no more than five binary decisions; however, in the conventional decimal format, two tenfold decisions are required. Although it is both practical and efficient to carry out the actual calculations in binary format, input and output are commonly translated into the decimal system. Some simple illustrations of binary mathematics are given in Figure 87. The capacity of

Binary numbers

0	=		0	=	00000
1	=		1	=	00001
2	=		$2^1 + 0$	=	00010
3	=		$2^1 + 1$	=	00011
4	=		$2^2 + 0$	=	00100
5	=		$2^2 + 1$	=	00101
6	=		$2^2 + 2^1$	=	00110
7	=		$2^2 + 2^1 + 1$	—	00111

Addition			*Multiplication*	
7	=	00111	00111	
12	=	01100 +	01100 ×	
19	=	10011	11100	
			111000	
			1010100	$= 2^6 + 2^4 + 2^2$

Figure 87. Some illustrations of binary logic.

the storage registers ranges from about 100 words in the smallest desk-type computers, to six million or more words in the latest third generation machines. If a small computer is to be used, it is particularly important to check that the combined requirements for *instruction* and *data storage* registers do not exceed the capacity of the memory. Much ingenuity of programming may be needed to bring a problem within the scope of the apparatus.

Instructions are essentially commands to carry out a logical sequence of operations such as the algorithm of Figure 85. In the desk calculator, the ordering of events is regulated by the brain of the human operator. In the modern computer, control instructions are inserted as an initial sequence of cards or as a length of magnetic tape. Having received such control input, the computer then transfers the data to be analyzed from the input store to the data storage portion of the memory. It subsequently carries out an orderly sequence of mathematical operations, as in the simple example of Figure 88. Each instruction location normally contains the binary equivalent of a twelve-decimal digit. The first two digits specify the operating code, and two subsequent sequences of five digits identify the address. A total of some thirty-five instructions such as add, subtract, multiply, divide, jump, jump if accumulator is greater than zero, and shift decimal point are sufficient

Instruction	Operating code	Address	Interpretation
01	2	00	Clear accumulator
02	5	00	Read number (a) from input into accumulator
03	4	95	Store in data location 95
04	5	00	Read number (b) from input into accumulator
05	4	96	Store in data location 96
06	0	95	Add number from location 95 to accumulator
07	0	96	Add number from location 96 to accumulator
08	6	00	Print contents of accumulator.

Figure 88. Example of simple digital computer program (see also Figs. 85 and 86).

to describe and carry out most conceivable mathematical operations. In small computers, such instructions are given individually, but if a long program has to be arranged, this can be quite time consuming, with numerous possibilities for errors of both logic and execution. Larger computers thus store packaged sequences of operations that can be called into play through codes such as the Fortran language. In some smaller computers, prepackaged programs can again be inserted via magnetic cards or tape cassettes. Theoretically, a program should function immediately after it has been written, but in practice the inexperienced programmer often omits a crucial step in the logical sequence, and many patient hours may be necessary to *debug* a complex program. The *data storage* portion of the memory also contains the binary equivalent of twelve decimal digits. At a first glance, this seems an excellent level of precision for most purposes. However, if the computer is required to carry out lengthy iterative procedures, significance rapidly disappears from the terminal digits.

TABLE VIII. APPROXIMATE COST OF OPERATION FOR DIFFERENT SCALES OF DIGITAL COMPUTER

Scale	Time for single multiplication	Rent per hour	Cost for 1×10^6 multiplications
Human brain	1 min	$6	$100,000
Desk calculator plus clerk	10 sec	$6.50	$18,000
IBM®7094 plus technical support staff	25 sec	$300	$2.50
Third generation computer	60 sec	$600	$0.01

The relative costs of differing scales of digital calculation are explored in Table VIII. If an extensive sequence of calculations is planned, the large digital computer is many orders cheaper than the simple desk calculator or the unassisted human brain. However, if the theoretical economies are to be realized in practice, it is vital to ensure that the large computer is restricted to large calculations, and that it is kept fully occupied twenty-four hours per day by suitable time-sharing arrangements. Too rarely is this achieved (Rowley, 1969). Further, it must be stressed that the stated costs make no allowance for expenditures on updating programs and software, with a small computer, this is a simple matter, but on a large machine, much costly time can be lost from this cause. With older computers as much as 10 to 20 percent of potential operating time was lost through mechanical failure, but current computers have much greater reliability, operating to at least 98 percent of their potential. This is partly a consequence of improved design (including arrangements for preventive maintenance) and partly because the latest types of computer are programmed to carry out periodic internal checks on their circuitry, halting operations whenever a serious fault is detected.

THE ANALOGUE COMPUTER

The analogue computer deals with continuous variation rather than discreet integer or digital variations. A mechanical, pneumatic or electrical model may simulate a process and study its behavior. Operations such as addition, subtraction, multiplication, division, integration and differentiation can also be carried out.

The familiar *slide rule* is a simple mechanical analogue. Here, the operator is provided with two scales, each calibrated logarithmically, and multiplication is carried out by adding the scale readings:

$$\text{Log } (ab) = \text{Log } a + \text{Log } b$$

The *planimeter* is a mechanical integrator used to find the area under a curve such as that produced by a physiological chart recorder. The integration may be carried out manually after an experiment is completed, or may proceed during recording, using a ball and disc system

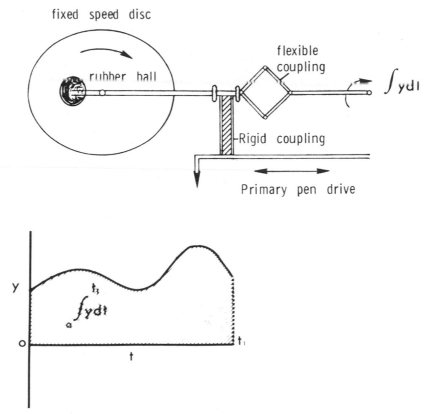

Figure 89. Schematic diagram of ball and disc integrator.

(Fig. 89). In the latter case the primary pen drive is rigidly coupled to a freely rotating rod that carries at its tip a rubber ball. The ball presses firmly upon a disc that rotates at a fixed speed, and the proportion of the angular velocity transmitted to the ball depends upon the displacement of the latter induced by the pen linkage. When at the center of the disc, the ball speed is zero, and if it moves to the right of center, the direction of rotation is reversed. The angular movement of the ball shaft may operate a system of dials or it may drive a second, reciprocating pen.

Scaling can be achieved mechanically by a suitable arrangement of sprockets. Thus, to calculate $W_2 = \dfrac{(n_1)}{(n_2)} W_1$, the variable is presented as an angular velocity W_1; after passing from a sprocket with n_1 teeth to a second sprocket with n_2 teeth, the angular velocity W_2 is derived. *Addition* and *subtraction* can be carried out using a differential gear of the type found on a car rear axle (Fig. 90). Note that the angular velocity of the *drive shaft* is proportional to the algebraic sum of speeds

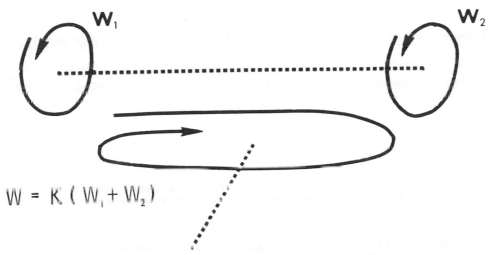

$$W = K (W_1 + W_2)$$

Figure 90. The use of a differential gear for addition and subtraction.

for the two *road wheels*, so that if desired one of the latter may have a negative sign.

The usual type of analogue computer is electrical, being based upon a bank of from twenty to several hundred highly stable direct current *operational amplifiers*. Such amplifiers have a high gain and draw no significant current. The variables under study are represented by proportional voltages. The basic operations (Fig. 91) are much the same as in older mechanical systems, including addition and subtraction, multiplication and division, integration and differentiation, scaling, and change of sign. The input to the computer could theoretically be presented as a sequence of discreet digital voltages, but more usually it is derived on a continuously varying basis from such sources as a physiological transducer, an electrical signal generator, a mechanical/electrical converter, or a preceding operational amplifier. The output, also, can be read from a digital voltmeter, but is more usually presented as a continuous, time-related curve on an x-y plotter. Thus, in order to plot the distance S traveled by a falling object in time t, and the corresponding velocity V, a voltage proportional to the gravitational acceleration G is passed through two integrating circuits (Fig. 92), and the output is plotted against time.

RELATIVE MERITS OF ANALOGUE AND DIGITAL SYSTEMS

The main advantages of the analogue computer are a relatively low cost, simplicity and a potential for continuous as opposed to stepwise analysis of the data. Where input variables are derived from physiological transducers that are somewhat out of phase, appropriate delay cir-

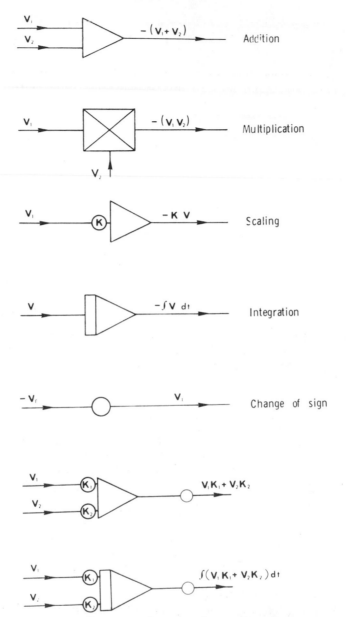

Figure 91. Operations carried out by electrical analogue computers.

cuits can also be introduced to permit *on-line* computation of results.
The main problem with the analogue computer is its limited accuracy.
At best, individual operational amplifiers have an accuracy of ± 0.5
percent. Furthermore, for this theoretical accuracy to be realized, voltages must approximate but not exceed the maximum permitted input
(commonly 10 volts). It is almost impossible to arrange this when carry-

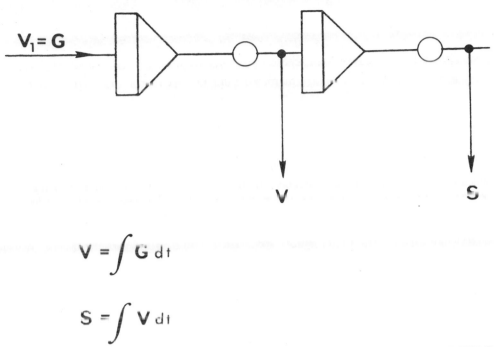

$$V = \int G\, dt$$

$$S = \int V\, dt$$

Figure 92. Arrangement of an analogue computer to calculate the distance traveled by a falling object (S) and its velocity (V).

ing out a sequence of multiplications and divisions, so that effective accuracy may drop from 0.5 percent to 5.0 percent per amplifier. If data must be processed through a bank of two hundred amplifiers, it can be appreciated that the end result is often no more than qualitative.

Programming of individual problems is also cumbersome and time consuming. Analogue computers are thus best reserved for relatively constant and simple patterns of calculation, where limited capacity and accuracy are acceptable, and a continuously varying *on-line* solution is required. Specific applications of analogue computers to the simulation of respiratory processes are discussed in Chapter 6. Problems concerning the uptake and distribution of materials through body compartments, and the simulations of biochemical reactions involving three to seven enzyme systems are two further examples of questions suitable for solution by analogue techniques (Chance *et al.*, 1962).

HYBRID COMPUTERS

Hybrid computers combine the various functions of analogue and digital machines, interconnections between analogue and digital portions of the apparatus being made through appropriate analogue $\rightleftarrows$ digi-

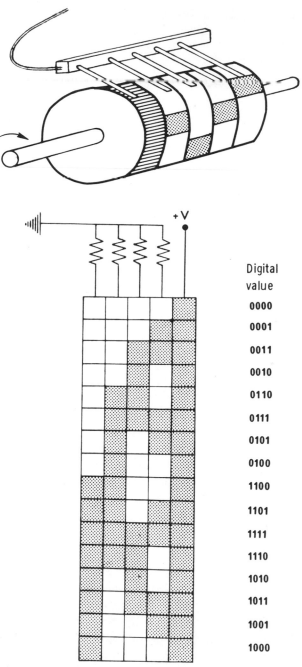

Figure 93. An example of an electromechanical analogue into digital converter.

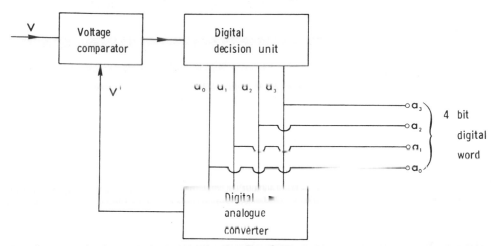

Figure 94. Example of electronic feedback technique for conversion of analogue into digital signal.

tal transducers. In some older machines, conversion was electro-mechanical in type (Fig. 93). An analogue signal was applied to a shaft carrying a switch patchwork that generated four or more binary voltages. Digital to analogue conversion was carried out in the converse manner, a comparator and servomotor driving the switch patchwork to a position matching a series of binary voltages.

Conversion is now carried out electronically. With the *feedback method* (Fig. 94), a digital *decision unit* is initially set to one volt. It is then compared with the analogue input. The feedback voltage V′ is increased progressively following each of a sequence of comparisons until it is equal to the input V. The digital *word* is then displayed as an appropriate series of binary voltages. Alternatively, a *voltage doubler* may be used (Fig. 95). A standard voltage V_s, half of the full permitted range (for example, 7.5 volts with a 4-bit system) is compared with the analogue input V_i. If V_s is greater than V_i, a zero is registered as the corresponding binary voltage. If V_i exceeds V_s, a *shift register* control subtracts a sequence of 0, 1, 2 and 4 volts from V_i as comparison proceeds. *Digital into analogue* conversion is carried out very simply with a constant voltage source, an appropriate resistance network and a voltage adder (Fig. 96).

One important current application of the hybrid computer is in the automatic processing of laboratory data. A batch of urine samples are passed automatically into a spectrophotometer, and the corresponding optical densities are transcribed directly onto magnetic tape or punch cards. At a more sophisticated level, measurements may be made of respiratory gas volume, expired gas composition and end-tidal composition before and after CO_2 rebreathing. Voltages from the physiological transducers are converted to the corresponding digital values, and after

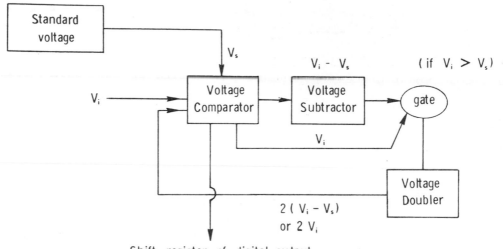

Figure 95. Analogue to digital conversion by use of a voltage doubler. A voltage of 7.0 becomes 0111, while a voltage of 14.0 becomes 1110 (See Text).

Examples:

(a) V_s = 7.5 volts
    V_i = 7.0 volts

Step	Bit	Old V_i	New V_i
1	0	7	14
2	1	14	13
3	1	13	11
4	1	11	—

(b) V_s = 7.5 volts
    V_i = 14.0 volts

Step	Bit	Old V_i	New V_i
1	1	14	13
2	1	13	11
3	1	11	7
4	0	7	—

due processing the oxygen consumption and cardiac output are reported by a high-speed data printer. Such an arrangement is expensive to install and requires skilled maintenance, but can be justified if there is a need for many repetitions of a rather inflexible process. The main advantages are (1) the output is free of observer bias and (2) subsequent tests can be modified in the light of results. On the other hand, unless the program is carefully arranged to signal and reject poor data, malfunction of the detectors can continue in a way that would not be possible with close human control of the primary detectors. Furthermore, although claims are made for economies in technician time, convincing statistics of a greater patient through-put per dollar of capital and operating costs have yet to be presented for most proposed applications.

HUMAN PROBLEMS OF COMPUTER USAGE

It seems widely agreed that there are ever-expanding opportunities for the computer. In the foreseeable future, it will dominate almost every sphere of human endeavor, from school learning and the ordering

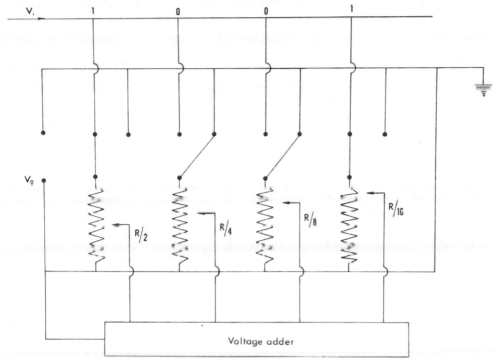

Figure 96. Electronic digital into analogue conversion.

of the weekly groceries to international decision making. It is thus surprising that little attention has been directed to the human problems involved in matching many levels of computer sophistication with a world population equally diverse in personality, experience and intelligence. Rowley (1969) stated that 70 percent of computer installations were barely cost-efficient, and to the present author this seems a charitable estimate of the current situation.

Possible reasons why the ergonomic study of computers has been unpopular (Nickerson, 1969; Shackel, 1969) include:

1. a disparity between the time required for convincing formal research and the rapidity of modifications in computer design,

2. the prohibitive cost of using a large computer exclusively for ergonomic research, and the attendant user dissatisfaction when a service installation diverts its energies to a research problem,

3. the problem of obtaining broadly applicable results in the face of wide variations in computer hardware and software,

4. an intuitive suspicion that *standard* ergonomic research can be applied to problems of computer design.

With regard to the last point, there is certainly scope to apply current knowledge of control theory and console design to the computer field, but at the same time the modern computer presents many unique problems associated with its capacity for *dialogue* with the human operator.

Linguistic Problems

It is usually more important for the operator to be familiar with machine language than for him to understand all the intricacies of computer hardware.

Problems of communication can arise from the diversity of design and language, but the greatest source of difficulty is the wide range of experience and interests in the user population. Even given a common language, the *grammar* and *syntax* required for a business application differs markedly from that needed by the research scientist. Simple list processing, for example, is hard to reconcile with a demand for complex algebraic calculations.

There is a regrettable tendency for individual users to proliferate local machine *dialects*. In the interests of effective communication and wide applicability of programs, the manager of a computer operation should resist this trend. Suggested dialects must be examined very critically to assess whether they are helpful in capitalizing upon the idiosyncracies of a particular computer or whether they are inconsequential for efficiency but have an adverse impact upon man/machine interactions.

Several levels of vocabulary may be necessary to allow for differences in skill between individual users of a facility. Most users are irritated when a machine presents information that seems redundant to them, although there have been no formal studies on the influence of such redundancies upon learning patterns and other aspects of efficiency. One possible solution to this problem is for the user to specify his level of competence prior to entering a dialogue; however, in practice, this fails to resolve the dilemma, since individual aspects of personal skill are developed to differing extents. A more effective alternative is to have several levels of information available within the computer, detailed statements being presented in response to specific requests.

User Behavior

More attention should be directed to the impact of the computer upon user behavior. In a general sense, man is resistant to change, and acceptance of the behavior patterns essential to computer processing of such items as monthly bank statements may depend upon careful preparation of attitudes in the user population.

Patterns of computer usage vary not only with instrument characteristics but also with management policies. If the input and output mechanisms are cumbersome and create delays, then users tend to adopt long and involved programs to minimize the time spent at input and output stations. They also check rather carefully to ensure that each intervention brings them significantly nearer to their objective. The frustration encountered in the operation of a conversational system varies not only with mean delay times, but also with the variability about mean values (Neisser, 1965). Delays of more than two seconds lead to loss of efficiency, and delays of more than fifteen seconds are generally regarded as intolerable (Miller, 1968). Color signals that indicate an operation is proceeding or that an error has occurred reduce the uncertainty of the operator and make a given delay more acceptable. Thus, the flashing green light of the slowly moving Olivetti® 101 gives the user the comforting impression that the machine is lumbering towards a solution of his problem. In some large, time-sharing computers a fixed delay is imposed, so that all instructions are treated as though waiting in a queue with three or four other potential users.

The debate over the most appropriate *billing* of computer costs has yet to be resolved. Should costing reflect the period of on-line time, the percentage of the memory utilized, or the number of *bits* processed? By how much should charges be reduced if a user is prepared to accept other than prime time or a low priority in a queue? Does an over-tight control lead to ultimate inefficiency because a user prefers to operate with a known but unsatisfactory program rather than run-up a substantial bill designing something better? It is important to resolve such questions, for as Scherr (1965) has demonstrated, billing procedures have a significant impact upon patterns of user behavior. Institution of a penalty charge for large programs produces an almost immediate reduction in average program size. The user tries to beat the system, and in the process is brought closer to the computer manager's image of an ideal user.

Bottleneck Identification

The development of an appropriate descriptive terminology is a necessary first step towards the location and elimination of bottlenecks within a computing facility. Note must be kept of the percentage of total time allocated to the various phases of computer activity—dead, dormant, working, awaiting command, awaiting input and output (Scherr, 1965). Such data has more bearing upon the efficiency of operation than do the traditional criteria of storage capacity, word length and processor speed. The responses of a given computer may be evaluated by running a test program, and noting the percentage of the total potential time the computer allocates to this specific program, a measure of queue length within the machine. Alternatively, a checklist may examine such items as the percentage saturation of various operational stages, average response times, reliability, accessibility, and capital and operating costs.

Continuing improvement of computer design is likely, and the time will soon arrive when the main bottleneck is not within the machine, but rather in the human operator and his interaction with the terminal facility. Operator thinking time will thus become an increasingly important variable to include in the evaluation of new equipment.

Interface Design

Since man is destined to become the limiting factor in computer operation, more attention must be paid to interface design. Although a typewriter keyboard has traditionally been used for data input, there is no strong evidence that such an arrangement is preferable to a pattern recognition device (voice or letters) or the use of carbon-sensitized data sheets. The optimum combination of speed and accuracy is obtained when data are entered by an individual personally concerned with the answer received. However, lack of skill on the part of a potential user may force the interposition of a *buffer* (such as the airline reservation clerk). It is also possible that some potential users (such as high-salaried physicians) may lack the time to learn the necessary techniques of data entry and carry out the mechanics of data input. However, even if it is decided to relegate the bulk of data entry to a statistical clerk, critical and potentially dangerous information such as details of drug dosage should be entered by the appropriate specialist. Where reliance is placed upon carbon-sensitized data sheets or the transcription of questionnaire responses by a keypunch operator, the incidence of errors can be greatly reduced if careful thought is given to the layout of the questionnaire.

The environment of the average computer terminal offers the operator a comfortable combination of temperature and humidity, since controlled thermal conditions facilitate the stability of instrument electronics. However, lighting may be less satisfactory, and the noise from teleprinters often rises above an acceptable level. Furthermore, the operating console may lack needed work space for data sheets and reference books. Improvement of the immediate working environment can often reduce the incidence of minor but annoying errors.

The output of a conversational machine is speeded if the teleprinter is replaced by an alpha-numeric cathode ray tube display. Typically, this has no more than thirteen lines, with forty to eighty characters per line—less than a third of the normal book page. It is also somewhat more difficult to *turn the page* than would be the case in a conventional book. Presumably, these factors diminish the efficiency of reading, although formal studies are lacking. If a picture (such as a circuit diagram or an architect's plan) is to be displayed, the limitations of size and resolution make it difficult to display the entire drawing at one time; there is thus scope to develop an effective image enlarger. For some purposes, speed of output is not important, and a teleprinter may then be the most useful arrangement. However, there is still scope for improvements in efficiency. Current printouts often use vast quantities of paper where cost and convenience would be better served by a more compact format. Attention to details of tabulation can also facilitate the transfer of information to the user, very different modes of presentation may be needed by the business accountant and the research scientist.

Other Human Problems

There remain many fascinating human problems of computer operation, ranging from the practical to the esoteric (Voas, 1961; Grodsky, 1967; Mudd, 1968; Shackel *et al*, 1967) What types of grammar and *syntax* are most readily graded to meet varying levels of operator skills? What forms of language permit maximum speed of user operation with a minimum of errors? How may such items of software as programs, records of dialogue, and service instructions be improved to facilitate understanding? How may hardware be redesigned to facilitate location of faults and replacement of damaged parts?

The incidence of personnel problems in computer facilities is often high. This difficulty may be traced in part to the relative youth and immaturity of many staff, and in part to the pressures of operating with high load factors on a twenty-four hour basis. More thought should be devoted to methods of selecting programmers and computer technicians, with particular emphasis upon their ability to withstand shift work (page 275), and to handle customer enquiries and complaints (particularly those relating to the apparent loss of vital data!).

From the managerial viewpoint, there is also a need for clearer decisions on the cost/effectiveness of different computing styles, including the generally accepted time-sharing concepts versus batch processing. Consideration must be given not only to the cost per computer hour, but also to the cost per user hour, and the financial implications of implementing user preferences.

SIMULATION TECHNIQUES

The present section examines the principles governing process simulation (Parton and Roberts, 1959) and discusses a limited number of specific examples. Other applications—particularly those involving the use of analogue computers—will be found in Chapter 6.

In essence, *simulation* involves the operation of a model that represents the parts of a system and their interactions with one another. It may be physical or mathematical in type, and it is used in preference to more direct experimentation when the real process is difficult, dangerous or expensive. The objectives of the modeler may be to establish estimates and guidelines (system capacity, response time, weaknesses and vulnerability to damage), to seek modifications of design that will improve the system, and to obtain a detailed evaluation of performance. Often, the mere formulation of a model helps to clarify the events in a particular process, forcing the modeler to place individual occurrences in an orderly and logical sequence, and clarifying areas that need further analysis.

For some applications in the physical sciences, a *deterministic* model (error-free) is appropriate. Most biological problems call for a *stochastic* model, where provision is made for description of error. The analogue computer is effective for a deterministic problem (such as the speed of a falling object, Fig. 92), but it has difficulty in handling models that cannot be described by rigid logical and mathematical statements. Digital treatment is mandatory for stochastic simulation. Thus, if there is a need to analyze traffic patterns (page 364), the basic question of queuing is susceptible to analogue treatment, but if a range of associated variables such as pedestrian density and weather conditions are to be incorporated into the model, a simple mathematical formulation is no longer appropriate.

As in a more direct laboratory experiments, simulation requires meticulous planning, design and subsequent analysis of data. *Initial conditions* must define the starting state of the model, and *boundary conditions* must specify environmental restrictions upon the operation of the system (page 15). The change in state of the model must be recorded after each operation, so that a *state history* is progressively described. Any description of the system is necessarily less than perfect, and checks of adequacy and validity of the model are thus necessary. The state history of the model is compared with the corresponding behavior of the real system, and an assessment is made of the importance of any phenomena not represented in the model. Although of equal importance in a mathematical formulation, this particular point is conveniently illustrated by electromechanical training devices. In a road vehicle simulator, how necessary is it to match the film speed and thus the apparent vehicle speed with throttle position? In a Link®-type aircraft trainer, how important to validity are such sensory cues as the bump of a bad landing, the screech of tires and the smell of burning rubber?

Erroneous conclusions may be drawn from modeling if there is an unsuspected *bias* in the assumptions made (for instance, population behavior may be predicted from data collected only on young male subjects). The model may lack *generality*, and it is necessary to give careful consideration to both the type and the spread of variables that are used. The results obtained may also lack *precision*, due to uncertainty regarding input data, or the excessive use of iterative procedures (leading to cumulative *rounding errors*).

Assuming that such technical problems are successfully overcome, the model can be used to assess *performance* (how well is a system serving its *purpose*, page 8), *efficiency* (how completely are the various components of the system being utilized?), economy (what is the unit cost of operating the system?) and *effectiveness* (how well is the system matched to the environment in which it must operate?).

Physical Models

Digital computer simulation has certain advantages. It is completely repeatable, and free from any physical constraints. Quantitative evaluations are also possible as simulation proceeds. Nevertheless, it can be a lengthy and inflexible process, and some problems are more conveniently represented by physical models. Thus, the airflow over a car or an aircraft frame is very easily studied by Schlierren photography of a scale model inserted into a wind tunnel. Again, hydrodynamic problems such as flood control and the design of canals can usefully be examined in scale models. However, there may be physical reasons why a valid scale model cannot be prepared. Thus, some hydrodynamic models cannot be reduced to manageable size without giving disproportionate importance to the surface tension of water.

Monte Carlo Methods

Monte Carlo methods exploit the capacity of large digital computers to store and/or generate random numbers. Let us suppose that we wish to study the dynamics of tumor growth. Individual *cells* are represented by a single binary digit. Cell division is required after a given time, with a finite probability of *death* in the daughter cells. Variance in the time to cell division and the likelihood of cell death is specified by the size and shape of suitable distribution curves. The computer can then *decide* by reference to random numbers, the time of the next cell division, and the likelihood that the daughter cells will live. After allowing the computer an appropriate period of freedom—perhaps the equivalent of three months real time—the total number of living cells and thus the tumor size may be determined. Proposed treatments can then be examined in terms of their impact upon the time to cell division and the likelihood of life in daughter cells.

A second example with ergonomic relevance is a learning model. Let us suppose that a subject is confronted by two keys, left (L) and right (R). If he presses the R key a green light shines, but the L key illuminates a red light. The initial probability of choosing R is close to 0.5, but there is a finite increment in this probability with each correct choice. Ultimately, the probability of choosing R reaches a limit of 0.7 to 1.0 (depending upon the intelligence of the subject and the effectiveness of any immediate reward). The entire learning process can be mirrored on the computer, with an initial probability of 0.5, and modification of this probability with each trial of the two keys.

Missing Data

If information has been lost through the carelessness of the investigator, it is usually desirable to repeat the experiment. However, in some circumstances, such as the anthropologist who discovers a few fragments of a skull, there may be no possibility of obtaining more complete data.

The available options are (1) to make an initial guess at the missing values, and allow the computer to reach the most likely answer by a series of successive approximations, and (2) to test the difference in the final result, given an initial selection of discreet digital values ranging from one to ten.

Unfortunately, the mathematical complexity of the solution varies as the power of the number of missing results. Thus, if n figures are missing, we are confronted by 2^n possible solutions. Where data are obtained in a logical sequence, the process of convergence can sometimes be helped by fractional replication. Thus, if a sample is one eighth of the size, the number of missing data may be reduced from sixteen to two, and the search process will correspondingly be reduced from 2^{16} to 2^2 possibilities. Alternatively, it may be possible to reach an arbitrary judgment regarding the relative importance of the missing items, and to concentrate the search upon one or two of the more significant variables.

War Games

In times of peace, when military equipment and supplies are limited, army chiefs of staff often test tactics and strategy by computer-based war games. These have well-defined rules, such as the number of troops killed by the dropping of a 2000-pound high explosive bomb. Validity is naturally limited by the accuracy of such estimates, although unfortunately independent confirmation of data is sometimes obtained through subsequent warfare.

Comparable games are now played by the politician and the businessman to test the impact of such imponderables as a presidential assassination upon the political stability of a Latin American republic, and an increase in production or a decrease in product cost upon the competitive position of a giant manufacturer. Let us examine some of the limitations of the latter model. Our assumptions are as follows: (1)

both the company under consideration and its major competitor are profit maximizers (the second company may be guilty of an unsuspected and irrational philanthropic streak!); (2) both organizations make their decisions on the *rate* of production (whereas the competitor may prefer to emphasize the quality of its product); (3) the overall demand for the product is stable and known to both manufacturers (whereas in practice the demand for most consumer durables varies rather unpredictably with the buoyancy of the national economy); (4) costs vary stochastically with the total production of the two companies, (5) if one company modifies production the output of the competitor remains unchanged; (6) the user population has a fixed quantity of money available for purchase of the product, unit costs at a given rate of production being fixed and equal for the competing firms; (7) the products are uniform and homogenous; (8) there are no physical limitations (such as factory size and availability of land or labor) limiting the potential production of either company. It is immediately obvious that in most situations we are unlikely to satisfy all of these assumptions. Equally, the design of a more realistic model is prohibitively expensive. It is thus hardly surprising that *duopology*, or economic modeling remains an imprecise science.

Pattern Recognition

The potential for the feedback of game results into real life is readily appreciated. Sometimes a model may have more immediate ergonomic application. Thus, one current bottleneck in automation is the problem of visual and speech pattern recognition. A device that could read writing such as signatures would have immediate application in commerce, and at the same time would add greatly to the happiness and effectiveness of a blind person (page 292). The basic problem is that the eye can recognize much finer detail than might be anticipated from the number of primary image receptors (the cone cells of the retina). The explanation apparently lies in a complex pattern of neural intercommunication (page 151), and by formulating a model (Fig. 97) it is possible to test which arrangement gives the closest approach to normal pattern recognition.

The initial layer of the model corresponds to the image receptors. It is presented with a suitable pattern of binary information (0 or 1). At the second level of organization, the computing units of the ganglion cells, the input is weighted; thus, the output of the ganglion cells may be arranged to read zero unless the input from at least three of the primary receptors is equal to one. Other *structured* requirements are laid upon the cognitive and decision-taking levels of organization in the model. Finally, by making systematic and progressive changes

Image Units (Rods and Cones)

Computing Units (ganglion cells)
                  of retina

Decision Units

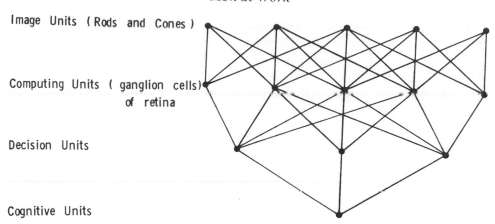

Cognitive Units

Figure 97. Suggested model of pattern recognition in the human eye.

in the computing rules, the process of pattern learning can be simulated. Unfortunately, this type of modeling makes prohibitive demands upon computer time, and to date the output of useful information on either pattern recognition or adaptive behavior has been slight.

Queuing Models

The relevance of queuing models to computer design has already been noted (page 343). Depending upon the arrangement of the computer, individual transactions may be subject to a fixed delay (e.g. 3 seconds) or a delay with a statistically specified variability (3 ± 1 sec). The probability of a *competitor* arriving to use the computer is initially close to the chance figure (P = 0.5), but as a specific operation continues, the probability of competition increases progressively towards unity. The objective of the modeler may be to modify computer and/or user tactics so that the rate of servicing of data is faster than the average rate at which new information is arriving. If the probability of competition is kept close to 0.5, then service will be good. If the probability rises to 0.9, long queues may be anticipated. There is naturally a *trade-off* between fast service and cost. To avoid long queues, 10 to 20 percent of computer capacity must be idle for much of the day. Such losses can be minimized if the computer maintains a file of low priority *busy work* that can be undertaken when the probability of competition drops below an economically acceptable figure.

Similar questions may be analyzed for many human situations, ranging from the checkout counter of the local supermarket to the provision of emergency services for a city or a military regiment. In general, a customer is unlikely to patronize a service station if more than three cars are already waiting at the pump islands. Nor will he shop at a supermarket that has an average of more than three or four customers waiting to pay for their groceries. The manager of such a facility must control average and maximum queue lengths, average and maximum waiting times per customer, and the percentage of the total working day for which his staff are idle. Flexibility can be built into the system by opening further checkout counters or adding packaging staff when an unacceptable queue length is approached. Larger queues may be acceptable at a discount warehouse, where customer expectations of service are modified by a change of price structure.

An interesting digital study of military casualty evacuation was carried out by the U.S. Army in 1960. The objectives were to estimate the optimum number and size of ambulances required by a medical aid center serving five platoons. *Casualties* were created in the computer model at statistically predetermined intervals. If an *ambulance* was available at the *medical aid center*, this was dispatched immediately to the casualty; alternatively, if space was available in an ambulance already at platoon level, this was sent to collect the injured soldier. If no transport was available, the casualty was entered on a list (request for ambulance queue). When the duties of the ambulance were completed, it was returned to the vehicle pool (ambulance availability queue). The model tested the effects of rate and variability of casualty production, and the speed, variability of speed and capacity of individual ambulances upon the adequacy and efficiency of the medical service provided. Because random numbers were used to play the game, a series of plays were undertaken to establish the mean performance of the system and its variability. Criteria of adequate service included the evacuation time, and the number of casualties waiting in the ambulance request queue. Efficiency was judged from the number of ambulances standing in the ambulance availability queue. One limitation of the model was uncertainty regarding the importance of the medical service to the larger objectives of the system (page 8). What was the purpose of the system—to win the war or to provide good treatment of injured soldiers? And if the former, would the war be won faster if the entire medical service was disbanded and the staff transferred to the front line? This question assumed particular pertinence in that one proposed solution for casualty evacuation (the use of helicopters) required the periodic halting of any artillery barrage.

The general result from the simulation was that evacuation would be speeded by use of a large number of small ambulances. The ambulance request queue was also shortened more by vehicle speed than by vehicle capacity, although the model did not assess the influence of speed upon the condition of the injured patient. This led to the conclusion that if capital cost was not a major consideration, and the prime objective of the system was to treat the casualty, then evacuation requirements were best satisfied by a fleet of small helicopters.

APPLICATIONS TO LITERATURE
SURVEYS AND DIAGNOSIS

Literature Surveys

A logarithmic growth in the volume of scientific literature has created an enormous practical problem for the earnest investigator who seeks to remain abreast of current developments in his field (Barber *et al.*, 1972). How far can computers help in surveys of world literature?

The complexity of many proposed search schemes has led to the channeling of searches through a trained librarian or technical editor. Inevitably, difficulties have arisen from the use of an intermediary. He is somewhat uncertain of objectives, and cannot state with authority which references should be accepted, and which deleted. Further, he may lack the background to modify search instructions in the light of obvious deficiencies in early output data; indeed, procedural delays commonly preclude any rapid answers to queries.

The current emphasis is upon provision of conversational-type terminals where the concerned research worker can formulate his search request, and modify input instructions on the basis of the early output of information. In the Medlars System, the first step is to insert a subject title approved by the computer, normally one of 8000 headings used by the *Index Medicus*. The machine responds by indicating the number of citations available, together with a vocabulary of related terms. One subheading of interest may be selected and printed out in greater detail. The operator then decides whether to progress further within this tree, or move laterally to a related tree. Having reviewed several trees, he may instruct the computer to concentrate on a combination of items located in two or more trees, and to print out a complete list of relevant references meeting combined criteria such as "ventilation/perfusion ratio" and "respiration."

Validation of the search procedure and an assessment of its efficiency can be carried out by comparing the tactics and strategy of the man/machine system with procedures used by an unaided research worker.

Diagnostic Reasoning

There is increasing evidence that a diagnostic computer program can carry out the routine clinical questioning of patients more efficiently and more objectively than the average physician. In a large and cosmopolitan city, there is also the significant advantage that multilingual conversation is possible. A necessary preliminary to computer diagnosis is to formulate a logical sequence of questions with a rigidity foreign to the intuitive thinking of many physicians (Barnett, 1968). Several schemes of reasoning can be identified (Fig. 98). The *key* is widely used as a means of species identification in zoology, botany and bac-

(a) *Key*
    1. Glucose positive: 2
    1. Glucose negative: *N. catarrhalis*
        2. Maltose positive: 3
        2. Maltose negative: *N. gonorrhoea*
            3. Fructose positive: 4
            3. Fructose negative: *N. meningitidis*
                4. Sucrose positive: *N. sicca*
                4. Sucrose negative: *N. flava*

(b) *Decision table*

Bacterium	Carbohydrate metabolism			
	glucose	maltose	fructose	sucrose
N. catarrhalis	0	0	0	0
N. gonorrhoea	+	0	0	0
N. meningitidis	+	+	0	0
N. flava	+	+	+	+

(c) *Flow chart*
    Neisseria
    Glucose positive?   No = *N. catarrhalis*
    Maltose positive?   No = *N. gonorrhoea*
    Fructose positive?   No = *N. meningitidis*
    Sucrose positive?   No = *N. flava*
    Yes = *N. sicca*

Figure 68. Schemes of reasoning for the identification of *Neisseria* (based on Ames, 1971).

teriology. As an alternative, the same information can be presented as a *decision table*. The *flow chart* and the *logic tree* are minor variants of the key. The flow chart, which presents all questions in yes/no form, is most readily adapted to the digital computer.

The ideal starting point of medical diagnosis is the presenting sign or symptom. Thereafter, the classification must be arranged as an all-inclusive series of binary decisions. Unfortunately, the demands made upon the computer are extreme, and as yet there are no published programs that will carry a person with a complaint such as dizziness through all possible diagnoses. More success has been attained with classifications that begin from a stated or implied diagnosis. The computer can also play a useful role in indicating possible areas where a physician should concentrate additional questioning. The design of a flow chart sharpens the logic of the clinician and ensures that critical points in the diagnosis of a particular disease entity have been given adequate consideration.

The current dilemma of the clinician is well illustrated by a quotation from Asher (1959):

> Formal logic is based largely on statements of this kind: All X is Y. All Y is Z. Therefore all X is Z. For instance: All men are mortal. Socrates is a man. Therefore Socrates is mortal. Medical facts are rarely suitable for this

discipline. They are more of this kind: Some X is Y, but I'm not really sure of it. Possibly all Y may be Z (as Smith claims to show), but Brown affirms that no Y is ever the slightest Z, and Robinson too has strong views, but I find them hard to understand because he uses X to mean both X and Z, and he calls Y what I call X.

Hopefully, the increasing use of the computer will force physicians and other biological scientists to a more rigid terminology (page 9) and thus more logical thought patterns. This should yield important dividends, allowing not only an automation of practical diagnostic procedures but a clearer understanding of the mode of operation of the underlying system.

CHAPTER 19

ERGONOMICS OF TRAFFIC SYSTEMS

A T VARIOUS POINTS IN THIS BOOK we have made reference to ergonomic problems encountered in the design and control of both vehicles and traffic systems. A brief integrating chapter seems desirable, partly because of the medical and economic importance of system failures, and partly because a well-designed traffic system provides a good example of the practical ramifications of ergonomics.

MEDICAL AND ECONOMIC IMPORTANCE

Western society is heavily dependent upon its systems of transportation. A national airline, rail or bus strike can rapidly lead to economic disaster. A further medical and economic burden is imposed by the ever-rising total of road accidents. World fatalities now total 250,000 per year, with a further ten million serious injuries (W.H.O., 1970).

Some twenty years ago, a high proportion of the casualties were elderly pedestrians, but currently the majority of those killed are vehicle occupants. From the economic viewpoint, young casualties are of particular significance. Some 50 percent of male deaths in the age decade 15 to 24 years are presently attributable to road accidents, and it has been estimated that elimination of such fatalities would increase the life expectancy of the young adult male by two years, a larger gain than could be obtained from the elimination of all infectious disease. Further drains upon the economy result from (1) continuing care of crash survivors, particularly patients with irremediable brain and spinal injuries, (2) court costs, and (3) property damage.

Insipient system failure is reflected in the ever-increasing congestion and pollution of large urban centers. A typical city dweller may spend two hours of each day in his car or public transportation, and there would be a substantial benefit to the economy if even a half of this time could be put to productive use.

TRAFFIC ACCIDENTS

In industry, it has been possible to reduce the incidence of accidents and associated economic losses by a careful review of critical incidents, not only actual accidents, but also dangerous situations with a significant potential for mishap.

Unfortunately, traffic accidents have so far resisted such an analytic approach. The main problems seem (1) the absence of a simple,

355

stereotyped hazard for which an appropriate countermeasure can be devised, and (2) a high proportion of incidents in which the use of alcohol or other drugs is a contributing factor. Nevertheless, the study of road accidents continues, since most laboratory simulations of driving conditions lack realism.

The present discussion is restricted to traffic accidents. Nevertheless, the principles considered are generally relevant to the causation and prevention of accidents elsewhere in industry and the home.

Causative Factors

Mackay attempted to apportion blame for accidents between the driver, the vehicle and his environment. His conclusions can be summarized in a format similar to an analysis of variance:

Source	Percent of accidents
Driver	12.4
Vehicle	4.8
Environment	4.6
Driver/vehicle interaction	7.2
Driver/environment	48.4
Vehicle/environment	4.8
Driver/vehicle/environment	16.4

The driver was totally or partially at fault almost 85 percent of the time. Specific problems of mental and physical health accounted for only 12.4 percent of accidents, and 72.4 percent were attributable to a mismatching between the traffic system and the decision-making process of the individual.

Accident Proneness

Some drivers can operate a vehicle for as long as forty years without incident, but others—apparently *accident prone*—are involved in frequent accidents. On theoretical grounds, one might anticipate a bell-shaped distribution of skill among a population of road users, with a corresponding *normal curve* reflecting the varying demands of the road system (Fig. 99); accidents would then become probable when the demands of the system exceeded the available skills.

Unfortunately, this simple relationship is not observed in practice. Often, it is the apparently skillful driver who is involved in an accident. Part of the problem is an undue emphasis upon control skills in the training and evaluation of the driver. A newly qualified road user may have justifiable confidence in his ability to handle a vehicle, but overlooks his deficiency in the more important skills of decision making. Moreover, driving skill is not a stable criterion; accidents frequently happen to the apparently experienced individual during a momentary

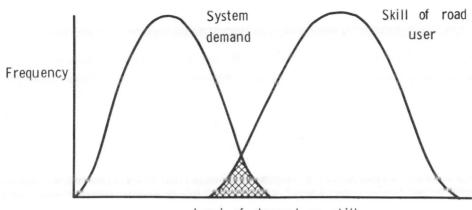

Figure 99. The theoretical relationship between the demands of a road system and the skill of a population of road users. Accidents might be anticipated in the cross-hatched area where the two curves overlap.

aberration (page 272). Presumably, the accident prone driver suffers more aberrations than the person with an accident-free record.

The youthfulness of the typical casualty is a further paradox. The drivers in the age decade fifteen to twenty-four years sustain eight times as many accidents as those who are fifty-five years of age and older. This may be partly a reflection of the relative mileage covered by the two age groups, but it also highlights the fallacy of medical examinations that stress vision, hearing and reaction time. All of these functions are near their peak in the average twenty-year-old man; his deficiencies are in the areas of attitude and behavior. The young tend to be aggressive, with an instinct for territorial defense, and are chance rather than skill oriented. Quenault noted a high proportion of extroverts among the more careless drivers; it may be presumed that such individuals are easily bored when sustained vigilance is required, and they do not readily accept social pressures towards safe and courteous motoring.

Fatigue

Accidents can undoubtedly arise from fatigue and loss of vigilance, particularly when a solo driver is operating under conditions that do not favor cerebral arousal (page 215). In this context, circadian rhythms of wakefulness (page 274) may be an important secondary cause of disaster.

Nevertheless, other psychological factors often outweigh fatigue. McFarland (1958) studied near accidents in truck drivers making a nine-hour journey, and noted that incidents occurred most frequently in the first two hours of travel. He postulated that to such men, initial lack of rest and domestic anxieties were more disturbing than the cumulative effects of a long drive. It is uncertain whether such findings can be extrapolated to the amateur driver who undertakes a marathon journey prohibited to the professional trucker by both unions and road safety legislation.

Deteriorations of performance can be documented as an increase in the number of gross control movements observed under comparable traffic conditions. It is also possible to demonstrate increased muscular activity at electromyography, and an increase in the total energy cost of driving. Exhaustion occurs most quickly on crowded routes where overtaking is difficult and potentially dangerous. Under such conditions, measurements of heart rate and palmar sweating suggest that the driver is frequently under emotional stress. Performance may also deteriorate due to group interaction with other bad and aggressive drivers. Self-discipline is important to accident prevention, and as in so many problems of human psychology behavior is more important than knowledge or attitudes.

Task Difficulty

The probability of an accident normally increases with the demands made upon the driver. In driving as in flying, an increase of speed reduces the time available for critical decisions (page 6). Rapid movement also makes the accurate judgment of speed difficult, and increases the number of interactions with trucks and other slowly moving vehicles.

Other factors increasing task difficulty include congested roads, overtaking, the use of a rear-view mirror, and the navigation of unfamiliar territory. Task difficulty can be judged from assessments of *spare* mental capacity (page 87), and there is scope to apply this technique to road safety research.

Crawford (1963) studied problems of overtaking at the British Road Research Laboratory. He noted that the time needed to pass a slow vehicle increased from 7½ seconds when it was traveling at 25 mph to 9½ seconds at 50 mph. Over the same speed range, the safe gap between overtaken and oncoming traffic increased from 200 to 400 yards. When a driver was presented with a safe gap, the decision to overtake was made in about 1.5 seconds, but as the gap became shorter, the typical driver wasted precious moments agonizing over the decision whether to pass or not.

A typical glance in a rear-view mirror occupies 0.8 seconds (1.0 seconds if a curved mirror is used). In expressway driving, a car can easily move 100 feet forwards during this time. The typical rearward glance can be reduced to 0.6 seconds by a gross enlargement of the mirror, and (perhaps more important to accident prevention) *long* glances (> 1.5 seconds) also diminish from 10 to 2 or 3 percent of the total. The use of rear-view mirrors is facilitated by a reduction in the size of roof pillars, but a finite limit is set to such improvements by the need for rigid roof support in case the car should roll over.

The principles governing the design of road signs seem simple, but the navigation of an unfamiliar expressway remains a difficult and hazardous undertaking. Letter size, contrast and illumination (page 153) must take account of the brief time available for viewing and interpretation. A sign that takes two seconds to read and understand must be clearly visible, both day and night, at a distance of several hundred feet. The height and positioning of the sign should be such that the driver's eye is not drawn from the road. Expectancy or *set* (page 219) influences the speed of interpretation. Thus, the exit ramp of a North American expressway is anticipated in the right-hand lane, and if a left-hand exit is unavoidable, additional warning is required. Diagrammatic guide signs have yet to prove their worth in minimizing erratic maneuvers at complex intersections. To be effective, they must be simple, internationally standardized, and introduced after a suitable program of driver education.

Night Vision

There is a marked increase in the frequency of accidents at night, particularly when this is related to the number of vehicles in operation. Consumption of alcohol and diminished wakefulness (page 215) are important considerations, but difficulties also arise from twilight (page 158), glare (page 152) and other problems of night vision. Most motorists overdrive the range of visibility afforded by their headlamps. A typical low-beam installation offers 360 feet of vision, permitting a safe speed of 50 mph under good road conditions. A triple low-beam system increases the range to 430 feet, and a triple mid-beam offers 470 feet of visibility. However, the safe stopping distance on a 70-mph expressway is nearer 560 feet, and the only practical method of attaining this range without dazzling oncoming traffic is to use high-intensity polarized light.

ACCIDENT PREVENTION

Any remedies proposed to diminish the frequency of traffic accidents must be both practical and socially acceptable. From MacKay's analysis (page 356), it might seem that the best solution (in terms of cost/effectiveness) would be to modify the interface between driver and traffic system through selection, education, legislation and improvements in hardware and software.

Selection

Formal selection procedures can be applied to bus and train drivers, where there may be as many as forty applicants for a single vacancy. However, a car is so essential to current Western society that it is socially unacceptable to demand more than rather mediocre driving ability as a condition of vehicle operation. Tests to distinguish adverse traits such as extreme extraversion, psychopathic tendencies and overt aggression may usefully be applied to drivers involved in several accidents, with a view to selecting those in need of remedial education.

Medical examination should take particular note of the subtle and fluctuating changes in decision-making capability that accompany ischemic cerebrovascular disease. Elderly drivers may suffer from several minor defects such as visual and hearing problems. Although not individually serious, together they affect both reaction times and driving ability. On the other hand, it may be socially unacceptable to deny a license to an elderly person in a large North American city where a car is an unfortunate necessity of normal life. The Province of British Columbia has introduced conditional licenses as a solution to this dilemma. Such licenses prohibit rush-hour, freeway and night driving.

There are presently no formal minimum standards of intelligence for drivers, although initial multiple-choice questionnaires may eliminate those who are pathologically stupid. Given patient training by a skilled instructor, many individuals of below average intelligence can serve their community well as short-haul commercial drivers.

Education

The average driver has not mastered the more advanced skills of motoring such as the control of a skidding vehicle. Through frequent repetition, the required movements can be developed to the point where they become *automatic* (page 87). However, the main emphasis of educational programs should be upon the improvement of judgment rather than control skills. The student should learn tricks for sustaining vigilance and acquire an ability to make decisions as an exponent rather than as a linear function of vehicle speed.

Unfortunately, attitudes towards instruction are often poor. Whereas a potential submariner or airline pilot accepts his professional preparation as essential to survival, the majority of aspirant drivers fail to recognize the hostility of the environment for which they are being prepared. Indeed, many of the younger generation regard road safety legislation as one more piece of oppression by *the establishment,* to be ignored or openly violated as opportunity presents itself.

It is difficult to measure the success of advanced driver education programs. Certainly, the accident record of graduates is good, but the population concerned is highly selected. Many of the pupils are women, and in terms of both mileage driven and system demands, the average woman is not at high risk. Many, also, are older drivers, and there is a natural tendency for the individual's driving record to improve with experience and maturity. Certainly, behavioral modification is difficult to achieve at any age, and an adverse trait may remain fundamentally unaltered despite a marked change of expressed attitudes (page 223). There is plainly a need for a broad program of public education aimed at making fast and aggressive driving a socially unacceptable habit. Possibly, there may be scope for operant conditioning techniques to improve the behavior of the recalcitrant driver (page 207).

Legislation

The most important area for the legislator is the control of alcohol and drug usage by the driver. It is also possible to specify minimum levels of knowledge, skill, judgment and health as a condition of vehicle operation, and to deny motor vehicle licenses to those who repeatedly infringe traffic regulations, perhaps linking this penalty with compulsory attendance at a driver re-education course. The effectiveness of court supervised education is difficult to evaluate. Those attending such courses certainly drive better in subsequent months, but the mere summons to appear before a court official often has a salutary effect upon driving habits.

Hardware and Software

Although vehicle and road together account for a relatively small proportion of traffic accidents, it is worthwhile eliminating undesirable features, as this has an immediate effect upon accident statistics. Changes in hardware and software may be addressed to both the prevention of accidents, and a minimization of the effects of collision.

Road improvements range from the simple introduction of nonslip surfaces and appropriately cambered bends with signs indicating safe speeds to the radical redesign of highway networks with the separation of pedestrians, slow moving and fast vehicles and the introduction of expressways having limited access via well-designed cloverleaf intersections. Bends, gradients and street lighting on more traditional highways can be adjusted to provide a better matching of forward visibility and permitted vehicle speeds. Navigational aids can be improved with regard to placement, legibility and intelligibility, and warning signs can incorporate such ergonomic principles as intermittency (page 94) and shape/color reinforcement (page 22). Road surfaces that emit a warning noise when a vehicle wanders from its lane may be practical in some situations, but more advanced concepts such as

partial or fully automated control of vehicles on busy highways are
as yet little more than the dreams of the visionary.

Within the vehicle, it may be useful to reallocate function between
the conflicting demands of status, comfort and the effective operation
of displays and controls. What proportion of the total vibrant energy
is transmitted to the driver? Are road noise and vibration contributing
to fatigue? Are soft forms of springing increasing the liability to motion
sickness (page 138)? Could reaction times in an unfamiliar rental vehicle
be improved through international standardization of control location,
display and coding (page 29). Often the vehicle designer has given
too little thought to controllability under typical and atypical road conditions. Can the driver manipulate all controls while wearing a lap belt
and shoulder harness? Can pedal force be modulated to allow both
gentle braking and the force needed in an emergency stop? Can overall
visibility be improved by such measures as reshaping the hood, or
displacement of windshield wipers and roof pillars (page 317). Are
laboratory measurements of a driver's field of vision realistic when
posture is altered by the tension of anxiety or the cumulative effects
of fatigue? Can the glare of night driving be reduced by an alteration
in texture of painted surfaces and the range of visibility increased
by the use of more powerful but polarized lights?

Road changes that reduce the severity of injuries following collision
include the positioning of bridge buttresses and other immovable
objects behind guardrails, the use of light metal collapsible lamp standards and other street furniture, and adequate separation of opposing
traffic streams, either by decelerating (*chicken wire*) barriers or physical
distance.

The most promising addition to the vehicle is an impact-inflated
bumper. Inside the car, injuries can be minimized by adequate padding
of the dashboard, with provision of a collapsible steering wheel, swivel
rear-view mirror, penetration-resistant windshield, and interlocking
door catches. The extent of damage to a car is influenced by its inertia
relative to that of an opposing vehicle; it is thus particularly hazardous
to drive a small imported car on North American roads. Personal injuries
are greatly reduced by the wearing of lap and shoulder harnesses. Unfortunately, current usage of the lap belt in North America is estimated
to be no more than 30 percent, and very few drivers accept a shoulder
belt. Possible alternative methods of increasing the use of safety
restraints include (1) the design of a car that will not start with unfas-
tened harnesses, (2) legislation prohibiting vehicle operation without
wearing appropriate safety belts, and (3) a massive program of driver
education. In Australia, legislation is claimed to produce 70 percent
usage of harnesses, with a 16 percent reduction of fatalities.

Mortality rates may also be reduced by improvements in health care delivery systems (page 16). Developments could logically include emergency first aid training for the entire adult population, higher standards of education for ambulance drivers (coupled with specific training in cardiac massage) and production of a simple vehicle-mounted sensor that would emit a warning radio beam if an accident occurred in a remote part of the country.

SPECIAL CLASSES OF VEHICLE OPERATOR

The Disabled Driver

The design of vehicles for patients with physical handicaps has been discussed elsewhere (page 291).

The Cyclist

The bicycle is currently enjoying a resurgence of popularity in North America. Large increases of domestic sales were reported in each of the years 1970, 1971 and 1972. There have also been an alarming increase in the number of accidents involving cyclists. Despite the apparent danger from cars on congested city highways, the most frequent reason for injury has been the hitting of some stationary object. Victims are usually boys under twenty years of age, and the majority of accidents occur during daylight hours. However, these statistics may reflect mainly patterns of bicycle usage; certainly, the poor lighting and reflectors of the average bicycle place the nocturnal rider at considerable risk, particularly when traveling on unlighted country roads.

Practical measures to increase the safety of cycling are similar to those proposed for the motorist. Educational programs are preferably concentrated in schools. Legislation may include the licensing of cyclists and statutory minimum standards for lighting and reflectors. Improvements of hardware range from the provision of specific cycle tracks to the elimination of unstable designs such as the notorious *high-rise* cycle. Injuries may be minimized if guards keep the feet from the wheels and the genital organs from impaction upon gear levers.

The Motorcyclist

The motorcycle is a particularly lethal form of transportation for the teenager. Education and legislation to ban excessively powerful machines may reduce the toll of accidents. The severity of injuries can be reduced by appropriate protective equipment, including well-designed crash helmets and crash bars that protect the legs if a machine skids on a loose gravel surface. Padding of the handlebars or the rider's clothing reduces the likelihood of pelvic injuries during forward ejection of the rider.

The Train Driver

Retrospective analysis of railway accidents is important, for although incidents are infrequent, the number of fatalities in a single crash may be large. The results of such enquiries also have general interest, for while there is scope for disagreement upon the definition of dangerous driving, there can be little argument that the train driver who passes a signal set at danger has committed a hazardous act.

The British Medical Research Council has reviewed a series of episodes where signals were passed (Edholm, 1967). As in the motorist, medical problems were rarely responsible for the error, and incidents bore little relation to such variables as visual acuity or the accuracy of color perception. Age, also, was unimportant; gains in experience and maturity apparently counterbalanced any decline in acuity of the physical senses. Somewhat surprisingly, the incident rate per train mile was no higher at night; apparently, drivers were very successful in adapting circadian rhythms of arousal (page 274) to the unusual demands of their work. Recurring factors were lack of experience, operation over an unfamiliar route, and approach to an unusual or complicated array of signals. Certain men were apparently accident prone; the individuals concerned not only had a poor work record, but also a history of playing truant at school, a high divorce rate, and a bad credit rating. In short, they had poorly developed social consciences.

THE AUTOMATION OF TRAFFIC CONTROL

Traffic control measures are necessary not only to reduce the frequency of accidents, but also to ensure the smooth flow of vehicles through congested areas of a city. In the case of road traffic, this involves supplementing the traditional tools of the traffic engineer—parking regulations, pavement markings and one-way streets—by an optimization of both vehicle speed and intersection control.

Safe following distances increase exponentially with speed, and for this reason there is a U-shaped relationship between average vehicle speed and the carrying capacity of a given road system. The optimum speed varies with the experience of the drivers and the nature of the vehicle controls (for instance, power brakes and automatic transmission). Maximum flow is commonly achieved at about 40 mph.

Control systems for both road and rail traffic were initially entrusted to the human operator. The police officer at a busy intersection became quite adept at his task as he accumulated experience of traffic patterns. However, such intuitive learning is necessarily inefficient. Further, the number of controlled intersections in large cities is now numbered in thousands, and the cost of manual control is prohibitively large.

Fortunately, the operations required at the majority of intersections are repetitive and well suited to automation.

One possibility explored in the 1930's was the traffic circle. Land requirements for a safe circle are large, and traffic capacity is rather limited. Further, there is a serious risk of accident when two equally aggressive drivers seek to cross paths.

A more practical alternative is the familiar traffic light. This has already profited from some ergonomic study, but more investigations are needed. What is the optimum distance between controlled intersections? Expediency, often political in type, demands lights every 600 feet or closer, but the distance that will establish a good traffic flow is probably two or three times this figure. If the interval between lights is too long, vehicles fail to form convoys, and drivers react more slowly because they are not anticipating a traffic light. Other questions remain. What is the optimum height and distance from the sidewalk? Can a placing be chosen that will secure maximum visibility without drawing the driver's eyes from the road? Can the relative positions of the three lights be standardized on an international basis? What power of illumination is needed in fog? Is there any practical advantage in using a larger light for the red than for the green signal? What legislation is needed to control advertising signs that could be confused with traffic lights?

Filtration arrows or the equivalent flashing green signals speed flow when there is much turning traffic. Several advantages are claimed for the flashing signal: (1) versatility—it can readily be omitted at periods of the day when traffic patterns differ, (2) visibility—the larger size of the signal and its intermittency help make it obvious in fog and (3) rapid perception—acceleration is much faster for a flashing signal than for an arrow. The duration of the signal can be adjusted according to the line of waiting traffic; two to three seconds must be allowed between cessation of the flash and a green signal for opposing traffic in order that the final two to three cars may clear the intersection.

The basic box-mounted control system for traffic lights uses a series of rotating cams and switches, so that specified green periods are allocated to each of the intersecting routes in turn. An amber period allows turning traffic to clear the intersection, and a brief *all red* phase is desirable for safety. Experience suggests allocating four seconds of amber and two seconds of all-red per cycle. Further time is lost due to the reaction period of the column of waiting motorists; this varies from three to four seconds, depending on whether drivers can see the opposing signals or not. Finally, the new packet of moving vehicles must accelerate to permitted city speeds. A maximum traffic flow is achieved by allowing a relatively long green period (60 to 100 seconds) for the major route. Under conditions of recent snowfall, acceleration is slow, and both green and amber phases require extension. Alternatively, signals can be switched by police or computer to a flashing amber (caution) or flashing red (halt) mode. Special needs of the pedestrian may be a further consideration influencing the timing of a given cycle. With an average walking speed of 4 feet per second, and a 100-foot arterial highway, it is necessary to obstruct the major route for about thirty seconds if pedestrians are to cross safely. In the vicinity of an old people's home, it may be desirable to allow for an even slower pace of walking. Even if pedestrians are few and the major route is narrow, the time allowed to the side street cannot usefully drop below ten seconds. With shorter intervals, drivers become confused and may even miss the green signal.

The main advantage of the fixed and locally operated control is its simplicity. The main disadvantage is that it is insensitive to diurnal fluctuations of traffic density. The advantages of local control can be combined with greater flexibility by the use of three or four switching circuits, possibly allowing the major route a 100-second green phase in the morning rush hour, an 80-second green phase in normal daylight hours, a 60-second green phase in the evening rush hour and a flashing amber signal from midnight to 5 AM. Selection of the appropriate switching circuit is controlled by a clock, with the possibility of manual override to cover such contingencies as a heavy snowfall, a football game, or even the imminent appearance of an ambulance or a fire truck.

A further stage of sophistication is to link the response of a series of traffic signals, so that if a vehicle proceeds at a specified pace (e.g. 32 mph) it will regularly encounter the green phase. The desired results can be achieved in a single street through the careful synchronization of individual control clocks, but if the objective is to maximize traffic flow patterns over a network of several thousand signals, the problem is better handled by an electronic computer.

Fixed phase relationships between intersections are ineffective if the average speed and density of traffic depart from anticipated levels. An alternative approach is to make a continuous count of vehicles approaching an intersection. In the United Kingdom, rubber strip electrical contacts have been used for this purpose for many years. They are reasonably satisfactory in light to moderate traffic, but fail completely if a vehicle is stationary over the tread. They are also impractical in areas where there are frequent snowfalls. An alternative type of detector, popular in North America, is a metal loop buried some two inches below the road surface 300 feet away from the intersection.

If a loop is crossed by a metallic mass such as a car, there is a change of inductance in the electrical circuit to which it is connected. This activates a counter, and by noting the frequency of signals not only traffic density but also vehicle speed may be estimated. Traffic responsive signals are particularly useful for the control of vehicles emerging from infrequently used side streets. If a vehicle is detected in the side street, this information can operate the signal either directly or through a computer; in the latter case, the green phase for the side street can be adjusted to optimize conditions in the main traffic stream. Normally, interruption of the main stream is brief (10 seconds), but an extension is allowed if a second vehicle appears in the side street. This is an important provision, for otherwise the second vehicle might remain stranded between the control strip and the light.

The most sophisticated control systems, such as that operated by Metropolitan Toronto (Hodges and Whitehead, 1969; Cass, 1970), use a relatively large electronic computer. The advantages claimed for such a system include

1. flexibility of signal phasing (since centrally located computer programs are more easily modified than peripheral signaling equipment),
2. compatibility of the controller with a wide range of detectors and signaling equipment,
3. versatility, including a potential to operate a variety of signaling plans over a short space of time,
4. judgment, involving an ability to make on-line measurements of traffic flow through a complex network, to devise an optimum theoretical solution at any given point in real time, and to check the resultant movement of vehicles to ensure that the chosen solution is appropriate,
5. memory, including the collection and cataloguing of objective data for the future planning of the traffic system.

Detailed computer programming for a large traffic network requires careful study of each major intersection over the course of several typical days. Account must be taken not only of vehicle counts on the major routes, but also any unusual causes of delay such as turning traffic, pedestrians crossing intersections, and pedestrians boarding streetcars. Where possible, detectors should be arranged so that the signals remain uninfluenced by right-turning traffic, allowing full use to be made of the possibility of *right turn on red* or an equivalent filtration arrow.

How practical is a fully computerized traffic control system? In the City of Toronto, the most vulnerable feature has been the metal loop detector, often inadvertently damaged by street repair crews or heavy vehicles. Currently, a little over 1000 signals are linked to the computer. Of these, about 180 are traffic-responsive signals controlling egress from side streets, 70 are traffic-responsive signals at major intersections, 50 are adjacent signals linked to responses at the major intersections, and the remaining 700 as yet operate to a diurnal schedule regulated by the computer. Traffic flow has certainly been speeded by the control system. Average journey times have increased 40 to 50 percent on the rare days when the computer has been out of service. However, the true advantage gained from the computer is smaller than 40 to 50 percent, since traffic flow could also be improved by local regulation of signals with three series of cams and a clock.

Let us assume that an average journey of thirty minutes is speeded by 10 percent twice in each working day. The annual saving is some twenty hours per vehicle, or with 500,000 vehicles entering the city each day 10 million vehicle hours per year. The capital cost of the Toronto System was about 5 million dollars (2 million dollars for the computer, 1½ million dollars for the detectors, and 1½ million dollars for other peripheral hardware). Operating costs are currently about 1 million dollars per year. Such figures are small relative to snow clearance (which costs Toronto 20 to 30 million dollars per year). In effect, ten hours of free time for a vehicle and driver is purchased for one dollar. Traffic lights also reduce road accidents by 10 to 20 percent, equivalent in Toronto to an annual saving of 1.5 million dollars property damage, plus court and medical costs. The total estimated commercial savings from the computer system are as large as 20 million dollars per annum (Cass, 1970). Unfortunately, not all of these gains are realized. In particular, much of the time lost in driving is that of office workers returning to their homes, and they may be surprisingly reluctant to agree to additional taxation in order to spend a few extra minutes each day by their firesides.

TOWARDS AN INTEGRATED SCHEME OF CITY TRANSIT

Freeway networks in a large city must be related to the overall purpose of the system in order to avoid unrealistic solutions (page 13). Unfortunately there is no unanimity upon the purpose of a city or its associated methods of transit, but many would consider objectives to include a reasonably rapid and safe interchange of workers between home and office or factory, while preserving the quality of the environment in domestic, commercial and industrial areas.

Where a metropolis is planned and built *ab initio,* as in certain of the British *new towns,* clustering of factories, commercial centers and residential districts may permit most journeys to be made both quickly and safely by foot or by bicycle. However, the average city planner must cope with the concepts and the mistakes of his predecessors, entombed in brick and mortar. Most cities were built before the coming of the motor car, with population, commerce and industry concentrated along public transit routes. Motor vehicles have now removed traditional restraints upon urban growth, and the periphery of an expanding North American city is characterized by low densities of both residential and industrial development. It is financially difficult to maintain an acceptable frequency of public transportation to such areas, and in many large cities more than 95 percent of all passenger journeys are made by private car. At the same time, increasing congestion and air pollution suggest this is an unacceptable ultimate solution of urban transit problems.

Long-term plans to maintain and develop public transportation networks must include (1) encouragement of high-density downtown growth, (2) support of travel corridors with a high frequency of public service vehicles, and (3) provision of facilities that compete with the private car in terms of speed, convenience and cost.

Rail transit systems are only economic when they can carry heavy peak hour loads (for example, 50,000 passengers per hour in the more busy direction of travel). Conditions that will support a rail network are unlikely in the absence of either a high central population density ($> 20,000/mile^2$) or the entry of at least 300,000 people per day into the city core (Smith, 1969).

An intermediate solution, currently being tried in several North American cities, is the provision of specific bus lanes on arterial roads and freeways. These enable the bus to travel at a speed competitive with that of a car, and thereby encourage the development of a densely populated transportation corridor that will later support the construction and operation of a subway route. Initially, the frequency of the bus service may not justify exclusive use of a freeway lane, and one possibility is to program the traffic computer so that it will meter a limited number of private cars into the bus lane, while permitting the bus to travel at the maximum legal speed. A further possible development is a device that will change traffic lights in favor of a bus.

A number of communities are currently seeking to attract passengers through the convenience of minibuses, directed to individual homes by radio telephone. But for the majority of commuters, the most effective argument is cost. If it is accepted that an increase of public transportation is important to the viability of a large city, then the relative usage of bus and car can be adjusted quite quickly by selective taxation of private vehicles entering the city core. Unfortunately, the proposed solution carries long-term rather than immediate rewards, and for this reason lacks the popular appeal that would encourage its early implementation by the city fathers.

BIBLIOGRAPHY

Agate, J.N., and Druett, H.A.: A study of portable vibrating tools in relation to the clinical effects which they produce. *Br J Ind Med, 4:*141–163, 1947.

Allnisi, E.A., and Chiles, W.D.: Sustained performance, work-test scheduling and diurnal rhythms in man. *Acta Psychol, 27:*436–442, 1967.

Allport, G.W., Vernon, P.E., and Lindzey, G.: *Study of Values*. Boston, Houghton Mifflin, 1951.

Andersen, K.L., Shephard, R.J., Denolin, H., Varnauskas, E., and Masironi, R.: Fundamentals of exercise testing. World Health Organization, Geneva, 1971.

Aschoff, J.: *Circadian Clocks*. Amsterdam, North Holland, 1965.

Asher, R.: Talking sense. *Lancet, ii:*417–419, 1959.

Asmussen, E.. In Jokl, E., and Simon, H.E.: *International Research in Sport and Physical Education*. Springfield, Ill., Thomas, 1964.

———, and Poulsen, E.: Energy expenditure in light industry. Its relation to age, sex, and aerobic capacity. *Comm Test Obs Inst Hellerup, 13:*1–13, 1963.

———, Poulsen, E., and Bogh, H.E.: Measurements of the muscle strength necessary for driving a motor car. *Comm Test Obs Inst Hellerup, 19:*1–12, 1964.

———, and Poulsen, E.: Disability as a handicap relative to occupation. *Comm Test Obs Inst Hellerup, 24:*1–16, 1966.

———, and Poulsen, E.: On the role of intra-abdominal pressure in relieving the back muscles while holding weights in a forward inclined position. *Comm Test Obs Inst Hellerup, 28,* 1968.

Astrand, I.: Aerobic work capacity in men and women with special reference to age. *Acta Physiol Scand (Suppl), 49(169):*1–92, 1960.

———: Degree of strain during building work as related to individual aerobic work capacity. *Ergonomics, 10:*293–303, 1967.

———: Circulatory responses to arm exercise in different work positions. *Scand J Clin Lab Invest, 27:*293–297, 1971.

———, Astrand, P.O., Christensen, E.H., and Hedman, R.: Intermittent muscular work. *Acta Physiol Scand, 48:*448–53, April 1960.

Astrand, P.O., and Rodahl, K.: *Textbook of Work Physiology*. New York, McGraw Hill, 1970.

Atkinson, J.W.: Motivational determinants of risk-taking behaviour. *Psychol Rev, 64:*359–372, 1957.

Auchinschloss, J.H., Gilbert, R., and Baule, G.H.: Unsteady-state measurement of oxygen transfer during treadmill exercise. *J Appl Physiol, 25:*283–293, 1968.

Baddely, A.D.: Influence of depth on the manual dexterity of free divers: a comparison between open sea and pressure chamber testing. *J Appl Psychol, 50:*81–85, 1966.

Bakan, P., Belton, J.A., and Toth, J.C.: Extraversion-introversion and decrement in an auditory vigilance task. In Buckner, and McGrath (Eds.): *Vigilance: A Symposium*. New York, McGraw Hill, 1960.

Baker, C.H.: Three minor studies of vigilance. D.R.B. Rep. 234–2. Defence Research Medical Laboratories, Canada, 1959.

Banister, E.W.: Energetics of muscular contraction. In Shephard, R.J. (Ed.): *Frontiers of Fitness*. Springfield, Ill., Thomas, 1971.

Barber, A.S., Barraclough, E.D., and Gray, W.A.: Closing the gap between the medical researcher and the literature. *Br Med J, i:*367–370, 1972.

Barnett, G.O.: Computers in patient care. *N Engl J Med, 279*:1321–7, 1968.

Barter, J.T.: Estimation of the mass of body segments. *WADC Tech Rep, 57*:260, 1957.

Bassett, C.A.L.: Effect of force on skeletal tissues. In Downey, J.A., and Darling, R.C. (Eds.): Physiological basis of rehabilitation medicine. Philadelphia, Saunders, 1971.

Battig, W.F., Voss, J.F., and Brogden, W.J.: Effect of frequency of target intermittence upon tracking. *J Exp Psychol, 49*:244–248, 1955.

Baumrind, D.: Some thoughts on ethics of research. *Am Psychol, 19*:421–423, 1964.

Beckett, R., and Chang, K.: An evaluation of the kinematics of gait by minimum energy. In Bootzin, D., and Muffley, H.C. (Eds.): *Biomechanics.* New York, Plenum Press, 1969.

Beckman, E.L., Duane, T.D., Ziegler, J.E., and Hunter, H.N.: Some observations on human tolerance to accelerative stress. Phase IV. Human tolerance to high positive G applied at a rate of 5–10G per second. *J Aviat Med, 25*:50–66, 1954.

Beckman, E.L., McNutt, D.C., and Rawlins, J.S.P.: An investigation into the feasability of using the standard Martin-Baker ejection seat systems for underwater escape from ditched aircraft. *Aerosp Med, 31*:715–732, 1960.

Bedale, E.M.: *Industrial Fatigue.* U.K. Research Board Rept. 29, 1924.

Bedford, T., and Warner, C.G.: The energy expended while walking in stooping postures. *Br J Ind Med, 12*:290–295, 1955.

Begbie, G.H., Gainford, J., Mansfield, P., Stirling J.M.M., and Walsh, E.G.: Head and eye movement during rail travel. *J Physiol, 165*,72P-73P, 1963.

Bekesy, G. Von: *Akush Z, 4*:316, 1939 (quoted by Guignard, 1965).

Belbin, E., Belbin, R.M., and Hill, F.: A comparison between the results of three different methods of operator training. *Ergonomics, 1*:39–50, 1957.

Belbin, E., and Down, S.: Interference effects from new learning: their relevance to the design of adult training programs. *J Gerontol, 20*:154–159, 1965.

Belleville, J.W., and Seed, J.C.: Respiratory carbon dioxide response curve computer. *Science, 130*:1079–1083, 1959a.

Belleville, J.W., and Seed, J.C.: The use of an analogue computer for measurement of respiratory depression. *N.Y. Acad Sci, 22*:34–43, 1959b.

Bennett, P.C., and Elliott, D.H.: *The Physiology and Medicine of Diving.* London, Baillière, Tindall & Cassell, 1969.

Berkun, M.M., Bialek, H.M., Kern, R.P., and Yagi, K.: Experimental studies of psychological stress in man. *Psychol Monogr, 76*(15) **534**– 1–39, 1962.

Bernard, C.: *Leçons sur les Phenomènes de la Vie.* Paris, Baillière, 1878.

Bertelson, P., Boons, J.P., and Renkin, A.: Vitesse libre et vitesse imposée dans une tache simulant le trimécanique de la correspondance. *Ergonomics, 8*:3–22, 1965.

Beswick, R.E.: Ilio-psoas fatigue in car-drivers. *Practitioner, 196*:688–9, 1966.

Bexton, W.H., Heron, W., and Scott, T.H.: Effects of decreased variation in the sensory environment. *Can J Psychol, 8*:70–76, 1954.

Bilodeau, E.A., and Bilodeau, I. McD.: Motor skills learning. *Ann Rev Psychol, 42*:243–280, 1961.

Birren, J.E., Butler, R.N., Greenhouse, S.W., Sokdoff, L., and Yarrow, M.R.: In Birren, J.E. (Ed.): *Human Ageing.* Bethesda, Md., U.S. Government Printing Office, 1963.

Bogh, H., and Poulsen, E.: Investigations of the demand/ability relation in handicapped motorists. *Comm Test Obs Inst Hellerup, 26*:1–16, 1967.

Bonjer, F.H.: Relationship between working time, physical working capacity, and allowable caloric expenditure. In Rohmert, W. (Ed.): *Muskelarbeit und Muskeltraining.* Gentner, Verlag, 1968.

Borg, G.: The perception of physical performance. In Shephard, R.J. (Ed.): *Frontiers of Fitness.* Springfield, Ill., Thomas, 1971.

Brand, J.J., Colquhoun, W.P., Gould, A.A., and Perry, W.L.M.: l-hyoscine and cyclizine as motion sickness remedies. *Br J Pharmacol Chemother, 30:*463–469, 1967.

Brener, J., and Kleinman, R.A.: Learned control of decrease in systolic blood pressure. *Nature, 226:*1063–1064, 1970.

Brent, F.N.: Pulse rate studies for evaluation of physical demand of manual work. Brouha Symposium, Toronto, Ont., 1968.

British Standards Institution: Specification for office desks, tables, and seating. B.S. *3893,* 1965.

Broadbent, D.E.: Classical conditioning and human watchkeeping. *Psychol Rev, 60:*331–339, 1953.

———: Effects of noises of high and low frequency on behaviour. *Ergonomics, 1:*21–29, 1957.

———. *Perception and Communication.* Oxford, Pergamon, 1958.

———: Vigilance. *Br Med Bull, 20:*17–20, 1964.

Brooks, C. McC., Kao, F.F., and Lloyd, B.B.: *Cerebrospinal Fluid and the Regulation of Ventilation.* Oxford, Blackwell, 1965.

Brouha, L.: *Physiology in Industry.* Oxford, Pergamon Press, 1960.

Brown, I.D., and Poulton, E.C.: Measuring the spare "mental capacity" of car drivers by a subsidiary task. *Ergonomics, 4:*35–40, 1961.

Brown, J.L.: Orientation of the vertical during water immersion. *Aerosp Med, 32:*209–217, 1961.

Brown, J.R.: Standards and scales of energy requirements for industrial workers. *Med Serv J (Can), 20:*365–373, 1964a.

———: Industrial fatigue. *Med Serv J (Can), 20:*221–231, 1964b.

———: The metabolic cost of industrial activity in relation to weight. *Med Serv J (Can), 22:*262–272, 1966.

———: Lifting as an industrial hazard. Labour Safety Council of Ontario, Dept. of Labour, Ontario, 1971.

———: Manual lifting and related fields. An annotated bibliography. Labour Safety Council, Dept. of Labour, Ontario, 1972.

———, and Shephard, R.J.: Some measurements of fitness in older female employees of a Toronto department store. *Can Med Assoc J, 97:*1208–1213, 1967.

Brunnstrom, S.: *Clinical Kinesiology,* 3rd ed. Philadelphia, F.A. Davis, 1972.

Buckley, W.: *Sociology and Modern Systems Theory.* Englewood, N.Y., Prentice Hall, 1967.

Buller, A.J.: The muscle spindle and the control of movement. In Howell, J.B.L., and Campbell, E.J.M. (Eds.): *Breathlessness.* Oxford, Blackwell, 1966.

Bienning, E.: *The Physiological Clock.* Berlin, Springer-Verlag, 1964.

Burger, G.C.E.: Permissible load and optimal adaptation. *Ergonomics, 7:*397–417, 1964.

Byford, G.H.: Eye movements and the optogyral illusion. *Aerosp Med, 34:*119–123, 1963.

Campbell, E.J.M., and Matthews, C.M.E.: The use of computers to simulate the physiology of respiration. *Br Med Bull, 24:*249–252, 1968.

Cannon, W.B.: Organization for physiological homeostasis. *Physiol Rev, 9:*399–431, 1929.

Cappon, D., and Banks, R.: Studies in perceptual distortion. *Arch Gen Psychiatry, 2:*346–349, 1960.

Cardozo, B.L., and Leopold, F.F.: Human code transmission. Letters and digits compared on the basis of immediate memory rates. *Ergonomics, 6:*133–141, 1963.

Carlston, A.: Influence of leg varicosities on the physical work performance. In Cumming, G.R., Snidal, D., and Taylor, A.W. (Eds.): Environmental effects on work performance. *Can Assoc Sports Sci.,* 207–214, 1972.

Cartmel, J.L., and Banister, E.W.: The physical working capacity of blind and deaf schoolchildren. *Can J Physiol Pharmacol, 47*:833–836, 1968.

Cass, S.: A computer controls traffic signals. Paper presented at the VIth World Highway Conference, Montreal, October, 4–10, 1970.

Cattell, W.: 16 P.F. test. Champaign, Illinois, Institute for Personality and Ability Testing, 1954

Chaffin, D.B., and Moulis, E.J.: An empirical investigation of low back strains and vertebral geometry. Unpublished Rept., Western Electric Company, U.S., 1968.

Chambers, R.M.: Operator performance in acceleration environments. In Burns, N.M., Chambers, R.M., and Heindler, E. (Eds.): *Unusual Environments and Human Behaviour.* New York, MacMillan, 1963.

Chance, B., Higgins, J., and Garfinkel, D.: Analogue and digital computer representations of biochemical processes. *Fed Proc, 21*:75–85, 1962.

C.I.S.: International Occupational Safety and Health Information Centre, I.L.O., Information Sheet 3. Geneva, 1962.

Clark, B., and Nicholson, M.A.: Aviator's vertigo, a cause of pilot error in naval aviation students. *J Aviat Med, 25*:171–179, 1954.

Clark, G.E.: A chronocyclegraph that will help you improve methods. *Factory, 112*:124–125, 1954.

Clark, K.: *Civilization.* London, B.B.C., Murray, 1969.

Cobb, P.W., and Moss, F.K.: The four variables of the visual threshold. *J Franklin Inst, 205*:831–847, 1928.

Coermann, R.: Untersuchungen uber die Einwirkung von Schwingungen auf den menschlichen Organismus. *Luftfahrtmedizin, 4*:73–117, 1940.

———: Mechanical vibrations. In *Ergonomics and Physical Environmental Factors.* I.L.O. Geneva, 1970.

Cohen, L.A.: Humanoid cover seeking and obstacle-avoidance functions. In Bootzun, D., and Muffley, H.C. (Eds.): *Biomechanics.* New York, Plenum, 1969.

Colquhoun, W.P., Blake, M.J.F., and Edwards, R.S.: Experimental studies of shift work. I. A comparison of "rotating" and "stabilized" 4-hour shift systems. II. Stabilized 8-hour shift systems. *Ergonomics, 11*:437–453, 527–556, 1968a, b.

Conrad, R.: Experimental psychology in the field of telecommunications. *Ergonomics, 3*:289–296, 1960.

Conrad, R.: Designing postal codes for public use. In Singleton, W.T., Easterby, R.S., and Whitfield, D. (Eds.): *The Human Operator in Complex Systems.* London, Taylor & Francis, 1967.

Cooper, J.M.: *Biomechanics.* Chicago Athletic Institute, 1971.

Corcoran, D.W.J.: Noise and loss of sleep. *Q J Exp Psychol, 14*:178–182, 1962.

———: Personality and the inverted-U relation. *Brt J Psychol, 52*:267–273, 1965.

———: Changes in heart rate and performance as a result of loss of sleep. *Br J Psychol, 55*:307–314, 1966.

Corcoran, P.J., Jebsen, R.H., Brengelmann, G.L., and Simons, B.C.: Effects of plastic and metal leg braces on speed and energy cost of hemiparetic ambulation. *Arch Phys Med Rehabil, 51*:69–77, 1970.

Cotes, J.E.: I.L.O. Occupational Safety and Health Series, 6, I.L.O. Geneva, 1966.

Cratty, B.J.: *Movement Behaviour and Motor Learning.* Philadelphia; Lea & Febiger, 1967.

———, and Sage, J.N.: The effects of primary and secondary group interaction upon improvement in a complex movement task. *Res Q, 35*:265–274, 1964.

Crawford, A.: The overtaking driver. *Ergonomics, 6*:153–170, 1963.

Croker, E.C., and Henderson, L.F.: Analysis and classification of odors. *Am Perfumessent Oil Rev, 22*:325–327, 1927.

Crossman: cited by Edholm, 1967.

Cumming, G.R.: Current levels of fitness. In Proceedings of Int. Symp. on Physical Activity and Cardiovasc. Health. *Can Med Assoc J, 96:*868–877, 1967.

———, Goulding, D., and Baggley, G.: Working capacity of deaf and visually and mentally handicapped children. *Arch Dis Child, 46:*490–494, 1971.

Dart, E.E.: Effects of high speed vibrating tools on operators engaged in the airplane industry. *Occup Med, 1:*515–550, 1946.

Davies, C.N.: *Design and Use of Respirators.* New York, MacMillan, 1962.

Davies, C.T.M.: Cardiac frequency in relation to aerobic capacity for work.*Ergonomics, 11:*511–520, 1970.

Davis, P.R., Troup, J.D.C., and Burnard, J.H.: Movements of the thoracic and lumbar spine when lifting; a chrono-cyclo-photographic study. *J Anat (Lond), 99:*13–26, 1965.

Deese, J.: Some problems in the theory of vigilance. *Psychol Rev, 62:*359–368, 1955.

———, and Ormond, E.: Studies of detectability during continuous visual search. U.S. Air Force W.A.D.C. Tech. Rept. TR 53:8, 1953.

Defares, J.G.. On the use of mathematical models in the analysis of the respiratory control system. In Lloyd, B.B. (Ed.): The regulation of human respiration. Oxford, Blackwell, 1963.

de Jong: cited by Edholm, 1967.

Del Veechio, V., and Mammarella, L.: Odours: Physiological and psychological aspects, detection, evaluation, control. *Ergonomics and physical environmental factors.* I.L.O., Geneva, 1970.

Dement, W.: The effect of dream deprivation. *Science, 131:*1705–1707, 1960.

Dement, W.C., and Kleitman, N.: Cyclic variations in EEG during sleep and their relation to eye movements, body motility and dreaming. *Electroencephalogr Clin Neurophysiol, 9:*673–690, 1957.

Dieckmann, D.: Einfluss vertikaler mechanischer Schwingungen auf den Menschen; Mechanische Modelle für den vertikal schwingenden menschlichen korper; Einfluss horizontaler mechanischer Schwingungen auf den Menschen. *Int Z Angew Arbeitsphysiol, 16:*519–564; *17:*67–82; *17:*83–100, 1959.

Dickson, J.F., and Brown, J.H.V.: *Future Goals of Engineering in Biology and Medicine.* New York, Academic Press, 1969.

di Prampero, P.E.: Anaerobic capacity and power. In Shephard, R.J. (Ed): *Frontiers of Fitness.* Springfield, Ill., Thomas, 1971.

Doane, B.K., Mahatoo, W., Heron, W., and Scott, T.H.: Changes in perceptual function after isolation. *Can J Psychol, 13:*210–219, 1959.

Dravineks, A., and Krotoszynsky, B.K.: Collection and processing of airborne chemical information.*J Gas Chrom, 4:*367–370, 1966.

Dreyer, G.: *The Assessment of Physical Fitness.* London, Cassell, 1920.

Droese, W., Kofranyi, E., Kraut, H., and Wildemann, L.: Energetische Untersuchung der Hausfrauenarbeit. *Arbeitsphysiol, 14:*63–81, 1949.

DuBois, E.F.: *Basal Metabolism in Health and Disease.* Philadelphia, Lea and Febiger, 1927.

Duffin, J.: The chemical regulation of ventilation. *Anaesthesia, 26:*142–156, 1971.

Dupertuis, C.W., and Emmanel, I.: A statistical comparison of the body typing methods of Hooton & Sheldon, U.S.A.F., W.A.D.C. *Tech Rep, 56:*366, 1956.

Durnin, J.V.G.A., and Passmore, R.: *Energy, Work and Leisure.* London, Heinemann, 1967.

Duvelleroy, M.: Simulation des phenomènes physiologiques. *Rev Inform Med, 3:*213–232, 1970.

Dyson, G.H.G.: *The Mechanics of Athletics,* 5th ed. London, University Press, 1970.

Eberhardt, H.D.: Fundamental studies of human locomotion and other information relating to design of artificial limbs. Report to U.S. Nat. Res. Council, Committee on Artificial Limbs, Berkeley, Calif., 1947.

Edholm, O.G.: *The Biology of Work.* New York, McGraw Hill, 1967.

Ernsting, J., and Gabb, J.E.: Problems of communication in breathing equipment. In Davies, C.N. (Ed.): *Design and Use of Respirators.* Oxford, Pergamon Press, 1962.

Eysenck, H.J.: *Manual of the Maudsley Personality Inventory.* London, University Press, 1959.

Faulkner, T.W.: Variability of performance in a vigilance task. *J Appl Psychol, 46:*325–328, 1962.

Ferster, C.B., and Perrott, M.C.: *Behaviour Principles.* New York, Appleton Century Crofts, 1968.

Fincham, E.F.: The accommodation reflex and its stimulus. *Br J Ophthalmol, 35:*381–393, 1951.

Fish, H.: Raising productivity in Israel. *Int Lab Rev, 68,* 1953 (quoted by Mundel, 1970).

Fitts, P.M., and Posner, M.I.: *Human Performance.* Belmont, Calif., Brooks & Cole, 1967.

Fletcher, H.: The nature of speech and its interpretation. *J Frankl Inst, 193:*729–747, 1922.

Flint, M.M.: Effect of increasing back and abdominal muscle strength on low back pain. *Res Q, 29:*160–171, 1958.

Floyd, W.F., and Roberts, D.F.: Anatomical and physiological principles in chair and table design. *Ergonomics, 2:*1–16, 1958.

Floyd, W.F., and Silver, P.H.S.: The function of the erectores spinae muscles in certain movements and postures in man. *J Physiol, 129:*184–203, 1955.

Forbes, A.R.: Memo 105, F.P.R.C. U.K. Air Ministry, 1959.

Frankenhauser, M.: Effects of prolonged gravitational stress on performance. *Acta Psychol, 14:*92–108, 1958.

Fredrik, W.S.: Some aspects of human engineering. *Arch Environ Health, 21:*192–199, 1960.

Fried, T., and Shephard, R.J.: Deterioration and restoration of physical fitness after trauma. *Can Med Assoc J, 100:*831–837, 1969.

Gerathewohl, S.J., Strughold, H., and Stallings, H.D.: Sensomotor performance during weightlessness. *J Aviat Med, 28:*7–12, 1957.

Gilbreth, F.B., and Gilbreth, L.M.: *Motion Study for the Handicapped.* London, Routledge, 1920.

Gillies, J.A.: *A Textbook of Aviation Physiology.* Oxford, Pergamon Press, 1965.

Givoni, B., and Goldman, R.F.: Predicting metabolic energy cost. *J Appl Physiol, 30:*429–433, 1971.

Glaister, D.H.: The effects of accelerations of short duration. In Gillies, J.A. (Ed.): *Textbook of Aviation Physiology.* Oxford, Pergamon Press, 1965.

Goddard, R.F.: *The International Symposium on the Effects of Altitude on Physical Performance.* Chicago, Athletic Institute, 1967.

Godin, G., and Shephard, R.J.: Activity patterns of the Canadian Eskimo. In Edholm, O.G. (Ed.): *Human Polar Biology.* London, Heinemann, 1973.

Goldman, D.E.: *Handbook on Noise Control.* New York, McGraw Hill, 1957.

———, and Von Gierke, H.E.: The effects of shock and vibration on man. U.S. Naval Medical Research Inst. Lecture & Rev. Series, *60:*3, Bethesda, Md., 1960.

Grandjean, E.: *Sitting Posture*. London, Taylor & Francis, 1969.

——, Boni, A., and Kretschmer, H.: The development of a rest chair profile for healthy and notalgic people. In Grandjean, E.: *Sitting Posture*. London, Taylor & Francis, 1969.

Graybiel, A., and Clark, B.: Validity of the oculogravic illusion as a specific indicator of otolith function. *Aerosp Med*, 36:1173–1181, 1965.

Graybiel, A., and Kellogg, R.S.: Inversion illusion in parabolic flight: its probable dependence on otolith function. *Aerosp Med*, 38:1099–1103, 1967.

Grieve, D.W., and Gear, R.J.: The relationship between length of stride, step frequency, time of swing, and speed of walking for children and adults. *Ergonomics*, 5:379–399, 1962.

Griew, S.: Age, information transmission and the positional relationship between signals and responses in the performance of a choice task. *Ergonomics*, 7:267–277, 1964.

Grodins, F.S., Gray, J.S., Schroeder, K.R., Norins, A.L., and Jones, R.W.: Respiratory responses to CO_2 inhalation. A theoretical study of a non-linear biological regulator. *J Appl Physiol*, 7:283–308, 1954.

Grodsky, M.A.: The use of full-scale mission simulation for the assessment of complex operator performance. *Hum Factors*, 9:341–348, 1967.

Guignard, J.C.: Noise. In Gillies, J.A. (Ed.): *Textbook of Aviation Physiology*. Oxford, Pergamon Press, 1965.

Guthrie, D.I.: A new approach to handling in industry. A rational approach to the prevention of low back pain. *S Afr Med J*, 37:651–656, 1963.

Halberg, F., Nelson, W., Runge, W., and Schmitt, O.H.: Delay of circadian rhythm in rat temperature by phase-shift of lighting regimen is faster than advance. *Fed Proc*, 26:599, 1967.

Halberg, F., Vallbona, C., Dietlein, L.F., Rummell, J.A., Berry, C.A., Pitts, G.C., and Nunneley, S.A.: Human circadian rhythms during weightlessness in extra-terrestrial flight or bed rest with and without exercise. *Space Life Sci*, 2:18–32, 1970.

Harlow, H.F.: The nature of learning sets. *Psychol Rev*, 56:51–65, 1949.

Harrison, G.A., Weiner, J.S., Tanner, J.M., and Bernicot, N.A.: *Human Biology. An Introduction to Human Evolution Variation and Growth*. Oxford, Clarendon Press, 1964.

Hartman, B.O.: Field study of transport aircrew workload and rest. *J Aerosp Med*, 42:817–821, 1971.

Hashimoto, K., Kogi, K., and Grandjean, E.: *Methodology in Human Fatigue Assessment*. London, Taylor & Francis, 1971.

Hattem, D. Van: Studiegroep "Ergonomie." Ergonomische Vragenlijst voor de Beoordeling van Machines en Werktuigen. *Areeks*, 6, Nederlands Inst. voor Efficiencie, Den Haag, 1965.

Hauty, G.T., and Adams, T.: Phase shifts of the human circadian system and performance deficit during the periods of transition. I. East-West flight. II. West-East Flight. *U.S. Fed Aviat Admin Reps*, 65:28–29, 1965.

Hayden, F.J.: Physical activity for the severely retarded. *Can Assoc Retarded*, Annual Meeting, Halifax, N.S., 1962.

Hering, E., and Breuer, J.: Die Selbstuerung der Atmung durch den Nervus Vagus. *Akad Sitzungsb Wien*, 57:672–677; 58:909–937, 1868.

Hess, H.H.: Physiology in the space environment. *Natl Acad Sci*, Washington, D.C., Publication 1485B, 1967.

Hickish, D.E.: Industrial noise. In Permaggiani, L. (Ed.): *Ergonomics and Physical Environmental Factors*. I.L.O., Geneva, 1970.

Hill, J.: Care of the sea-sick. *Br Med J, ii:*802–807, 1936.

Hodges, J.D., and Whitehead, D.W.: Automatic traffic signal control systems—the metropolitan Toronto experience. *Am Fed Inf Proc Soc Joint Computer Conf Proc, 34:*529–535, 1969.

Holland, J.G.: Human vigilance. In Buchner, and McGrath (Eds.): *Symposium.* New York, McGraw Hill, 1960.

————: Human vigilance. *Science, 128:*61–67, 1958.

Hornick, R.J., Boettcher, C.A., and Simons, A.K.: *Final Report, Project TE1–1000.* Milwaukee, Bostrom Research Laboratories, 1966.

Hornick, R.J., and Lefritz, N.M.: A study and review of human response to prolonged random vibration. *Hum Factors, 8:*481–492, 1966.

Howard, P.: The physiology of positive acceleration. In Gillies, J.A. (Ed.): *A Textbook of Aviation Physiology.* Oxford, Pergamon Press, 1965.

Hughes, A.L., and Goldman, R.F.: Energy cost of hard work. *J Appl Physiol, 29:*570–572, 1970.

Hultman, E.: Muscle glycogen and prolonged exercise. In Shephard, R.J.(Ed.): *Frontiers of Fitness.* Springfield, Illinois, Thomas, 1971.

Hůzl, F., Stolaruk, R., Mainerová, J., Janková, J., and Sýkora, J.: Vibrationserkrankungen beim Holzeinschlag. In *Ergonomics and Physical Environmental Factors.* I.L.O. Geneva, 1970.

International Labour Organization: Maximum permissible weight to be carried by one worker. I.L.O. Report. *Occupational Safety & Health Series, 5,* Geneva, 1964.

Isherwood, P.A.: Ergonomics and the disabled. *Int Rehab Rev, 21:*9–11, 1970.

Jackson, A.S.: *Analog Computation.* New York, McGraw Hill, 1960.

Jacobson, E.: *Tension in Medicine.* Springfield, Illinois, Thomas, 1967.

Janeway, R.N.: Passenger vibration limits. *Soc Air Eng J, 56* (4):48–49, 1948.

Johansson, G., Bergstrom, S., Jonnson, G., Ottander, C., Rumar, K., and Ornberg, G.: Visible distances in simulated night driving conditions with full and dipped headlights. *Ergonomics, 6:*171–179, 1963.

Johnson, W.H., Stubbs, R.A., Kelk, G.F., and Franks, W.R.: Stimulus required to produce motion sickness. Restriction of head movements as a preventative of airsickness—field studies in airborne troops. *J Aviat Med, 22:*365–374, 1951.

Jokl, E., and Jokl, P.: *Exercise and Altitude.* Basel, Karger, 1968.

Jones, D.: *Safe Lifting.* Canadian Occupational Safety Association, Toronto, Canada, 1969.

Jones, G.M.: Pressure changes in the middle ear after simulated flight in a decompression chamber. *J Physiol, 147:*43P–44P, 1959.

Jones, J.C.: Method and results of seating research. In Grandjean, E. (Ed.): *Sitting Posture.* London, Taylor & Francis, 1969.

Jones, P.R.M., and Pearson, J.: Anthropometric determination of leg fat and muscle plus bone volumes in young male and female adults. *Proc Physiol Soc UK,* 11P–14P, July 1969.

Jorgensen, K.: Back muscle strength and body weight as limiting factors for work in the standing slightly stooped position. *Comm Test Obs Inst, Hellerup, 30:*1–12, 1970.

Kaiser, F.R.: Odor and its measurement. In Stern, A.C. (Ed.): *Air Pollution.* New York, Academic Press, 1962, Vol. I.

Kamon, E.: Negative and positive work in climbing a ladder mill. *J Appl Physiol, 29:*1–5, 1970.

Katz, D., and Kahn, R.L.: *The Social Psychology of Organizations.* New York, Wiley, 1966.

Kavanagh, T., and Shephard, R.J.: The immediate antecedents of myocardial infarction. *Can Med Assoc J, 109:*19–22, 1973a.

Kavanagh, T., and Shephard, R.J.: The application of exercise testing to the elderly amputee. *Can Med Assoc J, 108:*314–317, 1973b.

Kay, C., and Shephard, R.J.: On muscle strength and the threshold of anaerobic work. *Int Z Angew Physiol, 27:*311 328, 1969.

Kay, H., Dodd, B., and Sime, M.: *Teaching Machines and Programmed Instruction.* Baltimore, Penguin Books, 1968.

Kay, H., and Poulton, E.C.: Anticipation in memorizing. *Br J Psychol, 42:*34–41, 1951.

Kemp, W.B.: The flow of energy in a hunting society. *Sci Am, 225:*104–116, 1971.

Kerslake, D. McK.: *The Stress of Hot Environments.* Cambridge, University Press, 1972.

Keul, J., Doll, E., and Keppler, D.: *Energy Metabolism of Human Muscle.* Baltimore Univ. Pk. Press, 1972.

Klausen, K.: The form and function of the loaded human spine. *Acta Physiol Scand, 65:*176–190, 1965.

Klein, K.E., Wegman, H.M., and Bruner, H.: Circadian rhythms in indices of human performance, physical fitness, and stress resistance. *J Aerosp Med, 39:*512–518, 1968.

Kleitman, N.: *Sleep and Wakefulness.* Chicago, Univ. Chicago Press, 1939.

Kretschmer, E., and Enke, W.: *Die Personlichkeit der Athletiker.* Leipzig, Thieme, 1936.

Kryter, K.D.: Effects of high altitude on speech intelligibility. *J Appl Psychol, 32:*503–511, 1948.

Lambertsen, C.J.: *Underwater Physiology.* Philadelphia, Williams & Wilkins, 1967.

———, Gelfand, R., and Kemp, R.A.: Dynamic response characteristics of several CO_2 reactive components of the respiratory control system. In Brooks, McC., Kao, F.F., and Lloyd, B.B. (Eds.): *Cerebrospinal Fluid and the Regulation of Breathing.* Oxford, Blackwell, 1965.

———, and Gelfand, R.: Breath by breath measurement of respiratory functions: instrumentation and applications. *J Appl Physiol, 21:*282–290, 1966.

Langdon, D.E., and Hartman, B.: Performance upon sudden awakening. U.S. School of Aviat. Med. Rep. 62, 17. Brooks A.F.B., 1961.

Laties, V.G.: Modification of affect, social behaviour and performance by sleep deprivation and drugs. *J Psychiatr Res, 1:*12–25, 1961.

Lavery, J.J.: The effect of one-trial delay in knowledge of results on the acquisition and retention of a tossing skill. *Am J Psychol, 77:*427–443, 1964.

Lawley, D.N., and Maxwell, A.E.: *Factor Analysis as a Statistical Method.* London, Butterworth, 1963.

Lee, R.B.: ! Kung bushmen subsistence: an input-output analysis. In Vayda, A.P. (Ed.): *Environment and Cultural Behaviour.* New York, Natural History Press, 1969.

Lehman, H.C.: *Age and Achievement.* Princeton, University Press, 1953.

Leithead, C.S., and Lind, A.R.: *Heat Stress and Heat Disorders.* London, Cassell, 1964.

Leschly, V., Kjaer, A., and Kjaer, B.: General lines in designs of dwellings for handicapped confined to wheelchairs. *Comm Test Obs Inst Hellerup, 6:*1–68, 1960.

Levi, L.: The urinary output of adrenalin and noradrenalin during pleasant and unpleasant emotional states. *Psychosom Med, 27:*80–85, 1965.

Lewis, H.E., and Masterson, J.P.: Sleep and wakefulness in the Arctic. *Lancet, i:*1262–1266, 1957.

Lind, A.R., and McNicol, G.W.: Muscular factors which determine the cardiovascular responses to sustained and rhythmic exercise. *Can Med Assoc J, 96*:706–713, 1967.

Linden, V.: Absence from work and physical fitness. *Br J Ind Med, 26*:47–53, 1969.

Liston, R.A.: Walking machine studies and force-feedback controls. In Bootzin, D., and Muffley, H.C. (Eds.): *Biomechanics.* New York, Plenum Press, 1969.

Loach, J.C.: A new method of assessing the riding of vehicles and some results obtained. *J Inst Loco Eng, 48* (ii):183–240, 1958.

Lobban, M.C.: Time, light, and diurnal rhythms. In Edholm, O.G., and Bacharach, A.L. (Eds.): *The Physiology of Human Survival.* London, Academic Press, 1965.

Locke, E.A., and Bryan, J.F.: Cognitive aspects of psychomotor performance. The effects of performance goals on levels of performance. *J Appl Psychol, 50*:286–291, 1966.

Loveless, N.E.: Signal detection with simultaneous visual and auditory presentation. FPRC *1027*, U.K. Air Ministry, 1957.

Lovell, V.R.: The human use of personality tests: a dissenting view. *Am Psycholog, 22*:383–393, 1967.

Luby, E.D., Grisell, J.L., Frohman, C.E., Lees, H., Cohen, B.D., and Gottlieb, J.S.: Biochemical, psychological and behavioural responses to sleep deprivation. *Ann NY Acad Sci, 96*:71–79, 1962.

Ludvigh, E.: Direction sense of the eye. *Am J Ophthalmol, 36*:139–142, 1953.

Mackworth, N.H.: The breakdown of vigilance during prolonged visual search. *Q J Exp. Psychol, 1*:6–21, 1948.

———: Researches on the measurement of human performance. Med. Res. Council, U.K. *Spec Rep, 268*, 1950.

Malik, R.: Computer abuse confronts parliament. *Sci J, 5*:11–13, 1968.

Margaria, R.: *Exercise at Altitude.* Dordrecht, Netherlands, Excerpta Medica Foundation, 1967.

———: Positive and negative work performance and their efficiencies in human locomotion. *Int Z Angew Physiol, 25*:339–351, 1968.

———: Current concepts of walking and running. In Shephard, R.J. (Ed.): *Frontiers of Fitness.* Springfield, Illinois, Thomas, 1971.

———, Aghemo, P., and Rovelli, E.: Indirect determinations of maximal oxygen consumption in man. *J Appl Physiol, 20*:1070–1073, 1965.

Maritz, J.S., Morrison, J.F., Peter, J., Strydom, N.B., and Wyndham, C.H.: A practical method of estimating an individual's maximum oxygen intake. *Ergonomics, 4*:97–122, 1961.

Maslow, A.: *Motivation and Personality.* New York, Harper & Row, 1954.

———: *Towards a Psychology of Being.* Princeton, N.J., Van Nostrand, 1962.

Masterson, J.P.: Patterns of sleep. In Edholm, O.G., and Bacharach, A.L. (Eds.): *The Physiology of Survival.* London, Academic Press, 1965.

McCormick, E.J.: *Human Engineering.* New York, McGraw Hill, 1957.

McFarland, R.A.: *Human factors in Air Transportation.* New York:, McGraw Hill, 1953.

———: Human engineering and occupational safety. In Patty, F.A. (Ed.): Industrial Hygiene and Toxicology. New York, Interscience, 1958.

———: Experimental evidence of the relationship between ageing and oxygen want: in search of a theory of ageing. *Ergonomics, 6*:339–366, 1963.

McGeogh, J.A.: *Psychology of Human Learning.* London, Longmans Green, 1951.

———, Weinberger, N.M., and Whalen, R.E.: *Psychobiology: The Biological Bases of Behaviour.* San Francisco, Freeman, 1967.

McKenzie, R.E., and Elliott, L.L.: Effects of secobarbital and d-amphetamine on performance during a simulated air mission. *Aerosp Med, 36:*774–779, 1965.

Medford, G.: *Handbook of Human Engineering Data.* Boston, Tufts College Institute of Applied Experimental Psychology, 1951.

Meiry, J.L.: The vestibular system and human dynamic space orientation. Man-vehicle control laboratory, Report T-65–1. Cambridge, M.I.T., 1965 (quoted by Paulsen, 1970).

Melton, A.W.: Implications of short-term memory for a general theory of memory. *J Verb Learn Verb Behav, 2:*1–21, 1963.

Merkel, 1885, quoted by Woodworth, 1938.

Michal, E.D., Hutton, K.E., and Horvath, S.M.: Cardio-respiratory responses during prolonged exercise. *J Appl Physiol, 16:*997–1000, 1961.

Miles, S.: *Underwater Medicine.* London, Staples Press, 1962.

Milhorn, H.T.: *The Application of Control Theory to Physiological Systems* Philadelphia, Saunders, 1966.

———, Benton, R., Ross, R., and Guyton, A.C.: A mathematical model of the human respiratory control system *Biophys J, 5:*27–46, 1965a.

———, and Guyton, A.C.: An analog computer analysis of Cheyne-Stokes breathing. *J Appl Physiol, 20:*328–333, 1965b.

Miller, R.B.: Response time in man-computer transactions. *Amer Fed Inf Proc Soc Proc, 33* (1):267–277, 1968.

Mills, J.N.: Human circadian rhythms. *Physiol Rev, 46:*128–171, 1966.

Moores, B.: A comparison of work-load using physiological and time-study assessments. *Ergonomics, 13:*769–776, 1970.

Morant, G.M., and Gilson, J.C.: A report on a survey of body and clothing measurements of Royal Air Force Personnel, U.K. FPRC Rep. 633, 1945.

Morris, G.O., Williams, H.L., and Lubin, A.: Misperception and disorientation during sleep deprivation. *Arch Gen Psychiatr, 2:*247–254, 1960.

Morris, J.N., Kagan, A., Pattison, D.C., Gardner, M.J., and Raffle, P.A.B.: Incidence and prediction of ischaemic heart disease in London busmen. *Lancet, ii:*553–559, 1966.

Morrow, R.L.: *Time Study and Motion Economy.* New York, Ronald Press, 1946.

Mudd, S.: Assessment of the fidelity of dynamic flight simulators. *Hum Factors, 10:*351–358, 1968.

Mundel, M.E.: Memomotion. *Time and Motion Study, 7:*32–43, 1958.

———: *A Conceptual Framework for the Management Sciences,* New York, McGraw Hill, 1967.

———: *Motion and Time Study. Principles & Practices,* 4th ed. Englewood Cliffs, N.J., Prentice Hall, 1970.

Murdock, B.B.: Short-term memory and paired-associate learning. *J Verb Learn Verb Behav, 2:*320–328, 1963.

Murray, E.J. Schein, E.H., Erikson, K.T., Hill, W.F., and Cohen, M.: The effects of sleep deprivation on social behaviour. *J Soc Psychol, 49:*229–236, 1959.

Murray, M.P., Kory, R.C., and Sepic, S.B.: Walking patterns of normal women. *Arch Phys Med Rehab, 51:*637–650, 1970.

Murray, M.P., Seireg, A., and Scholz, R.C.: Center of gravity, center of pressure and supportive forces during human activities. *J Appl Physiol, 23:*831–838, 1967.

Neisser, U.: MAC and its users. Mass. Inst. Technol., Cambridge, Mass: Project MAC Memo 185, 1965.

Nicely, P.E., and Miller, G.A.: Some effects of unequal spatial distribution on the detectability of radar targets. *J Exp Psychol, 53:*195–198, 1957.

Nickerson, R.S.: Man-computer interaction: a challenge for human factors research. *Ergonomics, 12:*501–517, 1969.

Nuttall, J.B., and Sandford, W.E.: Publication M-27-56. Directorate of Flight Safety, U.S. Air Force, 1956.

Oxford, H.W.: Anthropometric data for educational chairs. In Grandjean, E. (Ed.): *Sitting Posture.* London, Taylor & Francis, 1969.

Padget, P.: The respiratory response to carbon dioxide. *Am J Physiol, 83:*384–394, 1927–8.

Parton, K.C., and Roberts, D.R.: Principles and applications of a control system simulator. *G.E.C. Journal, 26:*1–12, 1959.

Pascoe, C.: Towards a brighter tomorrow. *Can Doctor,* Sept:41–46, 1971.

Passmore, R., and Durnin, D.V.J.A.: Human energy expenditure. *Physiol Rev, 35:*801–840, 1955.

Pennock, G.A.: Investigation of rest periods, working conditions and other influences. *Personnel J, 8,* 1930 (quoted by Mundel, 1970).

Perrow, C.: An analysis of goals in a complex organization. *Appl Soc Rev, 26:*854–866, 1961.

Peterson, L.H.: Introduction to the principles of digital and analogue computers. *Fed Proc, 21:*69–74, 1962.

Peterson, L.R.: Short-term verbal memory and learning. *Psychol Rev, 73:*193–207, 1966.

Posner, I.: Immediate memory in sequential tasks. *Psychol Bull, 60:*333–349, 1963.

Post, W., and Gatty, H.: *Around the World in Eight Days.* London, Hamilton, 1931.

Postman, L.: Short-term memory and incidental learning. In Melton, A.W. (Ed.): *Categories of Human Learning.* New York, Academic Press, 1964.

Poulsen, E.: Analysis of job demands in rehabilitation of physically handicapped. *Comm Test Obs Inst Hellerup, 14:*1–12, 1963.

Poulton, E.C.: Simultaneous and alternate listening and speaking. *J Acoust Soc Am, 27:*1204–1207, 1955.

———: *Environment and Human Efficiency.* Springfield, Ill., Thomas, 1970.

———, Kendall, P.G., and Thomas, R.J.: Reading efficiency in flickering light. *Nature, 209:*1267–1268, 1966.

Rabbit, P.M.A.: Age and time for choice between stimuli and between responses. *J Gerontol, 19:*307–312, 1964.

Radloff, R., and Helmreich, R.: *Groups under Stress: Psychological Research in Sea Lab II.* New York, Appleton Century Crofts, 1968.

Ralston, H.J., and Lukin, L.: Energy levels of human body segments during level walking. *Ergonomics, 12:*39–46, 1969.

Rasch, P.J., and Burke, R.K.: *Kinesiology and Applied Anatomy. The Science of Human Movement,* 4th ed. Philadelphia, Lea & Febiger, 1971.

Raven, J.C.: Guide to the standard progressive matrices. Sets A,B,C,D,E. Revised. H.K. Lewis, London, 1956.

Rebiffe, R.: Le Siege du conducteur: son adaptation aux exigences fonctionelles et anthropometriques. In Grandjean, E. (Ed.): *Sitting Posture.* London, Taylor & Francis, 1969.

Rochmis, P., and Blackburn, H.: Exercise tests. A survey of procedures, safety and litigation experience in approximately 170,000 tests. *JAMA, 217:*1061–1066, 1971.

Rode, A., and Shephard, R.J.: Fitness for arctic life. The cardio-respiratory status of the Canadian Eskimo. In Edholm, O.G. (Ed.): *Human Polar Biology.* London, Heinemann, 1973a.

Rode, A., and Shephard, R.J.: Pulmonary function of Canadian Eskimos. *Scand J Resp Dis*, *54*:191–205, 1973b.

Rode, A., and Shephard, R.J.: Growth, development and fitness of the Canadian Eskimo. *Med Sci Sports*, 1973c, in press.

Roebuck, J.A.: N.A.S.A. study, presented to the AAHPER Council on Kinesiology, Chicago, Ill., 1966.

Rohmert, W., and Jenik, P.: Isometric muscular strength in women. In Shephard, R.J. (Ed.): *Frontiers of Fitness*. Springfield, Ill., Thomas, 1971.

Rosenblith, W.A., and Stevens, K.N.: U.S.A.F. Tech. Rep. 52, 204 W.A.D.C., 1953 (cited by Guignard, 1965).

Rowley, A.: High failure rate of computer projects. London, *The Times*, 21st March, 1969.

Sadoul, P., Heran, J., Arouette, A., and Grieco, B.: Valeur de la puissance maximale supportee, determinee par des exercises musculaires de 20 minutes pour évaluer la capacité fonctionelle des handicapes respiratoires. *Bull Physio Pathol Resp*, *2*:209–222, 1966.

Salisbury, L.L., and Colman, A.: A prosthetic device with automatic proportional control of grasp. In Bootzin, D., and Mufflcy, H.C. (Eds.): *Biomechanics*. New York, Plenum, 1969.

Saltin, B., Blomqvist, G., Mitchell, J.H., Johnson, R.L., Wildenthal, K., and Chapman, C.B.: Response to exercise after bed rest and after training. *Am Heart Assoc Monogr*, *23*, 1968.

Sauvy, A.: Demographic and economic aspects of the retirement problem. 1st International Course in Social Gerontology, Lisbon, 1970.

Schein, E.H.: *Organizational Psychology*. Englewood, N.J., Prentice Hall, 1965.

Scherr, A.L.: An analysis of time-shared computer systems. Ph.D. thesis, Massachusetts Institute of Technology, Cambridge, Massachusetts, quoted by Nickerson, 1969.

Schroter, G.: *Die Berufsschaden des Stutz und Bewengungsystems*. Leipzig, Barth, 1958.

Shackel, B.: Ergonomics in the design of a large digital computer console. *Ergonomics*, *5*:229–241, 1962.

———: Man-computer interaction—the contribution of the human sciences. *Ergonomics*, *12*:485–499, 1969.

———, Chidsey, K.D., and Shipley, P.: The assessment of chair comfort. In Grandjean, E. (Ed.): *Sitting Posture*. London, Taylor & Francis, 1969.

Sheldon, W.H.: *The Varieties of Human Physique. An Introduction to Constitutional Psychology*. New York, Hafner, 1963.

Shephard, R.J.: The ergonomics of the respirator. In Davies, C.N. (Ed.): *Design and Use of Respirators*. Oxford, Pergamon Press, 1962.

———: Dynamic characteristics of the human airway. *Aerosp Med*, *37*:1014–1025, 1966.

———: Devices for the teaching of expired air resuscitation. *Med Serv J* (Canad), *22*:273–284, 1966.

———: Ethical considerations in human experimentation. *J Can Assoc Phys Health Recr*, *33*:13–16, 1967.

———: Normal levels of activity in Canadian city dwellers. *Can Med Assoc J*, *97*:313–318, 1967.

———: The heart and circulation under stress of Olympic conditions. *JAMA 205*:150–155, 1968.

———: *Endurance Fitness*. Toronto, University Press, 1969a.

———: The working capacity of the older employee. *Arch Environ Health*, *18*:928–986, 1969b.

————: Computer programmes for solution of the Astrand nomogram. *J Sports Med Fitness, 10:*206–210, 1970a.

————: Prediction formulas and some normal lung volumes: man. In *Handbook of Circulation and Respiration.* Washington, D.C., Amer. Physiol. Soc., 1970b.

————: Comments on "Cardiac frequency in relation to aerobic capacity for work." *Ergonomioo, 13:*500 513, 1070c.

————: *Alive Man!* Springfield, Ill., Thomas, 1972a.

————: Opening lecture. Internat. Conference on Pediatric Work Physiology, Tel Aviv, 1972b.

————: Athletic performance at moderate altitudes. *Med del Sport (Torino),* 1973a, in press.

————: Future research on the quantifying of endurance training. Amer. Coll. Sports Med., Seattle, Wash., 1973b.

————, Allen, C., Benade, A.J.S., Davies, C.T.M., di Prampero, P.E., Hedman, R., Merriman, J.E., Myhre, K., and Simmons, R.: The maximum oxygen intake: an international reference standard of cardio-respiratory fitness. *Bull WHO, 38:*757–764, 1968a.

————, Allen, C., Benade, A.J.S., Davies, C.T.M., di Prampero, P.E., Hedman, R., Merriman, J.E. Myhre, K., and Simmons, R.: Standardization of sub-maximal exercise tests. *Bull WHO, 38:*765–775, 1968b.

————, Jones G., Ishii, K., Kaneko, M., and Olbrecht, A.J.: Factors affecting body density and thickness of subcutaneous fat. Data on 518 Canadian City Dwellers. *Am J Clin Nutr, 22:*1175–1189, 1969.

————, Kaneko, M., and Ishii, K.: Simple indices of obesity. *J Sports Med Phys Fitness, 11:*154–161, 1971.

Siegel, P.V., Gerathewohl, S.J., and Mohler, S.R.: Time zone effects. *US Fed Aviat Admin Reps, 69:*17, 1969.

Simonsen, E.: *Physiology of Work Capacity and Fatigue.* Springfield, Ill, Thomas, 1971.

Singleton, W.B., Easterby, R.S., and Whitfield, D.C.: *The Human Operator in Complex Systems.* London, Taylor & Francis, 1967.

Sippl, C.J.: *Computer Dictionary and Handbook.* Indianapolis, Sams, 1966.

Skinner, B.F.: Are theories of learning necessary? *Psychol Rev, 57:*193–216, 1950.

————: *Science and Human Behaviour.* New York, MacMillan, 1953.

Smith, W.: The potential for bus rapid transit. Detroit, Mich., Automobile Manufacturers' Association, 1969.

Smyllie, H.C., Taylor, M.P., and Cunningham-Green, R.A.: Acute myocardial infarction in Doncaster. *Br Med J, i:*31–34, 34–36, 1972.

Snook, S.H., and Irvine, C.H.: Maximum acceptable weight of lift. *Am Ind Hyg Assoc J, 28:*322–329, 1967.

Society of Actuaries: Build and Blood Pressure Study. Chicago, Ill., 1959.

Spicer, C.C. Computers in Medicine. *Brit Med Bull, 24:*187–276, 1968.

Stapp, J.P.: Effects of mechanical force on living tissue. 1. Abrupt deceleration and wind blast. *J Aviat Med, 26:*268–288, 1955a.

————: Tolerance to abrupt deceleration. In *Collected Papers on Aviation Medicine.* London, Butterworth, 1955b.

Stoll, A.M.: Human tolerance to positive G as determined by the physiological endpoints. *J Aviat Med, 27:*356–367, 1956.

Strughold, H.: The human time factor in flight. II. Chains of latencies in vision. *J Aviat Med, 22:*100–108, 1951.

————: Physiological day-night cycle in global flights. *J Aviat Med, 23:*464–473, 1952.

Swan, A.W.: Time study and statistics. *The Engineer,* November 1956.

Swearingen, J.J., McFadden, E.B., Garner, J.D., and Blethrow, J.G.: Human voluntary tolerance to vertical impact. *Aerosp Med, 31*:989–998, 1960.

Swearingen, J.J., Hasbrook, A.H., Snyder, R.G., and McFadden, E.B.: Kinematic behavior of the human body during deceleration. *Aerosp Med, 33*:188–197, 1962.

Tanner, W.R., and Swets, J.A.: A decision-making theory of visual detection. *Psychol Rev, 61*:101–100, 1054.

Thomas, E. Ll.: Man/Machine interface. *J Am Soc Agr Eng, 48*:495, 510–511, 1967.

Tinker, M.A.: *Legibility of Print.* Iowa, Ames, 1963.

Tippett, L.H.C.: Ratio-delay study. *J Text Inst Trans, 36*, 1935 (cited by Mundel, 1970).

Torrance, R.W.: *Arterial chemoreceptors.* Oxford, Blackwell, 1966.

Truax, C.B., and Carkhuff, R.R.: *Towards the Effective Counselling and Psychotherapy: Training and Practice.* Chicago, Aldine Publishing Co., 1967.

Tune, G.S.: Sleep and wakefulness in normal human adults. *Br Med J, II*:209–271, 1968.

Tyler, D.B.: The influence of a placebo, body position and medication on motion sickness. *Am J Physiol, 146*:450–456, 1946.

Ulrich, R., Stachnik, T., and Mabry, J.: *Control of Human Behaviour.* Glenview, Illinois, Scott Foresman, 1966.

Voas, R.B.: A description of the astronaut's task in Project Mercury. *Hum Factors, 3*:149–165, 1961.

Voigt, E.D., and Bahn, D.: Metabolism and pulse rate in physically handicapped when propelling a wheelchair up an incline. *Scand J Rehab Med, 1*:101–106, 1969.

Vos, H.W.: Human effort in the stacking of bales on a moving waggon. *J Agric Eng Res, 11*:238–242, 1966.

Vroom, V.H.: *Work and Motivation.* New York, Wiley, 1964.

Wald, G.: Macey Symposium on the Nervous Impulse, 1954 (cited by Keele, E., and Neil, E. in Samson Wright's *Applied Physiology,* 1964).

Wahlund, H.: Determination of the physical working capacity. *Acta Med Scand Suppl, 215*:1–78 plus unnumbered appendices, 1948.

Warner, W.K., and Havens, A.E.: Goal displacement and the intangibility of organizational goals. *Appl Soc Q, 12*:539–555, 1968.

Warren, B.H., Roman, J.A., and Graybiel, A.: Exclusion of angular accelerations as the principal cause of visual illusions during parabolic flight manoeuvres. *Aerosp Med, 35*:228–232, 1964.

Warren, R.M.: A new concept in industrial training. Paper presented to Can. Vocational Assoc, May 28th, 1968.

Weick, K.: *The Social Psychology of Organizing.* Reading, Mass., Addison Wesley, 1969.

Weiner, J.S., and Lourie, J.A.: *Human Biology—A Guide to Field Methods.* Oxford, U.K., Blackwell Scientific Publications, 1969.

Welford, A.T.: The "psychological refractory period" and the timing of high-speed performance—a review and a theory. *Br J Psychol, 43*:2–19, 1952.

West, J.C.: *Textbook of Servo-mechanisms.* London, Eng. Univ. Press, 1953.

White, R.W.: Motivation reconsidered—the concept of competence. *Psychol Rev, 66*:297–332, 1959.

Whiteside, T.C.D.: FPRC Rep., *724,* U.K. Air Ministry, 1949.

Whiteside, T.C.D.: *Problems of Vision in Flight at High Altitude.* London, Butterworth, 1957.

———: Hand-eye coordination in weightlessness. *Aerosp Med, 32*:719–725, 1961.

———: Motion sickness. In Gillies, J.A. (Ed.): *Textbook of Aviation Physiology.* Oxford, Pergamon Press, 1965.

Whittingham, H.E.: In *British Encyclopaedia of Medical Practice*, 2nd ed. London, Butterworth, 1950, p. 357, vol. 2.

Wiener, N.: *Cybernetics*. New York, Wiley, 1948.

Wilkinson, R.T.: Sleep deprivation. In Edholm, O.G., and Bacharach, A.L. (Eds.): *The Physiology of Human Survival*. London, Academic Press, 1965.

———: Sleep deprivation: performance tests for partial and selective sleep deprivation. In Abt, L.A., and Riess, B.F. (Eds.): *Progr Clin Psychol*, 7:28–43, 1969.

Woodhead, M.M.: Effect of a brief loud noise on decision making. *J Acoust Soc Am*, 31:1329–1331, 1959.

Woodson, W.E., and Conover, D.W.: *Human Engineering Guide for Equipment Designers*, 2nd ed. Berkeley, Univ. Calif. Press, 1964.

Woodworth, R.S.: *Experimental Psychology*. New York, Holt, 1938.

World Health Organization: *World Health Statistics Annual*. Geneva, Switzerland, 1970.

World Medical Association: Human experimentation. Code of ethics. *Br Med J, ii*: 177, 1964.

Wyndham, C.H.: An examination of the methods of physical classification of African labourers for manual work. *SA Med J*, 40:275–278, 1966.

———: Physical exercise at high temperatures. In Hollmann, W. (Ed.): *Sportmedizin*. New York, Springer, 1972.

———, Strydom, N.B., Morrison, J.F., Williams, C.G., Brendell, G., Peter, J., Cooke, H.M., and Joffe, A.: The influence of gross body weight on oxygen consumption and on physical working capacity of manual labourers. *Ergonomics*, 6:275–286, 1963.

Ziegarnik, B.: Uber das Behalten von erledigten und unerledigten Handlungen. *Psychol Forsch*, 9:1–85, 1927.

Zuckerman, M., and Cohen, N.: Sources of reports of visual and auditory sensations in perceptual-isolation experiments. *Psychol Bull*, 62:1–20, 1964.

INDEX

387